Introductory Graph Theory
with Applications

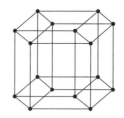

Introductory Graph Theory
with Applications

Fred Buckley
Baruch College (CUNY)

Marty Lewinter
Purchase College (SUNY)

Long Grove, Illinois

For information about this book, contact:
   Waveland Press, Inc.
   4180 IL Route 83, Suite 101
   Long Grove, IL 60047-9580
   (847) 634-0081
   info@waveland.com
   www.waveland.com

Copyright © 2003 by Fred Buckley and Marty Lewinter
Reissued 2013 by Waveland Press, Inc.
Previously published as *A Friendly Introduction to Graph Theory*

10-digit ISBN 1-4786-1175-8
13-digit ISBN 978-1-4786-1175-2

All rights reserved. No part of this book may be reproduced, stored in a retrieval system, or transmitted in any form or by any means without permission in writing from the publisher.

Printed in the United States of America

7   6   5   4   3   2   1

Dedicated to Frank Harary,
the lion of graph theory

Contents

Preface	ix
Notation	xiii

1. Introductory Concepts — 1
- 1.1 Mathematical Preliminaries — 1
- 1.2 Mathematical Induction — 15
- 1.3 Permutations and Combinations — 25
- 1.4 Pascal's Triangle and Combinatorial Identities — 36

2. Introduction to Graphs and their Uses — 47
- 2.1 Graphs as Models — 47
- 2.2 Subgraphs and Types of Graphs — 57
- 2.3 Isomorphic Graphs — 64
- 2.4 Graph Operations — 74

3. Trees and Bipartite Graphs — 85
- 3.1 Properties of Trees — 85
- 3.2 Minimum Spanning Trees — 93
- 3.3 A Characterization of Bipartite Graphs — 102
- 3.4 Matchings and Job Assignments — 107

4. Distance and Connectivity — 119
- 4.1 Distance in Graphs — 119
- 4.2 Connectivity Concepts — 129
- 4.3 Applications — 136

5. Eulerian and Hamiltonian Graphs — 143
- 5.1 Characterization of Eulerian Graphs — 143
- 5.2 Hamiltonicity — 151
- 5.3 Applications — 159

6. Graph Coloring — **165**
6.1 Vertex Coloring and Independent Sets — 165
6.2 Edge Coloring — 173
6.3 Applications of Graph Coloring — 182

7. Matrices — **191**
7.1 Review of Matrix Concepts — 191
7.2 The Adjacency Matrix — 199
7.3 The Distance Matrix — 207

8. Graph Algorithms — **215**
8.1 Graph Searching — 215
8.2 Graph Coloring Algorithms — 223
8.3 Tree Codes — 228

9. Planar Graphs — **237**
9.1 Planarity — 237
9.2 Planar Graphs, Graph Coloring, and Embedding — 246
9.3 Graph Duals and a Planar Graph Application — 251

10. Digraphs and Networks — **261**
10.1 Directed Graphs — 261
10.2 Networks — 271
10.3 The Critical Path Method — 286

11. Special Topics — **297**
11.1 Ramsey Theory — 297
11.2 Domination in Graphs — 304

Answers/Solutions to Selected Exercises — **313**

Index — **355**

Preface

Graph theory is a delightful subject with a host of applications in such fields as anthropology, computer science, chemistry, environmental conservation, fluid dynamics, psychology, sociology, traffic management, and telecommunications, among others. With the advent of operations research in the twentieth century, graph theory has risen several notches in the esteem in which it is held. Formally a branch of combinatorics, graph theory intersects topology, group theory, and number theory, to name just a few fields. While its theorems and proofs range from easy to almost incomprehensible, graph theory is, after all, the study of dots and lines!

Because of its wide range of applications, it is important that students in many diverse fields, not just mathematics and computer science, gain a foundation in the basics of graph theory. By doing so, they will have additional powerful tools at their disposal to analyze problems within their own area of study. We have written this text mainly with those students in mind and designed it to be easily accessible to undergraduate students as early as the sophomore level. We assume nothing more than a good grasp of algebra. Thus, *Introductory Graph Theory with Applications* provides early access to this wonderful and useful area of study for students in mathematics, computer science, the social sciences, business, engineering—wherever graph theory is needed.

The student will find this text quite readable. This book should be read with pencil and paper nearby—it is not bedtime reading. The exercises reflect the contents of the chapter and range from easy to hard. Answers, solutions, or hints are supplied in the back of the book for a wide selection of the exercises. It is important to realize that mathematics is not a spectator sport (although it can be fascinating to watch someone present a surprisingly simple proof of an interesting result). Do the exercises, or you will not retain the material of the nth chapter as you attempt the $(n+1)$-st. A bonus to the diligent students who work through the more challenging exercises is that those students will develop a strong foundation for research in graph theory and an ability to apply the concepts learned to a wide range of real-world problems.

Content

In Chapter 1, we discuss basic prerequisite concepts and ideas that the reader should be familiar with. Although it is meant as a review, a somewhat thorough treatment is given to provide the necessary tools for understanding the material that will be examined throughout the rest of the book. Thus we discuss topics such as sets, functions, parity, mathematical induction, proof techniques, counting techniques, permutations and combinations, Pascal's triangle, and combinatorial identities. For those readers unfamiliar with proof techniques, this introductory chapter will be particularly helpful. In a class of better-prepared students, the instructor may choose to skip some or all of the contents of Chapter 1.

As mentioned earlier, graphs have a wide variety of applications. In Chapter 2, we introduce the most basic concepts of graph theory and illustrate some of the areas where graphs are used. We will see many other applications throughout the book.

One class of graphs is so important that it deserves treatment in its own chapter. Because of their simple structure, trees are often used as a testing ground for possible new theorems. Trees arise in many applications, such as analyzing business hierarchies and determining minimum cost transportation networks, and are the basis of important data structures in computer science. A tree is a special type of graph in a larger class called bipartite graphs. In Chapter 3, we examine trees, bipartite graphs, and their uses. The chapter concludes with an application to job assignment problems.

The concept of distance is widely used throughout graph theory and its applications. Distance is used in various graph operations, in isomorphism testing, and in convexity problems, and is the basis of several graph symmetry concepts. Distance is used to define many graph centrality concepts, which in turn are useful in facility location problems. Numerous graph algorithms are distance related in that they search for paths of various lengths within the graph. Distance is an important factor in extremal problems in graph connectivity. Graph connectivity is important in its own right because of its strong relation to the reliability and vulnerability of computer networks. In Chapter 4, we discuss some of the many important concepts and results concerning distance and connectivity in graphs. We conclude with a discussion of a facility location problem and an introduction to the concept of reliability of computer networks.

Numerous theoretical and applied problems in graph theory require one to traverse a graph in a particular way. In some problems, the goal is to find a trail or circuit in order to pass through each edge exactly once. In other problems, one must find a path or cycle that includes each vertex exactly once. We discuss such problems in Chapter 5. We conclude with a discussion of two well-known related problems, namely, the Chinese postman problem, and the Traveling salesman problem.

Various real-world problems that can be modeled by graphs require that the vertex set or edge set be partitioned into disjoint sets such that items within a given set are mutually nonadjacent. Common problems include scheduling meetings or exams to avoid conflicts and storage of chemicals to prevent adverse interactions. These problems are related to graph coloring, the subject of Chapter 6.

A matrix is a rectangular table of numbers. One of the simplest ways of storing a graph in a computer is by using a matrix or its computer science counterpart, an array. Graph theory makes effective use of matrices as a tool in examining structural and other properties of graphs. In Chapter 7, we first review some basic properties of matrices and then examine some uses of matrices within graph theory.

There are many connections between graph theory and computer science. Thus it should not be surprising that algorithms have played a strong role in recent graph theory research, so much so that several books have been devoted to algorithmic graph theory. In Chapter 8, we give an introduction to graph algorithms and indicate some of their uses. We discuss the important breadth-first search and depth-first search algorithms and examine algorithms for graph coloring as well as for coding trees. Several other algorithms are presented throughout the text. We shall not examine efficiency of algorithms or NP-completeness here, but refer the interested reader to appropriate sources for discussions of those topics.

A graph is planar if it can be drawn in the plane with no crossing edges. Planar graphs have received a great deal of attention over the years because of a long-standing problem, the four-color conjecture, that took over one hundred years to prove. Planar graphs remain important today because of their applications. Planar graphs are important in facility layout problems within operations research, and they are crucial in the design of printed circuit boards in computer science. We discuss planar graphs and their properties in Chapter 9.

To model certain real-world problems, a structure more complex than a graph is needed. In Chapter 10, we consider some additional structures. Digraphs are similar to graphs except there are directions on the edges. Digraphs are used to model problems where the direction of flow of some quantity (information, traffic, liquid, electrons, and so on) is of importance. When limits are placed on how much of that quantity can flow through a given directed edge, we obtain a network. A special type of digraph with no directed cycles, called an activity digraph, has weights on the directed edges indicating the duration of a given activity. These weighted digraphs are used to aid in scheduling individual activities that compose a complex project. In Chapter 10, we explore various types of digraphs and their properties, study network flows, and present an algorithm to maximize total flow. We conclude the chapter with a discussion of activity digraphs and their uses.

In Chapter 11, we consider two additional topics, Ramsey theory and graph domination. The first is related to edge colorings in graphs, and the second is related to both distance and independence and has a wide range of applications. The one thing that these last two topics have in common is that they are lots of fun to work on.

We hope this book is as much fun for you to read and learn from as it was for us to write.

Acknowledgments

There are numerous people we must thank for their help in getting this book into its final form. First, we thank the various reviewers of the early versions of this text and Professor Michael Gargano of Pace University for suggestions at that stage. We especially thank Professor Nancy Eaton of the University of Rhode Island and Professor Renate Scheidler of the University of Calgary for extensive detailed insightful comments and suggestions on the penultimate version of the manuscript. We thank Professor Daphne Liu, California State University, Los Angeles, for pointing out an exercise that required fixing. We thank our editor, George Lobell, for his wise guidance in bringing this project to fruition. We also thank Bayani DeLeon for guiding the project through the production stage.

The first-named author did all of the typesetting using PCTeX4.01 and created all the figures using Adobe Illustrator 10.0. He thanks Professor Doug Howard of Baruch College (CUNY) for numerous extremely helpful suggestions concerning TeX. He also thanks Professor Nakhle Asmar of the University of Missouri, Columbia, for his generosity in sharing his knowledge on how to import Adobe Illustrator figures into LaTeX so that all items appear as intended, which is far from a trivial matter.

I would like to thank my wife, Wai Mui Choy, for her assistance in ensuring that I meet all deadlines and for her extreme patience while I was working on this text.—F. B.

I give special thanks to my friend and colleague, Timothy Bocchi, for his many valuable suggestions.—M. L.

Fred Buckley
Marty Lewinter

Notation

$\|A\|$	the cardinality of A
$A \cup B$	the union of sets A and B, or the union of graphs A and B
$A \cap B$	the intersection of sets A and B
$A \times B$	the cartesian product of sets A and B
A'	the complement of set A
$A \subseteq B$	A is contained in B, or A is a subset of B
$a \in A$	a is an element of A
$a \notin A$	a is not an element of A
a_{ij}	the (i,j)-entry of matrix A, that is, the entry in the row i, column j of A
$C(G)$	the center of G
C_n	the cycle of n vertices
$d(x,y)$	the distance from x to y
$\mathrm{diam}(G)$	the diameter of G
$E(G)$	the set of edges of G
$e(v)$	the eccentricity of vertex v
$\bar{G}$	the complement of graph G
G^*	the dual of G
$G + H$	the join of G and H
$G_1 + G_2 + \cdots + G_k$	the sequential join of graphs G_1 through G_k
$G \times H$	the Cartesian product of G and H
G-e	the graph resulting from removing edge e from G
G/e	the graph resulting from contracting edge e from G
G-v	the graph resulting from removing vertex v and all of its incident edges from G
I	the identity matrix
J	matrix of 1's
$K_{m,n}$	the complete bipartite graph with part of order m and n, respectively
K_n	the complete graph of order n
$L(G)$	the line graph of G
$M(n_1, n_2, \ldots, n_k)$	the n-mesh equivalent to the cartesian product $P_{n_1} \times P_{n_2} \times \cdots \times P_{n_k}$
M^t	the transpose of matrix M
N	the set of natural numbers; the positive integers

N_j	the jth distance set of the center: vertices at distance j from a closest central vertex
$N(v)$	the neighbors of v
P	$\{2, 3, 5, 7, 11, 13, 17, 19, 23, 29, 31, 37, \ldots\}$, the set of prime numbers
$P(G)$	the periphery of G
P_n	the path on n vertices
Q	the rational numbers
Q_n	the hypercube on 2^n vertices
$\Re$	the real numbers
$\mathrm{rad}(G)$	the radius of G
$S(G)$	the subdivision graph of G
$\langle S \rangle$	the induced subgraph on set S
t_n	the nth triangular number $n(n+1)/2$
$V(G)$	the set of vertices of G
$W_{1,n}$	the wheel with n spokes, equivalent to the join $K_1 + C_n$
$W_{m,n}$	the generalized wheel, equivalent to the join $\bar{K}_m + C_n$
$\lfloor x \rfloor$	the round down of x (often called "the *floor* of x")
$\lceil x \rceil$	the round up of x (often called "the *ceiling* of x")
Z	the integers
$\alpha(G)$	the vertex covering number of G
$\beta(G)$	the independence number of G
$\delta(G)$	the minimum degree of G
$\Delta(G)$	the maximum degree of G
$\gamma(G)$	the domination number of G
$\kappa(G)$	the vertex connectivity of G
$\lambda(G)$	the edge connectivity of G
$\chi(G)$	the chromatic number of G
$\chi_1(G)$	the edge-chromatic number of G

Introductory Graph Theory
with Applications

Chapter 1

Introductory Concepts

In this chapter, we discuss basic prerequisite concepts and ideas that the reader should be familiar with. Although it is meant as a review, a somewhat thorough treatment is given to provide the necessary tools for understanding the material that will be examined throughout the rest of the book. Thus we discuss such topics as sets, functions, parity, mathematical induction, proof techniques, counting techniques, permutations and combinations, Pascal's triangle, and combinatorial identities. For those readers unfamiliar with proof techniques, this introductory chapter will be particularly helpful.

1.1 Mathematical Preliminaries

In mathematics, theorems come in two basic forms. A theorem of the form "(statement A) if and only if (statement B)" means that A implies B *and* B implies A. We often write $\boldsymbol{A} \Leftrightarrow \boldsymbol{B}$, which is read "$A$ if and only if B." One simple theorem having that form is the following: Suppose that x is an integer. Then x^2 is even if and only if x is even. We will see why that statement is true shortly when we discuss parity. The second basic form of a theorem is "if A then B." Here we may write $\boldsymbol{A} \Rightarrow \boldsymbol{B}$, which means that "$A$ implies B," but no inference should be made that B implies A. This requires a bit of thought before it becomes second nature. Thus the statement "if you work hard you will be rich" doesn't negate the fact that some rich people inherited their wealth and did not work hard; that is, we cannot conclude that "if you are rich then you must have worked hard."

Rounding

Recall that an integer is an element of the set $Z = \{0, 1, -1, 2, -2, 3, -3, \ldots\}$. Many calculations in combinatorics require rounding a number up or down to the closest integer. Mathematicians have a convenient notation for these operations. $\lfloor x \rfloor$ represents the **round down** or **floor** of x and $\lceil x \rceil$ means **round up** or **ceiling** of x. Thus $\lfloor \sqrt{2} \rfloor = 1$, $\lfloor 5.98 \rfloor = 5$, and $\lfloor -3.1 \rfloor = -4$; whereas $\lceil \pi \rceil = 4$, $\lceil 5.38 \rceil = 6$, and $\lceil -6.7 \rceil = -6$.

Remember that when n is an integer, $\lceil n \rceil = \lfloor n \rfloor = n$. You could say that these two functions have no effect on integers. On the other hand, when n is not an integer, it follows that $\lceil n \rceil - \lfloor n \rfloor = 1$.

Many relationships that these rounding functions satisfy involve inequalities. Suppose we wish to compare $\lfloor 2x \rfloor$ and $2 \lfloor x \rfloor$. (We are using x here to represent any positive real number; n is usually reserved for integers.) In the first of these, we double x and then round down, while in the second, we first round down and then double. Changing order can be a tricky business. Sometimes we get the same answer and sometimes we don't. When $x = 3.2$, if we compute $\lfloor 2x \rfloor$, we get $\lfloor 6.4 \rfloor = 6$. If we compute $2 \lfloor x \rfloor$, we get $2 \lfloor 3.2 \rfloor = 6$—the same answer! Before you assert that $\lfloor 2x \rfloor = 2 \lfloor x \rfloor$, put $x = 3.6$ into this equation and see what happens. $\lfloor 2(3.6) \rfloor = \lfloor 7.2 \rfloor = 7$ does not equal $2 \lfloor (3.6) \rfloor = 2(3) = 6$. The correct relationship is

$$\lfloor 2x \rfloor \geq 2 \lfloor x \rfloor \tag{1.1}$$

Note that even dozens of examples supporting the truth of a given statement is no guarantee that the statement is actually true. To confirm that a statement is always true, we must *prove* that it is. Let's prove inequality (1.1). To do so, we need to be precise about what the floor function means. It is defined formally as follows.

Definition 1.1 Suppose that $x = n + \varepsilon$ (the letter ε, "epsilon" is commonly used in mathematics to represent small quantities), where n is an integer and $0 \leq \varepsilon < 1$. Then $\lfloor x \rfloor = n$.

Thus $\lfloor 2x \rfloor = \lfloor 2(n + \varepsilon) \rfloor = \lfloor 2n + 2\varepsilon \rfloor$. Note that if $\varepsilon < 1/2$, we will get $2n$, and if $1/2 \leq \varepsilon < 1$, we get $2n + 1$. Now let's compare that to the right side of inequality (1.1). Here we have $2 \lfloor x \rfloor = 2 \lfloor n + \varepsilon \rfloor = 2n$. Thus, no matter where ε is within the range $0 \leq \varepsilon < 1$, we always get $\lfloor 2x \rfloor \geq 2 \lfloor x \rfloor$.

Do you see why the inequality sign in (1.1) must be reversed when we consider the round-up function? The correct relationship for the ceiling function is

$$\lceil 2x \rceil \leq 2 \lceil x \rceil \tag{1.2}$$

Compare $\lceil 2(3.4) \rceil$ and $2 \lceil 3.4 \rceil$, and you might be convinced. However, to show that the relationship always holds, it must be proved formally. To do so, first note the following:

Definition 1.2 Suppose that $x = n + \varepsilon$, where n is an integer and $0 \leq \varepsilon < 1$. Then $\lceil x \rceil = n$ if $\varepsilon = 0$ and $\lceil x \rceil = n + 1$ when $0 < \varepsilon < 1$.

1.1 Mathematical Preliminaries

It is clear from Definition 1.2 that in order to prove inequality (1.2), two cases must be considered, namely, the case when $\varepsilon = 0$ and the case when $0 < \varepsilon < 1$. You will be asked in the Exercises to prove inequality (1.2).

Parity

Many questions in combinatorics, number theory, and other fields involve **parity**, that is, whether an integer is even or odd. It pays to remember that an even integer can be written in the form $2n$, while an odd integer can be written in the form $2n + 1$ (or $2n - 1$, whichever is more convenient), where n is an integer. Note that 0 is even because $0 = 2(0)$. Of course, if we wish to represent two even integers, we would call them $2n$ and $2m$ (or $2r$ and $2s$, etc.). For two odds, we use $2n + 1$ and $2m + 1$, whereas for an odd and an even integer, we would use $2n + 1$ and $2m$. Note that it is not appropriate to represent an even integer and an odd integer in a problem by $2n$ and $2n + 1$ unless we know that the integers are consecutive.

If j and k are integers, is $2jk$ even? Of course it is—since j and k are integers, the product jk is also an integer. Thus, if $n = jk$, we have $2jk = 2n$, which is even. What about $4rsp + 1$ when r, s, and p are integers? Let $n = 2rsp$; then $4rsp + 1$ is just $2n + 1$, which is odd.

What is the parity of $6m + 7$, where m is an integer? There are several approaches. If we let $n = 3m+3$, we have our old friend $2n+1$, which is odd. Alternatively, we might establish the principle, once and for all, that the sum of an even integer and an odd integer is odd. So $6m$ (even) plus 7 (odd) is odd. Denote the even integer, s, by $2n$ and the odd integer, t, by $2k + 1$, where n and k are integers. Then their sum is $s+t = 2n+2k+1 = 2(n+k)+1$. Since n and k are integers, $n+k$ is also an integer. Thus, $s+t = 2(n+k)+1$ is one more than twice an integer. Hence $s + t$ is one more than an even integer; so $s + t$ is odd. Therefore, the sum of an even integer and an odd integer is odd.

Parity is preserved when we square an integer. This proof is also quite easy. If s is even, then $s = 2n$ where n is an integer. Then $s^2 = (2n)^2 = 4n^2 = 2(2n^2)$. So s^2 is twice an integer, and therefore even. On the other hand, suppose that t is an odd integer. So $t = 2m+1$, where m is an integer. So $t^2 = (2m+1)^2 = 4m^2 + 4m + 1 = 2(2m^2 + 2m) + 1$. Thus t^2 is one more than twice an integer, so t^2 is odd. This proves the result we mentioned earlier, namely, that for an integer x, x^2 is even if and only if x is even.

Why is the product of an odd and an even integer even? Because when n and m are integers, $2n(2m+1) = 2[n(2m+1)]$, which is twice an integer, and therefore even. In fact, one even integer in a product of several integer factors guarantees that the product is even. This follows since if n and x_1 through x_k are all integers, then $2n(x_1 x_2 x_3 \cdots x_k) = 2(n x_1 x_2 x_3 \cdots x_k)$, which is even since it is twice an integer. What is the parity of the product of several integers if none of them are even? Start with two odds. $(2n+1)(2m+1) = 4nm+2n+2m+1 = 2(2nm+n+m)+1$, which is odd. Now use the associative law for three odd integers a, b, and c. Since $abc = (ab)c$, and since ab is odd (we just established that the product of two odd integers is odd) and c is odd, it follows that abc is odd. Say we bring in another odd factor, d. Then since $abcd = (abc)d$, we certainly have an odd product. A formal proof that the product of n odd integers is odd is done using mathematical induction, which will be discussed in Section 1.2.

To summarize our observations concerning parity, we have the following:

> **Note 1.1** A product of integers is even if and only if at least one of the factors is even.

Before we leave parity, suppose you are standing on one side of a river. If you cross the river n times, do you wind up on the side you started from or on the other side? The answer depends on the parity of n. If n is odd, you wind up on the other side, while if n is even, you wind up where you started. As mentioned earlier, zero is an even number. If you cross the river no times, you wind up, of course, where you started.

Finally, we note that the parity of an exponent on a negative number plays an important role. Consider, for example, $(-1)^n$, where n is an integer. As we consider successive positive integer values of n, we get $(-1)^1 = -1$, $(-1)^2 = 1$, $(-1)^3 = -1$, $(-1)^4 = 1$, and so on. We find that $(-1)^n = 1$ when n is even and $(-1)^n = -1$ when n is odd. In general, using simple factoring and the facts we have already established, it is not difficult to show the following general result.

> **Note 1.2** For any negative number x and integer n, x^n is positive when n is even, and x^n is negative when n is odd.

Sets

It is important to have a strong grasp of sets and operations on sets. Intuitively, a set is a collection of objects of any kind. Thus, we can speak of the set of all New York Stock Exchange listed stocks that reached new highs during 2002 (a rather small set) or the set of all students in your class who have gone skiing and who also possess a driver's license. To be useful mathematically, we must make the concept of a set precise. A **set** is a well-defined collection of distinct objects. An object in a set is said to be a member of the set or an **element** of the set. "Well-defined" means that there is a rule that enables us to unambiguously distinguish whether or not a particular object belongs to the set.

1.1 Mathematical Preliminaries

For example, we do not speak of the set of rich people. This is not a well-defined collection, because it is not clear which people belong to the set. If you have little money, you might consider someone having $100,000 rich, but a millionaire would not think of that person as being rich.

Generally, capital letters denote sets and lowercase letters denote elements. To indicate that x is an element of set A, we write $\boldsymbol{x \in A}$. The symbol "$\in$" is read "is an element of" or "belongs to" or "is in." To denote that element y "is not in A," we write $\boldsymbol{y \notin A}$. For example, if P is the set of prime numbers (a **prime number** is an integer greater than 1 that is exactly divisible only by 1 and itself), then $101 \in P$, while $91 \notin P$. There are various ways of describing a given set. The most common ways are as follows:

(1) listing the elements of the set,
(2) describing the elements of the set in terms of a property that characterizes them, or
(3) using unions, intersections, or complements of sets.

In the list format, separate the elements of the set by commas. An example is $\{1, 3, 8, 11, 15\}$. In general, the order in which we list the elements within a set does not matter. Thus $\{1, 3, 8, 11, 15\}$ is the same as the set $\{15, 3, 11, 1, 8\}$ When there are infinitely many elements or a large finite number of elements in the set, use an **ellipsis** "..." to indicate that the pattern of the listed elements continues, as in $\{5, 10, 15, 20, \ldots, 85\}$. When using the ellipsis, however, make sure that the pattern the elements follow is clear. In the property format, the set description usually begins with a variable, such as x or y, and then a colon, which is read "such that" or "where." (*Note:* Some authors use a vertical bar "|" instead of a colon.) The colon is followed by the property (or list of properties, separated by commas) that characterizes the element x or y. An example is

$$\{x : x \text{ is an even integer}\}$$

read "the set of all x such that x is an even integer." If there is a list of properties, the variable must satisfy all the properties. In this case, the separating commas are read "and." An example of this case is

$$\{y : y \text{ is a perfect square}, y \text{ is divisible by 3}\}$$

Example 1.1 Which of the following are sets? For those that are, describe them using the list format.

(a) The collection of positive integers greater than 18
(b) The collection of all smart students in the class
(c) The collection of letters in the word "possess"
(d) The set of odd integers divisible by 6

Solution

(a) Since we know exactly which integers belong to the collection, it is a set. This infinite set is $\{19, 20, 21, 22, \ldots\}$.

(b) This is not a set. The term "smart" is not specific enough to determine which students belong to the collection.

(c) This set is $\{p, o, s, e\} = \{e, o, p, s\}$.

(d) This is a set but happens to contain no elements. ◇

The set containing no elements encountered in Example 1.1(d) is an important one and is called the **empty set** (or **null set**) and is denoted by $\emptyset$. The set $\emptyset$ occurs often and can be described in many ways. For example, we can say that {humans who have run the mile in less than two minutes}= $\emptyset$. Five other important sets to know are the natural numbers, integers, rational numbers, real numbers, and prime numbers. These are commonly represented by the letters N, Z, Q, $\Re$, and P and are described in Note 1.3.

Note 1.3

The set of **natural numbers** is $N = \{1, 2, 3, 4, \ldots\}$.

The set of **integers** is $Z = \{0, 1, -1, 2, -2, 3, -3, \ldots\}$, which can also be written as $\{\ldots, -3, -2, -1, 0, 1, 2, 3, \ldots\}$.

The set of **rational numbers** is
$Q = \{x : x = a/b,\ a \in Z,\ b \in Z,\ b \neq 0\}$ (Q comes from "quotient.")

The set of **real numbers** is
$\Re = \{x : x \text{ corresponds to a point on the number line}\}$

The set of **prime numbers** is
$P = \{x : x \in N,\ x \geq 2, \text{ if } y \in N \text{ divides } x, \text{ then } y = 1 \text{ or } y = x\}$
$ = \{2, 3, 5, 7, 11, 13, 17, 19, 23, \ldots\}$

In our study of graph theory, we shall occasionally encounter the sets N and Z. The natural numbers are also referred to as the positive integers. Note that the order of the elements in the listing of a set is not important. A set is **finite** if we can count the number of elements that it contains. Otherwise, the set is **infinite**. Although order is not important for sets, if a set is finite with a large number of elements or if the set is infinite, the order could help the reader to detect the pattern of the elements. We usually include the beginning elements, and for a large finite set, the last one or two elements as well.

Example 1.2 Write the following sets using the list format but without writing all the elements.

(a) $A = \{x : x \text{ is an even integer}, -4 \leq x \leq 58\}$
(b) $B = \{x : x \in N, x \text{ is a perfect square}\}$
(c) $C = \{x : x \in N, x \text{ is a multiple of } 7\}$

Solution
(a) $A = \{-4, -2, 0, 2, 4, 6, \ldots, 58\}$
(b) $B = \{1, 4, 9, 16, 25, 36, \ldots\}$
(c) $C = \{7, 14, 21, 28, 35, 42, \ldots\}$ ◇

The **cardinality** of a finite set X is the number of elements contained in X and is denoted $|X|$. In Example 1.2, set A is finite, while sets B and C are infinite. The cardinality of set A is $|A| = 32$. Do you agree?

Subsets

Two sets, though described differently, may contain precisely the same elements. For two such sets A and B, we say that A and B are **equal** and write $A = B$. Set A is a **subset** of set B if every element of A is also an element of B. In this case, we write $A \subseteq B$. If $A \subseteq B$ and B has additional elements not in A, then A is a **proper subset** of B. In that case, we sometimes write $A \subset B$ rather than $A \subseteq B$. Also, for any set A, we have $\emptyset \subseteq A$.

> **Note 1.4** For sets A and B, $A = B$ if and only if $A \subseteq B$ and $B \subseteq A$.

Note 1.4 is important because it provides a technique for proving that two sets are equal. That is, to prove that $A = B$, we prove first that $A \subseteq B$ and then also prove that $B \subseteq A$.

Example 1.3 Which of the following are subsets of other sets that are described?
(a) $E = \{2, 4, 6, 8, 10, \ldots\}$
(b) $M = \{y : y \text{ is a positive multiple of } 8\}$
(c) $B = \{x : x \in N, x \text{ is a perfect square}\}$
(d) $T = \{16, 64, 196\}$

Solution
It is easy to verify the relations $M \subseteq E$, $T \subseteq E$, and $T \subseteq B$. There are no others involving distinct sets. ◇

Occasionally, we need to consider all possible subsets of a given set. For any set A, its **power set** $\wp(A)$ is the set of all subsets of A. So, for example, if $A = \{a, b\}$, then $\wp(A) = \{\emptyset, \{a\}, \{b\}, \{a, b\}\}$. It is an easy observation that if set A contains n elements, then $\wp(A)$ contains 2^n elements. This is because in generating each subset, we must decide for each element x of A whether x is in the subset or not. Thus there are two possibilities for each element, either in or out. There are then 2^n possible subsets; therefore, $\wp(A)$ contains

2^n elements. We shall prove this formally in Example 1.18. Note that all the elements of $\wp(A)$ are sets since they are subsets of A.

Set Operations

The union of two sets is the result of putting all the elements of the two sets into a single set. More formally, the **union** of sets A and B, denoted by $A \cup B$, is the set of all elements that belong to either A or B or both. Saying "or both" in the description means that elements contained in both A and B are included in $A \cup B$ along with the elements contained in only one of the two sets. In mathematics, "or" is always used in this inclusive sense. Using set notation,

$$A \cup B = \{x : x \in A \text{ or } x \in B\}$$

For the intersection of two sets, we want to find where the two sets intersect or overlap. The **intersection** of sets A and B, denoted by $A \cap B$, is the set of all elements that belong to both A and B. Using set notation,

$$A \cap B = \{x : x \in A, \ x \in B\}$$

Remember that the comma "," within a set is read "and."

In many problems involving sets, we restrict our attention to certain types of objects. For example, in set $C = \{x : x \in N, x \text{ is a multiple of } 7\}$, we restrict our attention to the set of natural numbers $N = \{1, 2, 3, 4, \ldots\}$. Set C is a particular subset of N consisting of those elements in N that are multiples of 7. The set of objects to which we restrict our attention is called the **universal set** and is denoted by U. Each set we consider contains elements selected from U. The **complement** of a set A is the set of all elements in U but not in A. We denote the complement of A by A . Thus,

$$A = \{x : x \in U, x \notin A\}$$

There are some occasions when working with a particular word problem, that one encounters "or'" used in an exclusive sense. Typically, such a problem involving sets A and B would use a phrase such as "either A or B but not both." This is translated into set theory notation as $(A \cup B) \cap (A \cap B)'$. Example 1.4 illustrates the operations of intersection, union, and complement.

Example 1.4 Suppose that $U = \{1, 2, 3, \ldots, 8\}$, $A = \{1, 7, 8\}$, $B = \{2, 6, 7\}$, and $C = \{1, 5, 6, 8\}$. Find each of the following sets:
(a) $A \cup B$ (b) $A \cap C$ (c) B' (d) $A \cup C'$ (e) $A' \cap B'$

Solution

(a) $A \cup B = \{1, 2, 6, 7, 8\}$. Although 7 is in both A and B, we list it only once in $A \cup B$.

(b) $A \cap C = \{1, 8\}$, the set of elements appearing simultaneously in both A and C.

1.1 Mathematical Preliminaries

(c) $B' = \{1, 3, 4, 5, 8\}$. B' is the set of elements in U but not in B.

(d) First find C', which is $\{2, 3, 4, 7\}$, and then find the union of this set with A to get $A \cup C' = \{1, 2, 3, 4, 7, 8\}$.

(e) A' is the set of elements in U but not in A. Thus, $A' = \{2, 3, 4, 5, 6\}$. In part (c), we found $B' = \{1, 3, 4, 5, 8\}$. Since $A' \cap B'$ is the set of elements that appear simultaneously in both A' and B', we find that $A' \cap B' = \{3, 4, 5\}$. $\diamond$

We note that the set $A' \cap B'$ found in part (e) of Example 1.4 is actually the complement of the set found in part (a). That is, $A' \cap B' = (A \cup B)'$. This is not a coincidence. That property holds no matter what the sets A and B are. Note 1.5 lists several properties of the intersection, union, and complement of sets.

Note 1.5 Properties of Set Operations
1a. (Associativity of Unions) $(A \cup B) \cup C = A \cup (B \cup C)$
1b. (Associativity of Intersections) $(A \cap B) \cap C = A \cap (B \cap C)$
2a. (Commutativity of Unions) $A \cup B = B \cup A$
2b. (Commutativity of Intersections) $A \cap B = B \cap A$
3. (Double Negation) $(A')' = A$
4a. (Distributive Law) $A \cap (B \cup C) = (A \cap B) \cup (A \cap C)$
4b. (Distributive Law) $A \cup (B \cap C) = (A \cup B) \cap (A \cup C)$
5a. (DeMorgan's Law for Sets) $(A \cup B)' = A' \cap B'$
5b. (DeMorgan's Law for Sets) $(A \cap B)' = A' \cup B'$

As an introduction to proof techniques, we present proofs of properties (4a) and (5a) of Note 1.5.

Example 1.5 Prove that $A \cap (B \cup C) = (A \cap B) \cup (A \cap C)$.

Solution
According to Note 1.4, in order to show that $A \cap (B \cup C) = (A \cap B) \cup (A \cap C)$, we must show that $A \cap (B \cup C) \subseteq (A \cap B) \cup (A \cap C)$ and then show that $(A \cap B) \cup (A \cap C) \subseteq A \cap (B \cup C)$. For the first part, suppose that $x \in A \cap (B \cup C)$. Then $x \in A$ and $x \in B \cup C$. If $x \in B \cup C$, then $x \in B$ or $x \in C$. But since $x \in A$, and x is in at least one of B or C, we have $x \in A \cap B$ (if x is in B) or $x \in A \cap C$ (if x is in C). Thus $x \in A \cap B$ or $x \in A \cap C$, so $x \in (A \cap B) \cup (A \cap C)$. Since an arbitrary element x in $A \cap (B \cup C)$ must be in $(A \cap B) \cup (A \cap C)$, we have $A \cap (B \cup C) \subseteq (A \cap B) \cup (A \cap C)$.

For the second part, suppose that $y \in (A \cap B) \cup (A \cap C)$. Then $y \in A \cap B$ or $y \in A \cap C$. Now we examine what causes y to be in at least one of those sets. First, y must definitely be in A, that is, $y \in A$. But additionally, it must be the case that y is in at least one of B or C in order for y to be in at least one of $A \cap B$ or $A \cap C$. Therefore, $y \in A$ and $y \in B \cup C$, so

$y \in A \cap (B \cup C)$. Since an arbitrary element y in $(A \cap B) \cup (A \cap C)$ must also be in $A \cap (B \cup C)$, we have $(A \cap B) \cup (A \cap C) \subseteq A \cap (B \cup C)$. Combining this with the first part, we get $A \cap (B \cup C) = (A \cap B) \cup (A \cap C)$, which proves Property 4a of Note 1.5. ◇

Example 1.6 Suppose that A is the set of all students currently attending the University of California at Los Angeles (UCLA), B is the set of all students who play on UCLA's varsity football team, and C is the set of all psychology majors. Interpret the equation $A \cap (B \cup C) = (A \cap B) \cup (A \cap C)$ from Example 1.5 in this context.

Solution

The equation says that the set of all students currently attending UCLA and who are on the varsity football team or are psychology majors is the same as the set of students who are attending UCLA and play on the varsity football team or are attending UCLA and are psychology majors. ◇

Example 1.7 Prove that $(A \cup B)' = A' \cap B'$.

Solution

We prove $(A \cup B)' = A' \cap B'$ in a slightly different way. Although the same type of subset argument given in Example 1.5 would work fine, it is more tedious than necessary. The set $(A \cup B)'$ is the set of all elements that are not in $A \cup B$, that is, the set of all elements that are not in at least one of A or B. But this corresponds to the set of elements that are in neither A nor B, which is the set of elements that are not in A and simultaneously not in B. This corresponds to the set of elements in both A' and B', which is $A' \cap B'$. Thus, $(A \cup B)' = A' \cap B'$, which proves Property 5a of Note 1.5. ◇

Example 1.8 Let $A = \{x : x \text{ is a red suit in a deck of playing cards}\}$, and let $B = \{y : y \text{ is an ace}\}$. Interpret the equation $(A \cup B)' = A' \cap B'$ of Example 1.7 in this context.

Solution

$(A \cup B)'$ is the set of playing cards that are neither red nor aces. The equation says that this set is the same as $A' \cap B'$, which is the set of playing cards that are not red and are not aces. ◇

Cartesian Product

As one last operation on sets, we consider the Cartesian product. The **Cartesian product**, $A \times B$, of sets A and B is the set of all *ordered* pairs (a, b), where $a \in A$ and $b \in B$. Items a and b are called **coordinates** of the ordered pair (a, b). In set notation, we have

$$A \times B = \{(a, b) : a \in A, b \in B\}$$

1.1 Mathematical Preliminaries

Example 1.9 Given that $A = \{1, 2, 3\}$ and $B = \{x, y\}$, find $A \times B$.
Solution
$A \times B = \{(1, x), (1, y), (2, x), (2, y), (3, x), (3, y)\}$ ◇

A **relation on set A** is a subset of $A \times A$. A **relation from set A to set B** is a subset of $A \times B$. So for sets A and B of Example 1.9, one possible relation is $C = \{(1, y), (2, x), (2, y)(3, x)\}$. Other examples are $D = \{(2, y), (3, y)\}$ and $E = \{(1, x), (2, y), (3, x)\}$.

A **function from A to B** is a relation in which each element of A appears as the first coordinate of precisely one ordered pair in the relation. Some notation that is useful to know is the following: $\boldsymbol{f : A \to B}$, which we read "f maps A into B. If our function f contains the ordered pairs $(1, x), (2, y)$, and $(3, x)$, we may write $f(1) = x$, $f(2) = y$, and $f(3) = x$. Most students should remember that last bit of notation from a course in intermediate algebra or precalculus. Sometimes we draw a picture ("mapping") of a function indicating which elements of the first set get mapped into which elements of the second set. Such a drawing is displayed in Figure 1.1 in which $f : X \to Y$, where $X = \{1, 2, 3, 4, 5\}$, $Y = \{a, b, c, d, e\}$, $f(1) = b$, $f(2) = c$, $f(3) = b$, $f(4) = e$, and $f(5) = a$.

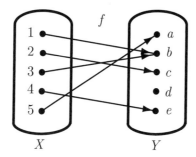

Figure 1.1 A function $f : X \to Y$.

A function f is **one-to-one** if the second coordinates are all distinct. A function (or relation) from A to B is **onto** if each element of B appears at least once as the second coordinate of some element of the function. Of the relations $C = \{(1, y), (2, x), (2, y)(3, x)\}$, $D = \{(2, y), (3, y)\}$, and $E = \{(1, x), (2, y), (3, x)\}$, the relations C and E are onto relations, but only E is an onto *function*. None are one-to-one. Note that for a finite set, we must first have $|B| \geq |A|$ for the possibility of a one-to-one function from A to B. A function that is both one-to-one and onto is called a **one-to-one onto function**. Such functions are very important in counting problems as well as in other mathematical problems and will be crucial for us in Section 2.3 when we study graph isomorphism. Some authors use alternate terminology, namely, *injection* for "one-to-one," *surjection* for "onto," and *bijection* for "one-to-one and onto."

Example 1.10 Let $X = \{\text{Al, Bob, Cara, Don}\}$, and let $Y = \{1, 2, 3, 4\}$. Find two different one-to-one onto functions from X to Y.

Solution
$f = \{(\text{Al},1), (\text{Bob},2), (\text{Cara},3), (\text{Don},4)\}$ and $g = \{(\text{Al},3), (\text{Bob},2), (\text{Cara},1), (\text{Don},4)\}$ are two such functions. We shall see in Section 1.3 that there are in fact 24 possible one-to-one functions from X to Y. Since X and Y have the same cardinality, all 24 of those functions are onto as well. ◇

Exercises 1.1

1. If x is a positive real number, what is the relationship between $\lfloor 3x \rfloor$ and $3\lfloor x \rfloor$? What about $\lceil 3x \rceil$ and $3\lceil x \rceil$?

2. Generalize the last exercise for $\lfloor nx \rfloor$ and $n\lfloor x \rfloor$, where n is a positive integer. Do the same for $\lceil nx \rceil$ and $n\lceil x \rceil$.

3. Show that $\lfloor x + y \rfloor \geq \lfloor x \rfloor + \lfloor y \rfloor$. Then use this to prove inequality (1.1) in this section.

4. Use one of the rounding functions to define $f(n)$ if the values of f when $n = 1, 2, 3, 4, 5, 6, 7, 8, 9$ are $1, 1, 1, 2, 2, 2, 3, 3, 3$, respectively. Notice that the function value jumps by 1 after each group of three integers, that is, $f = \{(1,1), (2,1), (3,1), (4,2), (5,2), (6,2), (7,3), (8,3), (9,3)\}$.

5. How far apart can $\lfloor 10x \rfloor$ and $10\lfloor x \rfloor$ be? In other words, what is the maximum value of $\lfloor 10x \rfloor - 10\lfloor x \rfloor$? *Hint:* First write x as $n+\varepsilon$, where n is the integer $\lfloor x \rfloor$ and ε satisfies $0 \leq \varepsilon < 1$. (For example, $6.7 = 6 + 0.7$, where $x = 6.7$, $n = 6$, and $\varepsilon = 0.7$.) Then observe that $10x = 10n + 10\varepsilon$. Now show that $\lfloor 10x \rfloor = 10n + \lfloor 10\varepsilon \rfloor$. Finally, figure out how big $\lfloor 10\varepsilon \rfloor$ can get.

6. Let us (temporarily) call all positive integers of the form $6k+1$ "awesome" numbers. Show that products of awesome numbers are awesome. Another way of stating this is that the set of numbers of the form $6k+1$ is **closed** under multiplication. Is the set closed under addition? Prove that the *sum* of seven awesome numbers is awesome. Under what condition is the sum of n awesome numbers awesome?

7. Give three examples of collections that are not well defined and, therefore, not sets.

8. Suggest how the descriptions of each of the following collections might be altered to describe well-defined sets.

(a) The collection of all elderly people

(b) The collection of all expensive cars

(c) The collection of all smart students in the class

Use a list format to describe each set in Exercises 9 through 12.

9. $\{x : x \in Z, x \text{ is a multiple of 3 or } x \text{ is a multiple of 7}\}$

10. $\{y : y \in N, \ y < 9\}$

11. $\{x : x \in N, \ 12 \leq x \leq 90\}$

12. $\{x : x = 5y + 1, \ y \in N, \ 7 \leq y < 13\}$

13. Suppose that $B = \{2, 4, 6, 8, ..., 90\}$. Express set B using a property format.

Express each set in Exercises 14–18 that uses a list format by using a property format, and vice versa.

14. $\{3, 4, 5, 6, 7, 8, 9, 10\}$

15. $\{x : x^3 = 64, \ x > 6\}$

16. $\{s : s \text{ is the name of a country in South America, } s \text{ begins with "U"}\}$

17. $\{x : x = 2k - 1, k \in Z\}$

18. $\{20, 40, 60, 80, \ldots\}$

19. Which of the sets in Exercises 14–18 are finite? Which are infinite?

20. Let $A = \{a, b, c\}$, $B = \{c, d\}$, $C = \{a, b\}$, and $D = \{c\}$. Which of the following are true? Explain each of your claims.

   a. $C \subseteq A$ b. $B \subseteq C$ c. $D \subseteq A$ d. $\emptyset \in B$ e. $C \in A$

21. Which of the following are subsets of the other sets? Are any of the sets equal to one another? Give proofs.

   a. $Y = \{y : y \text{ is prime}, y > 2\}$ b. $F = \{f : f = 2n - 1, n \in N\}$
   c. $D = \{x : x \in Z, x^2 < 0\}$ d. $W = \{1, 3, 5, 7, 9, \ldots\}$

22. Why is $\{3, 6, 15\}$ not a subset of $\{2, 15, 3, 10, 60\}$?

23. Is every subset of a finite set a finite set? Is every subset of an infinite set an infinite set? Explain.

24. Let $U = \{2, 4, 6, 8, 10, 12, 15\}$, $X = \{2, 6, 12\}$, $Y = \{6, 8, 12, 15\}$, and $Z = \{6, 8, 12, 15\}$. List the elements in each of the indicated sets.

    a. $X \cap Y$ b. $(Y \cup Z)'$ c. $(X \cup Y)' \cap Z'$ d. $((X \cup Y) \cup Y)'$

25. Prove $(A \cap B) \cap C = A \cap (B \cap C)$.

26. Prove $A \cup B = B \cup A$.

27. Prove $(A')' = A$.

28. Prove $A \cup (B \cap C) = (A \cup B) \cap (A \cup C)$.

29. Prove $(A \cap B)' = A' \cup B'$.

30. The elements of a set may be sets themselves. For example, consider $A = \{1, \{1\}, \{1, 2\}\}$. If $B = \{1\}$ and $C = \{2\}$, which of the following are true: $B \in A$, $B \subseteq A$, $C \in A$, $C \subseteq A$?

31. Give an example of a finite set S such that each element of S is also a subset of S. (*Hint*: Think small.)

32. Give an example of an infinite set S such that each element of S is also a subset of S.

33. Let $A = \{a, b, c\}$ and $B = \{c, d\}$. Find each of the following.

    a. $\wp(A)$ b. $A \times B$ c. $B \times A$ d. $A \times A$

34. Prove inequality (1.2), that is, prove that $\lceil 2x \rceil \leq 2 \lceil x \rceil$ for all real numbers x.

35. Given the function $f : Z \to N \cup \{0\}$ defined by $f(z) = z^2$,

    (a) Prove that $f(z)$ is an onto function.

    (b) Prove that $f(z)$ is not a one-to-one function.

36. Given the function $g : \Re \to \Re$ defined by $g(x) = x^2$, prove that $g(x)$ is not an onto function.

37. Given the function $h : \Re \to \Re$ defined by $h(x) = x^3$, prove that $h(x)$ is a one-to-one onto function.

38. Consider the ceiling function $c : \Re \to Z$ defined by $c(x) = \lceil x \rceil$.

    (a) Prove that $c(x)$ is an onto function.

    (b) Prove that $c(x)$ is not a one-to-one function.

1.2 Mathematical Induction

39. Suppose that X is a set where $|X| = n$ and $\wp(X)$ is the power set of X. Suppose that $f : \wp(X) \to Y$, where Y is some specific (yet unknown to us) set of integers. Give an example of a possible set Y such that

(a) $f(x)$ is an onto function, but not one-to-one.

(b) $f(x)$ is both one-to-one and onto.

1.2 Mathematical Induction

Mathematical induction is a powerful proof technique that can be used to verify equations, inequalities, and other mathematical truths. Let's begin with an important equation that gives a formula for the sum of the first n positive integers.

$$1 + 2 + 3 + \cdots + n = \frac{n(n+1)}{2} \tag{1.3}$$

Numbers of the form $\frac{n(n+1)}{2}$ are called **triangular numbers** and were originally studied in ancient Greece. While our interest in them stems from combinatorics, triangular numbers also play a prominent role in number theory. The nth triangular number is denoted t_n, so $t_n = \frac{n(n+1)}{2}$. The first few triangular numbers are 1, 3, 6, 10, 15, 21, and 28. In Example 1.11, $t_4 = 10$ and it is easy to see that such a triangle with 4 rows contains 10 dots.

Example 1.11 Find t_4.
Solution

$t_4 = 1 + 2 + 3 + 4 = \frac{4(4+1)}{2} = \frac{4(5)}{2} = 10$ ◇

The triangular numbers have many interesting properties. For example, the sum of two consecutive triangular numbers is a perfect square, that is, $t_{n-1} + t_n = n^2$. In Example 1.12, there are t_3 black dots and t_4 white dots that form the 4 by 4 square of 16 dots.

Example 1.12

○ ○ ○ ○
● ○ ○ ○
● ● ○ ○
● ● ● ○

$$\begin{aligned} t_{n-1} + t_n &= t_3 + t_4 \\ &= (1+2+3) + (1+2+3+4) = 6 + 10 = 16 = 4^2 \end{aligned} \quad \diamond$$

Karl Friedrich Gauss (1777–1855) was a German mathematician and physicist who made contributions in electricity and magnetism, in physics of fluids, and in astronomy. In statistics, Gauss did fundamental work on the normal distribution. In mathematics, Gauss developed the idea of complex numbers revolutionizing number theory and Euclidean geometry. In the nineteenth century, Gauss proved that every positive integer can be written as the sum of three or fewer triangular numbers! Equation (1.3) can be verified as follows. Let S denote the sum $1 + 2 + 3 + \cdots + n$. Then $S = 1 + 2 + 3 + \cdots + (n-1) + n$, and by reversing the terms on the right side of the equation we get an equivalent equation $S = n + (n-1) + \cdots + 3 + 2 + 1$. Now adding both sides of these equations yields

$$\begin{aligned} S &= 1 \;+\; 2 \;+\; \cdots \;+\; (n-1) \;+\; n \\ S &= n \;+\; (n-1) \;+\; \cdots \;+\; 2 \;+\; 1 \\ \hline 2S &= (n+1) + (n+1) + \cdots + (n+1) + (n+1) \end{aligned}$$

Note that there are n terms in the sum, so the last sum is $n(n+1)$. This quantity must be even because one of the two consecutive integers n and $n+1$ must be even. When we divide both sides of the last equation by 2, we obtain the result $S = \frac{n(n+1)}{2}$, which is an integer since $n(n+1)$ was even.

Example 1.13 Find the sum of the first hundred positive integers.

Solution
$1 + 2 + 3 + \cdots + 100 = \frac{100(100+1)}{2} = \frac{100(101)}{2} = 5050$ $\quad \diamond$

Equation (1.3), $1 + 2 + 3 + \cdots + n = \frac{n(n+1)}{2}$, can be more concisely expressed as

$$\sum_{i=1}^{n} i = \frac{n(n+1)}{2}$$

in summation notation. Σ, or *sigma*, is the Greek letter S (short for *sum*). As i runs through the values from 1 to n, the resulting numbers are added. Since many induction proofs involve summations, it is important to have a solid understanding of summation notation. Another example is

$$\sum_{i=3}^{7} i^2$$

read "the sum, as i goes from 3 to 7 of i squared," which means $3^2 + 4^2 + 5^2 + 6^2 + 7^2$. If we calculate this sum, we get $9 + 16 + 25 + 36 + 49 = 135$. Note that summation notation does not calculate the sum for us, it is simply a shorthand notation for the sum. The number 3 in the summation expression is the *initial value* of the variable i, and 7 is the *final value* of i that will be used in the function i^2. In many problems, particularly ones involving formulas, the final value is not a constant but an arbitrary integer n, as in the prior example.

Mathematical Induction

Now we discuss a powerful proof technique called **mathematical induction**. Suppose that we want to prove that a statement (proposition, equation, etc.) S_n is true for all positive integers. Mathematical induction consists of performing the following two steps.

Step 1 (Base step). Prove that the statement S_n is true for some starting value of n (usually $n = 0$ or $n = 1$).

Step 2 (Inductive step). *Assume* that the statement S_n is true when $n = k$ and then *prove* that it is also true when $n = k+1$. In proving the statement when $n = k+1$, we may use the assumption that S_n is true when $n = k$.

In this manner, the statement S_n is seen to be true for all positive integers. After all, it is true for $n = 1$ by step 1, so it must be true for $n = 2$ by step 2. Then since it is true for $n = 2$, it must be true for $n = 3$ by step 2. Then since it is true for $n = 3$, it must be true for $n = 4$, and so on for all positive integers. In step 2, the assumption that the statement S_n is true when $n = k$ is commonly called **the inductive hypothesis**. There we are assuming that the statement S_n is true for a *fixed* (that is, constant) value of the variable n. The challenge in step 2 is to prove that whenever S_n is true for the fixed value $n = k$, S_n is also true for the next value, namely, $n = k+1$.

Example 1.14 Using mathematical induction, prove that $7 + 13 + 20 + 27 + \cdots + (6n+1) = 3n^2 + 4n$, that is, $\sum_{i=1}^{n}(6i+1) = 3n^2 + 4n$, for all positive integers.

Solution

Step 1. The equation is certainly true when $n = 1$, since the left side is 7, while the right side, $3(1)^2 + 4(1)$, also equals 7.

Step 2. Now assume that the equation $\sum_{i=1}^{n}(6i+1) = 3n^2 + 4n$ is true for $n = k$, that is, $\sum_{i=1}^{k}(6i+1) = 3k^2 + 4k$.

We must show that equation is true when $n = k+1$. That is, we must prove that $\sum_{i=1}^{k+1}(6i+1) = 3(k+1)^2 + 4(k+1)$. Note that all that is done to write the last equation is to take the original summation in the example and replace each instance of n (there are three of them) by $k+1$. It is often useful to simplify the right side $[3(k+1)^2 + 4(k+1)$ in this example] of the equation. By expanding and combining like terms, we get

$$3(k+1)^2 + 4(k+1) = 3k^2 + 6k + 3 + 4k + 4 = 3k^2 + 10k + 7$$

We identify this modified form $3k^2 + 10k + 7$ of the right-hand side of the equation as our "goal". We then begin with the left side $\sum_{i=1}^{k+1}(6i+1) =$

$3(k+1)^2 + 4(k+1)$ and show that it is equal to the goal.

$$\sum_{i=1}^{k+1}(6i+1) = \underbrace{\sum_{i=1}^{k}(6i+1)}_{} + 6(k+1) + 1$$

$$\begin{aligned}&= (3k^2 + 4k) + (6(k+1) + 1) \quad \text{(by using step 2)}\\&= 3k^2 + 4k + 6k + 6 + 1\\&= 3k^2 + 10k + 7\end{aligned}$$

Since we have reached the goal, we can conclude, by the principle of mathematical induction, that the equation $\sum_{i=1}^{n}(6i+1) = 3n^2 + 4n$ is true for all natural numbers n. ◇

Note that after determining the goal in step 2 of Example 1.14, we explicitly separated out the last term $6(k+1)+1$ from the sum. Then by grouping the first k terms of the sum, we obtained $\sum_{i=1}^{k}(6i+1) + 6(k+1) + 1$. We were then able to make a replacement using the inductive hypothesis. That is typically what happens when doing mathematical induction proofs.

The proof given earlier to verify equation (1.3) employed a direct approach. Now let's prove the equation by using mathematical induction.

Example 1.15 Prove equation (1.3); that is, show that $1+2+3+\cdots+n = \frac{n(n+1)}{2}$ for all natural numbers n using mathematical induction.

Solution

Step 1. The equation is certainly true when $n=1$, since the left side is 1, while the right side, $\frac{1(2)}{2}$, also equals 1.

Step 2. Now assume that equation (1.3) is true for $n=k$, that is, assume that

$$1 + 2 + 3 + \cdots + k = \frac{k(k+1)}{2} \quad (1.4)$$

Now let's show that equation (1.3) is true when $n=k+1$. In other words, let's prove that $1 + 2 + 3 + \cdots + k + (k+1) = \frac{(k+1)[(k+1)+1]}{2} = \frac{(k+1)(k+2)}{2}$. So the "goal" is $\frac{(k+1)(k+2)}{2}$. . Thus we must prove that

$$1 + 2 + 3 + \cdots + k + (k+1) = \frac{(k+1)(k+2)}{2}. \quad (1.5)$$

Now replace the first k terms on the left side of (1.5) by the right side of (1.4). This gives us, after a bit of algebra, $1+2+3+\cdots+k+(k+1) = \frac{k(k+1)}{2} + (k+1)$. Factor out the common factor $(k+1)$ to get $(k+1)(\frac{k}{2}+1) = \frac{(k+1)(k+2)}{2}$. We have reached the goal. Thus, we conclude that $1+2+3+\cdots+n = \frac{n(n+1)}{2}$ for all natural numbers n. ◇

1.2 Mathematical Induction

It is nice, incidentally, to see two different proofs for the same fact. This is a common phenomenon in every branch and at every level of mathematics. The Pythagorean theorem, for example, has been proven in over 250 ways (See Loomis [1].) Note that "(see Loomis [1])" means see reference 1, a work by Loomis, at the end of this chapter. One of the great twentieth-century mathematicians, Paul Erdős (1913–1996), used to say that God has a "book" containing only the nicest proof of each theorem. If Erdős thought a proof of some theorem was simple, clever, and elegant, he would say, "That one is in the book!"

Before we consider another example, observe that in step 2 of an induction proof, we are using a hypothetical truth. We are not saying that we know the proposition to be true for $n = k$. We are, rather, *assuming* that it is and then showing that this would *imply* the truth of the proposition when $n = k+1$. This, in conjunction with the base step 1, proves the proposition for all positive integers. It is important to understand that the base step is crucial in this process. That is, we must first establish a "basis in fact" for making our assumption in step 2. To see why this is important, consider the following example.

Example 1.16 A careless student "proved" that the sum of the first n positive even integers is $n^2 + n - 2$, that is, $\sum_{i=1}^{n} 2i = n^2 + n - 2$ as follows. He assumed that $\sum_{i=1}^{k} 2i = k^2 + k - 2$. He then showed that when $n = k+1$, $\sum_{i=1}^{k+1} 2i = (k+1)^2 + (k+1) - 2$. He began by simplifying the right side to get $k^2 + 2k + 1 + k + 1 - 2 = k^2 + 3k$. His goal was $k^2 + 3k$. He then began with the left side of the equation and separated off the last term.

$$\sum_{i=1}^{k+1} 2i = \sum_{i=1}^{k} 2i + (2(k+1))$$

Using the inductive hypothesis, he replaced $\sum_{i=1}^{k} 2i$ by $k^2 + k - 2$ to get $k^2 + k - 2 + (2(k+1))$, which equals $k^2 + k - 2 + 2k + 2 = k^2 + 3k$. Since he reached the goal, he concluded that the formula $\sum_{i=1}^{n} 2i = n^2 + n - 2$ is true for all natural numbers n. What is wrong with the student's proof?

Solution
The student did not first verify the base case. If he had checked the base case, when $n = 1$, he would have found the left side to be $2(1) = 2$ and the right side to be $1^2 + 1 - 2 = 0$. The base case does not hold since $2 \neq 0$. Thus the statement

$$\sum_{i=1}^{n} 2i = n^2 + n - 2$$

is *not* true in general. ◇

Why does induction work? Imagine an infinite ladder with consecutive rungs labeled $1, 2, 3, \ldots$. Say you wish to climb the ladder and assume that you are given two permission slips to do so. The first allows you to get on the first rung, after which you can throw the first permission slip away since it is no longer needed. The second slip, which you better not throw away, gives you permission to advance from any rung to the next rung. This slip may be used as many times as you wish, and with it you can reach any higher rung (provided that you make it to the first rung).

Similarly, mathematical induction involves proving that a theorem is true for every integer from some initial point (the base point) and beyond. It requires a proof for the base case (usually $n = 1$), like the first permission slip, and a guarantee that the truth of the theorem travels from each integer to the next one (like the permission to climb the ladder, that is, to advance to the next rung).

Let's prove the following formula for the sum of the first n perfect squares (that is, numbers like $1^2 = 1$, $2^2 = 4$, $3^2 = 9$, $4^2 = 16$, $5^2 = 25$, and so on). Number theorists, by the way, have been fascinated by squares at least since the time of Pythagoras (540 b.c.). For example, Joseph Louis Lagrange (1736–1813), one of the greatest mathematicians of the eighteenth century, showed that every positive integer can be written as a sum of four or fewer perfect squares!

Example 1.17 Using mathematical induction, prove that

$$\sum_{i=1}^{n} i^2 = \frac{n(n+1)(2n+1)}{6} \tag{1.6}$$

is true for all positive integers.

Solution

Step 1. This formula is easily seen to be true for $n = 1$. The left side of the equation has the value $1^2 = 1$, while the right side becomes $\frac{1(1+1)(2+1)}{6} = \frac{1(2)(3)}{6} = 1$, so we get $1 = 1$.

Step 2. Now if formula (1.6) is assumed to be true when $n = k$, we get

$$\sum_{i=1}^{k} i^2 = \frac{k(k+1)(2k+1)}{6} \tag{1.7}$$

We must show that (1.6) is true when $n = k+1$. That is, we must prove that $\sum_{i=1}^{k+1} i^2 = \frac{(k+1)[(k+1)+1][2(k+1)+1]}{6} = \frac{(k+1)(k+2)(2k+3)}{6}$. Thus we must prove that

$$1^2 + 2^2 + 3^2 + \cdots + k^2 + (k+1)^2 = \frac{(k+1)(k+2)(2k+3)}{6} \tag{1.8}$$

Now replace the first k terms of the left side of (1.8) by the right side of (1.7). This gives us, after a bit of algebra, $[1^2 + 2^2 + 3^2 + \cdots + k^2] + (k+1)^2 = \frac{k(k+1)(2k+1)}{6} + (k+1)^2$. Now we factor out the common factor $(k+1)$ to

1.2 Mathematical Induction

get $(k+1)[\frac{k(2k+1)}{6} + \frac{6(k+1)}{6}] = \frac{(k+1)[k(2k+1)+6(k+1)]}{6} = \frac{(k+1)(2k^2+7k+6)}{6} = \frac{(k+1)(k+2)(2k+3)}{6}$. Again we have reached the goal, so the formula in (1.6) holds for all $n \in N$. ◇

Example 1.18 Using mathematical induction, prove that the product of n odd integers is an odd integer.

Solution
Since we need at least two numbers to form a product, we are only interested in values of n that are greater than 1. So we adjust our method by verifying the proposition for $n = 2$ in step 1 and proceeding as usual.

Step 1. The statement is certainly true when $n = 2$, since it was proved in Section 1.1, that is, $x_1 x_2$ is odd when x_1 and x_2 are both odd.

Step 2. Now assume that $x_1 x_2 \cdots x_k$ is odd when $x_1, x_2, \cdots, x_k$ are all odd. We must then show that $x_1 x_2 \cdots x_{k+1}$ is odd when $x_1, x_2, \cdots, x_{k+1}$ are all odd. Since $x_1 x_2 \cdots x_{k+1} = (x_1 x_2 \cdots x_k) x_{k+1}$ and the product in parentheses is odd by the assumption in step 2, $x_1 x_2 \cdots x_{k+1} = (x_1 x_2 \cdots x_k) x_{k+1}$ is the product of two odd integers, which was previously proven to be odd, and we are done. ◇

Example 1.19 Verify that $n < 3^n$ for all nonnegative integers $n = 0, 1, 2, 3, \cdots$ using mathematical induction.

Solution
In this problem, we are interested in integer values of n that are nonnegative. So we again adjust our method by verifying the proposition for $n = 0$ in step 1 and then proceeding as usual.

Step 1. The inequality is certainly true when $n = 0$, since $0 < 3^0 = 1$.

Step 2. Now assume that the inequality is true when $n = k$, that is, assume that $k < 3^k$ (the inductive hypothesis). We must show that the inequality is true when $n = k + 1$. In other words, we must prove that $k + 1 < 3^{k+1}$. It helps to think of 3^{k+1}, along with the "<" that precedes it, as our "goal" here. Replacing the value k on the left side of the last inequality using the inductive hypothesis, we get $k + 1 < 3^k + 1$. Aiming toward our goal of $< 3^{k+1}$, we want to modify the right side of $k + 1 < 3^k + 1$ to look like our goal. Noting that $1 < 2(3^k)$, we obtain $k+1 < 3^k+1 < 3^k+2(3^k) = 3(3^k) = 3^{k+1}$. If we follow this expression from left to right, we see that we have shown that $k + 1 < 3^{k+1}$, that is, we have achieved our goal. Thus, we conclude that $n < 3^n$ for all nonnegative integers $n = 0, 1, 2, 3, \cdots$. ◇

Example 1.20 Prove that if A is a finite set of cardinality n, then the power set $\wp(A)$ has cardinality 2^n for all nonnegative integers $n = 0, 1, 2, \ldots$ using mathematical induction.

Solution
Step 1. The statement is certainly true when $n = 0$, since the power set of the empty set is $\{\emptyset\}$, which has cardinality $1 = 2^0$.

Step 2. Now assume that the statement is true when $n = k$, that is, assume that the cardinality of the power set of a set with k elements is 2^k (this is the inductive hypothesis). Now we must prove that a set with $k + 1$ elements has a power set of cardinality 2^{k+1}. Let $A = \{a_1, a_2, \ldots, a_k, a_{k+1}\}$. In previous examples, we split off the last term of a sum; now we split off the last element a_{k+1} of a set, when the elements are listed in some specific order. Let $B = \{a_1, a_2, \ldots, a_k\}$. Then $A = B \cup \{a_{k+1}\}$. Since B contains k elements, the inductive hypothesis implies that $\wp(B)$ contains 2^k sets. For each set X in $\wp(B)$, there are precisely two sets in $\wp(A)$, namely, X and $X \cup \{a_{k+1}\}$. So there are twice as many sets in $\wp(A)$ as there are in $\wp(B)$, yielding $2(2^k) = 2^{k+1}$ sets in $\wp(A)$. Thus the cardinality of the power set for a set of n elements is 2^n for all nonnegative integers $n = 0, 1, 2, \ldots$. ◇

As we have seen, mathematical induction can be applied in a wide variety of interesting problems. Here's one more.

Example 1.21 Using mathematical induction, prove that

$$(A_1 \cup A_2 \cup \cdots \cup A_n)' = A_1' \cap A_2' \cap \cdots \cap A_n' \tag{1.9}$$

for all positive integers greater than 1, that is, for $n = 2, 3, 4, \ldots$. This is called the generalized DeMorgan's law.

Solution
Step 1. Equation (1.9) is certainly true when $n = 2$ since $(A_1 \cup A_2)' = A_1' \cap A_2'$ is DeMorgan's law (see Note 1.5 in Section 1.1).
Step 2. Now assume that (1.9) is true for $n = k$, that is, assume that

$$(A_1 \cup A_2 \cup \cdots \cup A_k)' = A_1' \cap A_2' \cap \cdots \cap A_k' \tag{1.10}$$

We must show that (1.9) is true when $n = k+1$, that is, we must prove that

$$(A_1 \cup A_2 \cup \cdots \cup A_k \cup A_{k+1})' = A_1' \cap A_2' \cap \cdots \cap A_k' \cap A_{k+1}'$$

By using the associative property and step 1, we obtain
$$\begin{aligned}(A_1 \cup A_2 \cup \cdots \cup A_k \cup A_{k+1})' &= ((A_1 \cup A_2 \cup \cdots \cup A_k) \cup A_{k+1})' \\ &= (A_1 \cup A_2 \cup \cdots \cup A_k)' \cap A_{k+1}'\end{aligned}$$
This becomes, using step 2, $A_1' \cap A_2' \cap \cdots \cap A_k' \cap A_{k+1}'$, and we are done. ◇

Example 1.22 Using mathematical induction, show that $n^3 - n$ is exactly divisible by 3 for all nonnegative integers $n = 0, 1, 2, 3, \ldots$.

Solution
Step 1. The statement is certainly true when $n = 0$, since $0^3 - 0 = 0$, which is exactly divisible by 3.
Step 2. Now assume that the statement is true for $n = k$, that is, assume that $k^3 - k$ is exactly divisible by 3. Now let's show that $n^3 - n$ is exactly divisible by 3 when $n = k + 1$, that is, $(k+1)^3 - (k+1)$ is exactly divisible by 3. So we must prove that $(k+1)^3 - (k+1) = k^3 + 3k^2 + 3k + 1 - k - 1 =$

$(k^3 - k) + (3k^2 + 3k) = (k^3 - k) + 3(k^2 + k)$ is exactly divisible by 3. But $(k^3 - k)$ is exactly divisible by 3 by our assumption in step 2, and obviously $3(k^2 + k)$ is exactly divisible by 3. Thus $n^3 - n$ is exactly divisible by 3 for all nonnegative integers $n = 0, 1, 2, 3, \ldots$. ◇

It is interesting to note that the statement proved in Example 1.22 is equivalent to saying that the product of any three consecutive integers is divisible by 3. To see why this is so, let the integers be $n - 1$, n, and $n + 1$. Then $(n - 1)(n)(n + 1) = n^3 - n$, and we have shown that 3 divides evenly into this quantity.

Strong Induction

Sometimes it is helpful to use an inductive hypothesis that appears somewhat stronger than that which we have been using. To prove a statement S_n by **strong induction**, we complete the following two steps:

Step 1: Prove that S_n is true for some initial integer value k.
Step 2: Assume that for some integer $t \geq k$, $S_k, S_{k+1}, S_{k+2}, \ldots, S_t$ are all true (this is the inductive hypothesis). Then prove that S_{t+1} is true.

By completing these two steps, we are assured that the statement S_n is true for all integers $n \geq k$. Although on the surface, this form of induction may appear to be stronger than that which we have been using (and the name seems to hint that it is), strong induction is actually logically equivalent to the form we have been using. This means that if we assume the original form, we can prove the strong form, and if we assume the strong form, we can prove the original form.

Exercises 1.2
Prove each of the following using mathematical induction. Unless otherwise stated, n is a positive integer.

1. The sum of the first n positive odd integers is n^2, that is, $1 + 3 + 5 + \cdots + (2n - 1) = n^2$, for all positive integers $n = 1, 2, 3, \ldots$.

2. The sum of the nonnegative powers of 2 up to n is $2^{n+1} - 1$, that is, $2^0 + 2^1 + 2^2 + \cdots + 2^n = 2^{n+1} - 1$ for all nonnegative integers $n = 0, 1, 2, 3, \ldots$.

3. $n < 2^n$ for all nonnegative integers $n = 0, 1, 2, 3, \ldots$.

4. $n^3 - n$ is exactly divisible by 6 for all nonnegative integers $n = 0, 1, 2, 3, \ldots$.

5. $2 + 5 + 8 + \cdots + (3n - 1) = \dfrac{n(3n + 1)}{2}$

6. $3 + 7 + 11 + \cdots + (4n-1) = n(2n+1)$

7. The sum of the first n cubes is $(t_n)^2$, that is, $1^3 + 2^3 + 3^3 + \cdots + n^3 = (t_n)^2 = \left(\frac{n(n+1)}{2}\right)^2$ for all positive integers $n = 1, 2, 3, \ldots$. Then show, using a little algebra, that $(t_n)^2 - (t_{n-1})^2 = n^3$, that is, the difference of the squares of two consecutive triangular numbers is a perfect cube.

8. $(A_1 \cap A_2 \cap \cdots \cap A_n)' = A_1' \cup A_2' \cup \cdots \cup A_n'$ for all positive integers greater than 1, that is, $n = 2, 3, 4, \ldots$ (Generalized DeMorgan law).

9. $2^2 + 4^2 + 6^2 + 8^2 + \cdots + (2n)^2 = \dfrac{2n(n+1)(2n+1)}{3}$

10. $1 + t + t^2 + t^3 + \cdots + t^{n-1} = \dfrac{1-t^n}{1-t}$, where $t \neq 1$

11. $\sum_{i=1}^{n} i(i+1) = \dfrac{n(n+1)(n+2)}{3}$

12. $1^2 + 3^2 + 5^2 + 7^2 + \cdots + (2n-1)^2 = \dfrac{n(2n-1)(2n+1)}{3}$

13. $\dfrac{1}{1 \cdot 2} + \dfrac{1}{2 \cdot 3} + \dfrac{1}{3 \cdot 4} + \dfrac{1}{4 \cdot 5} + \cdots + \dfrac{1}{n(n+1)} = \dfrac{n}{n+1}$

14. $\dfrac{1}{1 \cdot 3} + \dfrac{1}{3 \cdot 5} + \dfrac{1}{5 \cdot 7} + \dfrac{1}{7 \cdot 9} + \cdots + \dfrac{1}{(2n-1)(2n+1)} = \dfrac{n}{2n+1}$

15. $\sum_{i=1}^{n} (2i-1)^3 = n^2(2n^2 - 1)$

16. $6n < n^2$ for $n \geq 5$

17. $n^2 - 2n \geq n + 3$ for $n \geq 4$

18. $n^3 + 2n$ is exactly divisible by 3 for $n \geq 1$.

19. $7^n - 2^n$ is exactly divisible by 5 for $n \geq 1$.

20. $3^{2n} - 1$ is exactly divisible by 8 for $n \geq 1$.

21. $11^n - 6$ is exactly divisible by 5 for $n \geq 1$.

1.3 Permutations and Combinations

Counting is a very important activity in combinatorial mathematics. In graph theory, we are often concerned with determining the number of objects of a given type in particular graphs. In computer applications, determining the efficiency of algorithms requires some skill with counting. In this section, we shall examine some of the counting tools of combinatorics.

Permutations

A **permutation** is an arrangement of objects into a specific order. Suppose you were planning to read three different books during the summer. If we call the books A, B, and C, we have the six permutations ABC, ACB, BAC, BCA, CAB, and CBA of the orders in which you could read the books.

An alternative approach is illuminating. The letter A can be listed exactly once. If we add a B to this list, we get AB and BA, that is, we have a list of $2 \times 1 = 2$ permutations of A and B. If we now add a C, we use each permutation in the previous list (AB and BA) to generate permutations of A, B, and C as follows. Begin with AB and insert a C before AB (giving CAB), between A and B (giving ACB), and after AB (giving ABC). Then do the same to BA, yielding CBA, BCA, and BAC. This brings us to six permutations of A, B, and C. It also explains how we get the total of six. There are three ways to insert C into the 2×1 permutations of A and B, so we multiply by 3 to get $3 \times 2 \times 1 = 6$.

Now when we add D to the process, each permutation of A, B, and C can have D inserted in four ways (for example, ABC extends to $DABC$, $ADBC$, $ABDC$, and $ABCD$), in which case we multiply the number of permutations of A, B, and C by 4, yielding $4 \times 3 \times 2 \times 1 = 24$.

Using mathematical induction, we can extend this idea to any number of distinct letters. Imagine n consecutive slots in which we are to insert n distinct letters. The first slot can be filled in n ways. As the chosen letter selected for the first slot is no longer eligible for selection, the next slot can be filled in only $(n-1)$ ways. Continuing this line of thought, the third slot can be filled in only $(n-2)$ ways, and so on. These numbers are linked by the following fundamental counting principle.

> **Note 1.6 (The Multiplication Principle)**
> Suppose that a task consists of a sequence of n independent steps, where the first step can be performed in a_1 ways, the second step can be performed in a_2 ways, the third step can be performed in a_3 ways, and so on. Then the whole procedure can be completed in $a_1 \cdot a_2 \cdot a_3 \cdots a_n$ ways.

Example 1.23 Suppose that in making a car purchase, we can choose from three engine types, four model types, and three packages, and that each car comes in eight possible colors. How many different choices are available?

Solution
We have $a_1 = 3$, $a_2 = 4$, $a_3 = 3$, and $a_4 = 8$. By the multiplication principle, there are $a_1 \cdot a_2 \cdot a_3 \cdot a_4 = 3 \cdot 4 \cdot 3 \cdot 8 = 288$ ways in which we could make a purchase. $\diamond$

Incidentally, the multiplication principle itself is actually established by mathematical induction. Now back to our problem. Based on the multiplication principle, to find the number of ways that we can insert the n distinct letters into the n consecutive slots, we form the product of the numbers n, $(n-1)$, $(n-2)$, ..., 1 to get $n(n-1)(n-2) \cdots 1$ ways. This quantity comes up frequently in mathematics. Perhaps you have already encountered it. Because of its frequency of occurrence, the product is given a special name $n!$ ("n **factorial**"). That is,

Formula 1.1 $n! = n(n-1)(n-2) \cdots 1$

This yields the following basic result.

Note 1.7   The number of permutations of n distinct objects is $n!$.

The quantity $0!$ is defined to be 1. To help us understand the unusual fact that $0! = 1$, think about what $n!$ is counting. It is the number of ways of arranging n distinct object. There is just *one* way to arrange no objects—do nothing.

If we know the value of a particular factorial number, the next one is easy to get. For example, we already know that $5! = 120$. So to get $6!$, we simply multiply that quantity by 6. That is, $6! = 6 \cdot 5! = 6(120) = 720$. This makes sense because $6! = 6 \cdot 5 \cdot 4 \cdot 3 \cdot 2 \cdot 1 = 6 \cdot (5 \cdot 4 \cdot 3 \cdot 2 \cdot 1) = 6 \cdot 5!$. More generally, we have the following: For any natural number n, $(n+1)! = (n+1) \cdot n!$. This gives us a recursive definition of n factorial, once we define $1! = 1$. It says that the next factorial, that is, $(n+1)!$, is $(n+1) \cdot$ ["the previous factorial"]= $(n+1) \cdot n!$. So after we know that $1! = 1$, we get $2!$ by multiplying the previous factorial by 2, yielding $2 \cdot 1$. To get $3!$, we multiply the previous factorial by 3, yielding $3 \cdot 2 \cdot 1$, and so on.

Example 1.24 Prove the result in Note 1.7 by mathematical induction, that is, prove that the number of permutations of n distinct objects is $n!$ for all positive integers n.

Solution
When $n = 1$, there is just $1! = 1$ permutation of 1 object. This establishes the base case. Thus assume when $n = k$ that there are $k!$ permutations of

1.3 Permutations and Combinations

k distinct objects (this is the inductive hypothesis). We then must prove that when $n = k + 1$, there are $(k + 1)!$ permutations of $k + 1$ distinct objects. Thus suppose that $n = k + 1$. Then there are $k + 1$ possible choices of which object to place in position 1. Then we must arrange the remaining k distinct objects in the next k positions. But that is simply a permutation of k distinct objects, and by the inductive hypothesis, there are $k!$ such permutations. Thus, by the multiplication principle there are $(k + 1) \cdot k! = (k + 1)!$ permutations of $k + 1$ distinct objects. Therefore, by the principle of mathematical induction, the number of permutations of n distinct objects is $n!$ for all positive integers n. ◇

It is fortunate that scientific calculators have a factorial key, generally indicated as either $n!$ or $x!$. It is usually above some other key, so access requires the "shift" or "2nd function" key. So to calculate 7! on a calculator where $x!$ is above the number 3, we hit 7 $\boxed{\text{shift}}$ $\boxed{\substack{x! \\ 3}}$. Now we can do certain counting problems easily. Not only do we know how to determine the answer as a factorial, we can also calculate the numerical value instantly. Suppose we want to know the number of ways in which 10 rock enthusiasts can stand on line single file to buy tickets to a concert. We calculate 10! and are shocked to get an answer exceeding 3 million. Yes, 10! = 3,628,800, so there are 3,628,800 ways in which the rock enthusiasts could stand on line single file. In how many ways can we arrange five distinct math books on a shelf? The answer, in a flash, is 5!, which is 120. In how many ways can we arrange no books on a shelf? There is one way, namely, do nothing. Again, there's 0! = 1.

Now let's consider a slightly more challenging problem that will require using both the multiplication principle and permutations in a single problem.

Example 1.25 In how many ways can we arrange four mathematics books, five chemistry books, and six psychology books next to one another on a shelf if the books are all distinct and books of the same subject must remain together.

Solution

First arrange the subjects, that is, decide which subject's books will go toward the left on the shelf, which ones in the middle, and which ones toward the right, and then arrange the books within each subject. There are three subjects, so we can arrange them in 3! = 6 ways. Then we can arrange the four mathematics books in 4! = 24 ways, the five chemistry books in 5! = 120 ways, and the six psychology books in 6! = 720 ways. Since the process of arranging the books according to the given condition consists of four steps done in succession, the multiplication principle implies that the total number of arrangements is $6 \cdot 24 \cdot 120 \cdot 720 = 12{,}441{,}600$. ◇

Combinations

In each of the counting problems so far, the order of the items mattered. That is not always the case. A **combination** is an unordered selection or subset of items from a given set of items. To get an understanding of combinations, let us consider the following two related problems.

1. Given n distinct objects, in how many ways can we select k of them where the order of selection matters? An example is the selection of a president, vice president, and treasurer of a club with 10 members. This entails selecting an *ordered* trio of officials. The slate Joe for president, Susan for vice president, and Tom for treasurer is different from the slate Susan for president, Tom for vice president, and Joe for treasurer. It does not suffice merely to list the three candidates—we must *order* them. ◇

2. Given n distinct objects, in how many ways can we select k of them where the order does not matter? If you have three extra tickets to a football game and you have 10 friends, in how many ways can you select the lucky trio of friends who will receive free tickets? Here, order does not matter. As another example, the choice Newton, Archimedes, and Euclid is the same as the choice Euclid, Newton, and Archimedes. Either group represents the same trio of individuals named after famous mathematicians. ◇

Problem 1 involves *permutations* while Problem 2 involves *combinations*. The answers are related. They both begin with the first k factors of $n!$, namely, $n \cdot (n-1) \cdot (n-2) \cdots (n-k+1)$. This is because there are n ways to select the first object, $(n-1)$ ways to select the second object, and so on until the last choice (the kth object), which can be selected in $(n-k+1)$ ways. It is important to understand this last quantity. It is tempting to write $(n-k)$ instead. To see that $(n-k+1)$ is correct, start counting backward from, say, 10 and stop at the fourth number. The count is 10, 9, 8, 7—and the fourth number is 7. This is $10-4+1$. Similarly, the *first* number, n, can be written $n-0$. The *second* number is $n-1$, and the *third* number is $n-2$. Notice the staggering of the number after the minus sign. It is one less than the number that counts the factor's place in the product. So the kth number is not $n-k$. It is $n-(k-1)$, which equals $n-k+1$. Of course, when $k = n$, the final factor is $n-(n-1) = n-n+1 = 1$, that is, we simply get $n!$, as we would expect. Summarizing the ideas just discussed, we have the following:

> **Formula 1.2** The number of permutations of k out of n distinct objects is $\boldsymbol{n \cdot (n-1) \cdot (n-2) \cdots (n-k+1)}$.

1.3 Permutations and Combinations

The expression $n \cdot (n-1) \cdot (n-2) \cdots (n-k+1)$ gives too large a number to be the answer to the second problem, where order does not matter. To see this, consider the number of combinations of three objects from among seven distinct objects.

Example 1.26 Consider a baseball division with seven teams. At the end of the regular season, the top three teams in the division advance to the playoffs. How many possible outcomes are there for the three teams that will advance?

Solution

Let the seven teams be labeled by the letters A through G. The three-letter choice ACD, for example, should be counted just once, not six times as it would be for permutations. The order of the top three teams is not important here, just *which* three teams will advance. So for the six different permutations ACD, ADC, CDA, CAD, DCA, and DAC, we count just the one subset $\{A, C, D\}$ of teams that will advance. Similarly, for any other possible collection, such as CEF, we are just interested in the one subset $\{C, E, F\}$, not the six ways in which they could have placed first, second, and third. Then we must divide by 6—the number of permutations of three objects. Thus, we have $7 \cdot 6 \cdot 5 = 210$ ways to permute three out of seven objects, but since the order is unimportant, we must divide 210 by 6 to get 35, which is the number of possible outcomes for the three teams that will advance. ◊

The solution to Example 1.26 suggests the following procedure for the number of combinations of k objects chosen from among n distinct objects. Divide the product $n \cdot (n-1) \cdot (n-2) \cdots (n-k+1)$ by $k!$. In symbols, we have the following:

$$\text{Formula 1.3} \quad \binom{n}{k} = \frac{n \cdot (n-1) \cdot (n-2) \cdots (n-k+1)}{k!}$$

The strange-looking symbol on the left of Formula 1.3 is pronounced "n choose k." If we multiply the numerator and denominator of the fraction on the right side of Formula 1.3 by $(n-k)!$, we obtain the alternative formula

$$\text{Formula 1.4} \quad \binom{n}{k} = \frac{n!}{k!(n-k)!}$$

There is another way to understand the connection between permutations and combinations. Let's denote the number of permutations of k out of n distinct objects by $P(n, k)$. So $P(n, k) = n \cdot (n-1) \cdot (n-2) \cdots (n-k+1)$. Then we can form such a permutation in two successive steps: First, choose the k objects [we can do this in $\binom{n}{k}$ ways], and then arrange the k objects selected (we can do that in $k!$ ways). Then by the multiplication principle, $P(n,k) = \binom{n}{k} \cdot k!$. Equivalently, if we divide both sides of this equation by $k!$, we get $\frac{P(n,k)}{k!} = \binom{n}{k}$, which is equivalent to Formula 1.4.

The numbers k and $n-k$ in the denominator in Formula 1.4 add up to n. If we replace each k by $n-k$ in Formula 1.4, we obtain

> **Formula 1.5** $\qquad \binom{n}{n-k} = \binom{n}{k}$

This is logical. After all, choosing a meal of k distinct dishes from a menu of n distinct dishes is equivalent to indicating the $n-k$ dishes not chosen. Recalling Example 1.26, three teams out of the seven in the division advance to the playoffs. That means that four teams do not advance. Each choice of three teams that advance determines a unique collection of four teams that do not. So the number of ways of choosing three teams to advance is equal to the number of ways of choosing four teams to pack their bags and go home. That is, we have $\binom{7}{3} = \binom{7}{4}$.

An added bonus from Formula 1.5 is that to evaluate a quantity such as $\binom{100}{97}$, it is much easier to evaluate $\binom{100}{3} = \frac{100 \cdot 99 \cdot 98}{3 \cdot 2 \cdot 1}$. However, scientific calculators often have keys that will enable us to calculate $P(n,k)$ and $\binom{n}{k}$. Suppose that on a given calculator, $P(n,k)$ and $\binom{n}{k}$ are listed as ${}_nP_r$ and ${}_nC_r$, respectively. Suppose further that they appear above 1 and 2, respectively. Then on such a calculator we would calculate $P(10,3)$ by entering 10 $\boxed{\text{shift}}$ $\boxed{\overset{nPr}{1}}$ 3 $\boxed{=}$. Find where ${}_nP_r$ and ${}_nC_r$ are located on your own calculator and try it. You should get 720. This means that there are 720 permutations of three out of ten distinct objects. Similarly, to calculate $\binom{9}{4}$, enter 9 $\boxed{\text{shift}}$ $\boxed{\overset{nCr}{2}}$ 4 $\boxed{=}$. Try it. This time you should get 126. Thus there are 126 ways to select four distinct objects from a set of nine objects. Equivalently, a nine-element set has 126 four-element subsets.

Example 1.27 Determine the number of different five-member committees that can be formed from a set of 14 people.

Solution
Here we are simply choosing a five-element subset from a 14-element set. We can do this in $\binom{14}{5} = \frac{14!}{5! \cdot 9!} = \frac{14 \cdot 13 \cdot 12 \cdot 11 \cdot 10}{5 \cdot 4 \cdot 3 \cdot 2 \cdot 1}$ ways, which reduces to $14 \cdot 13 \cdot 11 = 2002$ ways. $\diamond$

Observe that if we calculate "n choose n" using Formula 1.3, we get $\binom{n}{n} = \frac{n!}{n! \cdot 0!}$. Since common sense dictates that "n choose n" must be 1— there is only one way to choose all n objects—it follows that the 0! in the denominator must be 1 and is defined to be so, as mentioned earlier. In

1.3 Permutations and Combinations

a similar manner, we find that $\binom{n}{0} = 1$. Another useful observation is that $\binom{n}{1} = n$. This could, of course, be verified without the help of Formula 1.3 by using common sense. There are n ways to choose one item among n distinct items; either choose the first, or the second, or the third,..., or the nth.

Exercises 1.3

1. How many permutations using two distinct elements from a, b, c, d, e are there? List them.

2. Calculate each of the following:

   a. 7!   b. $P(10, 4)$ c. $P(9, 3)$

3. Evaluate each quantity.

   a. $\binom{80}{76}$ b. $\binom{8}{2}$ c. $\binom{7}{3}$ c. $\dfrac{\binom{9}{2} \cdot \binom{14}{3}}{\binom{23}{5}}$

4. How many committees of five can be formed from 14 people?

5. Write the following using factorials:

   a. $19 \cdot 18 \cdot 17 \cdot 16$ b. $40 \cdot 39 \cdot 38$ c. $63 \cdot 62 \cdot 61 \cdots 35$

6. Simplify each of the following:

   a. $13!/9!$ b. $5! \cdot 10!/12!$ c. $P(10, 4)/4!$

7. The nine players on a baseball team will each bat in succession. In how many orders can the team's manager arrange them (that arrangement is known as the *batting order*)?

8. Imagine that you are positioned at the origin of the (x, y)-plane and wish to travel to the point (m, n) in the first quadrant in a sequence of special moves. Each such move consists of moving either one unit to the right or one unit up. Thus, if you are at the point (a, b), where a and b are positive integers, you may move to either $(a + 1, b)$ or $(a, b + 1)$. It should be clear that it will take a total of $m + n$ moves to get from $(0, 0)$ to (m, n). Explain why there are exactly $\binom{m+n}{m}$ ways in which to do this.

9. An ice cream parlor serves a sundae for which you can choose one of 20 different flavors, with one of seven different toppings. You can then choose to have whipped cream or not. How many different sundaes are possible?

10. In choosing a new car, you can select one of six different colors, three different transmissions, three types of interiors, and either get air conditioning or not. How many different car designs are available?

In Exercises 11–16, determine the number of possible license plates there are with the given conditions.

11. Two letters followed by four digits

12. Three letters followed by three digits

13. Four letters followed by two digits

14. Either one or two letters followed by three digits

15. Up to three letters followed by up to three digits. The plate might use no letters or it might use no digits, but it must contain at least one character.

16. Two distinct letters followed by three distinct digits

17. A witness to a hit-and-run accident tried to memorize the license plate of the car. She remembers that there were two letters followed by three numbers. The first letter was W and the second was either C, D, O, or Q. The last two numbers were both 7. How many different plate numbers would the police have to check given this information?

18. A briefcase has a combination lock with three labeled wheels, each labeled $0, 1, 2, 3, \cdots, 9$.

    (a) How many combinations are possible? (Be careful; a lock combination is not the same as a mathematical combination because order matters.)

    (b) Suppose that you forget the combination and try to discover it. Assuming that you can try one combination every five seconds, how long will it take you to find the combination if you try every possible combination before finding the correct one?

19. A midterm exam consists of four computer-generated essay questions. The computer extracts items from a list of comparable questions in a test bank. Suppose that there are six possible versions of the first questions, seven of the second, three of the third, and five of the fourth. How many different versions of the test are possible?

20. Suppose you plan to buy a stereo and a bookcase this year. You visit many stores and see six stereos at the first store, five at the second, ten

1.3 Permutations and Combinations

at the third, and seven at the fourth store you visit. You then visit two furniture stores and see five bookcases at one and three at the other. How many different stereo-bookcase purchases can you make?

21. A certain large city now has five area codes. A phone number consists of the area code followed by seven digits, the first of which cannot be zero or one. How many phone numbers are possible for that city?

22. If you decide to purchase one hundred shares of one of seven different stocks and fifty shares of one of five different bonds. How many stock-bond purchases are possible?

23. How many four-digit odd integers have all digits distinct (the first digit cannot be zero)?

24. A small country has license plates consisting of two letters followed by four nonzero digits. The country is switching to a system that will use three letters followed by three nonzero digits. How many additional license plate codes are possible with the new system than with the old?

25. How many n-bit binary strings have consecutive bits always distinct (in a binary string the only possible digits, called bits, are 0 and 1)?

26. How many six-letter strings from $\{a, b, c, \ldots, z\}$ contain exactly one of the vowels a, e, i, o, u (other letters may be repeated)?

27. How many three-digit numbers contain only distinct odd digits?

28. Two arrangements of elements around a circle are equivalent if one can be obtained from the other by rotation. Determine the number of possible arrangements of n distinct elements around a circle.

29. Determine the number of license plate codes consisting of three letters followed by three digits that contain a repeated letter or a repeated digit (or both).

30. Simplify $\frac{(n+2)!}{n!}$, $n \geq 0$.

31. Simplify each of the following:
    a. $P(n, n-2)$ b. $P(n+1, n-1)$ c. $P(n, 3) - P(n, 2)$

32. Find all $n \in N$ for which $P(n, 2) = 90$.

33. Find all $n \in N$ for which $P(n+1, 3) = n!$.

34. Show that $P(n, r+1) = nP(n-1, r)$ for all $n, r \in N$.

35. Find all $n \in N$ for which $7P(n,3) - 22P(n,2) = -14n$.

36. I plan to buy two ties from a group of eight and five shirts from a group of twelve. How many different selections are possible?

37. In the game of bridge, each player is dealt thirteen cards from the deck of 52 cards. How many different bridge hands are possible?

38. There are eight Democrats and four Republicans on a senate committee. They must form a subcommittee of five members for a project. How many possible subcommittees are there containing

  (a) exactly three Democrats?

  (b) at least one Republican?

39. A test has three parts, each containing six questions. You must answer five questions from part 1, three questions from part 2, and four from part 3. How many choices do you have?

40. How many boards of directors of six members can be selected from a group of 23 candidates?

41. A jar contains 10 red, 12 white, and 13 blue balls, all of which are distinct. In how many ways can seven balls be selected so there are at least two balls of each color?

42. There are 15 qualified applicants for five identical clerical positions in a company. If each position is filled by a different one of the applicants, how many different groups of five new employees are possible?

43. Repeat Exercise 42 if the clerical positions are not identical but instead all different.

44. In bowling there are 10 pins arranged in a triangular pattern, as in Figure 1.2. The bowler rolls a ball in hope of knocking over the pins. If not all pins are knocked over, the bowler gets a second try. One of the hardest second shots, called the 7-10 split, has only the 7 and 10 pins standing after the first roll.

  (a) How many different pin combinations are there with exactly two pins standing after the first roll?

  (b) How many different pin combinations are there with at most three pins standing after the first roll?

1.4 Combinatorial Identities

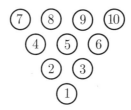

Figure 1.2

45. Prove that if $n \geq 4$, then $n! \geq 2^n$.

46. The fundamental theorem of arithmetic states that a positive integer can be factored down to primes in only one way, except for order. Thus, if the prime factors of n are p, q, r, s, for example, then we have $n = p^a q^b r^c s^d$, where a, b, c, d are each positive integers. Using the multiplication principle, show that the number of factors of n is given by $(a+1)(b+1)(c+1)(d+1)$. Generalize this to apply to any positive integer n.

47. How many ordered five-letter sequences can be made using the letters A, A, B, C, and D? An example of an ordered five-letter sequence using these letters is $BADAC$.

48. Generalize this problem to the case in which you are given n symbols and k of them are the same while the rest are all distinct. How many ordered n-letter sequences are there?

49. In how many different ways can we seat seven people at a circular table if all we care about is who is seated to each person's left and right?

50. Let A be a finite set. If $|A| = n$, determine

   (a) $|A \times A|$

   (b) The number of relations on A

51. Let A and B be finite sets. If $|A| = n$ and $|B| = m$, where $m \geq n$, determine

   (a) the number of functions from A to B

   (b) the number of one-to-one functions from A to B

1.4 Pascal's Triangle and Combinatorial Identities

There is a very ancient array of numbers called Pascal's triangle, named after the mathematician Blaise Pascal (1623–1662), who did pioneer work in probability theory as well as adding significantly to other areas of mathematics. Pascal's name is attached to this triangle because of the extensive work he did exploring its properties. However, the triangle seems to have been discovered independently by the Persians and the Chinese in the eleventh century. In fact, it is displayed in a Chinese book of 1303 a.d. by Chu Shih-Chieh, titled *Precious Mirror of the Four Elements.* Rows zero through seven of Pascal's triangle are shown in Figure 1.3. There are $n+1$ entries in row n, and we call the successive entries in row n, entries zero through n.

```
                        1                              Row 0
                     1     1                           Row 1
                  1     2     1                        Row 2
               1     3     3     1                     Row 3
            1     4     6     4     1                  Row 4
         1     5    10    10     5     1               Row 5
      1     6    [15]   [20]   15     6     1          Row 6
   1     7    21    [35]   35    21     7     1        Row 7
   .     .     .     .     .     .     .     .    .
```

Figure 1.3 Pascal's triangle (up to row 7).

Each entry, except for the 1's at the beginning and end of each row, is the sum of the two entries immediately above it to the right and left. Thus the first 35 in the seventh row is the sum of the 15 and 20 in the sixth row. You should have no trouble obtaining the entries of the eighth row using this method of generation.

Example 1.28 Determine row 8 of Pascal's triangle.

Solution
Entry zero is 1, as is entry eight. We obtain an internal entry in position k $(1 \leq k \leq 7)$ of row 8 by adding the entry in position $k-1$ of row 7 to the entry in position k of row 7. Thus, entry one is $1+7=8$, entry two is $7+21=28$, entry three is $21+35=56$, and so on. In this way we find that the entries of row 8 are 1, 8, 28, 56, 70, 56, 28, 8, 1. ◇

It is an amazing thing that the successive entries of row n of Pascal's triangle consist of the list of the $n+1$ numbers:

$$\binom{n}{0}, \binom{n}{1}, \binom{n}{2}, \ldots, \binom{n}{n-2}, \binom{n}{n-1}, \binom{n}{n}$$

1.4 Combinatorial Identities

It is also amazing that these numbers are the coefficients in the expansion of $(a+b)^n$. In fact, the numbers $\binom{n}{k}$ are commonly referred to as the **binomial coefficients**. The mathematical wonders of Pascal's triangle seem to be as endless as the triangle itself. Entry two (the third entry) of each row—that is, the number of the form $\binom{n}{2} = \frac{n(n-1)}{2}$—is the sums of the positive integers from 1 to $n-1$, as can be seen by using equation (1.3) in Section 1.2, after replacing n by $n-1$. In Pascal's triangle, suppose we travel along the elements in entry one and stop after six elements. We get $1, 2, 3, 4, 5, 6$. The sum of those entries is 21 and is found below and to the right of the last number, 6, that we encountered. It is true in general that if we stop after the kth of the entry-one items, the sum of those k integers appears as entry two (the third entry) in row $k+1$.

The important quantity $\binom{n}{2}$ counts the number of two-element subsets of an n-element set. Thus, it yields, for example, the total number of handshakes among n people, assuming that each person shakes hands with everyone else. This is, of course, "n choose 2." Another way to see this is as follows. The first person shakes hands with $n-1$ others (we assume he doesn't shake his own hand). The second person then shakes hands with $n-2$ people since he already shook hands with person number one. The third person then shakes hands with $n-3$ people since he already shook hands with person number one and person number two, and so on. We obtain the sum $(n-1) + (n-2) + (n-3) + \cdots + 2 + 1 + 0 = 1 + 2 + 3 + \cdots + (n-1)$, which by equation (1.3), with $n-1$ replacing n, is $\frac{n(n-1)}{2}$.

We shall be studying graphs in this text and will begin doing so in detail in Chapter 2. A **graph** is a mathematical structure consisting of a finite set of **vertices** together with a set of unordered pairs of distinct vertices called **edges**. A graph is typically represented by a drawing in which the vertices are depicted as points and edges as line segments joining those points. An important class of graphs we will encounter are the **complete graphs K_n**. These have n vertices (points) with every pair of vertices joined by an edge (line segment). Several small complete graphs are displayed in Figure 1.4. It is an easy observation that the number of edges in the graph K_n is $\frac{n(n-1)}{2}$. This makes sense since there are $\binom{n}{2}$ ways to select a pair of vertices that determine a particular edge of K_n.

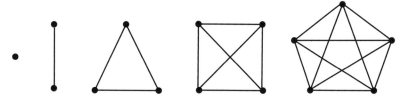

Figure 1.4 The complete graphs K_1, K_2, K_3, K_4, and K_5.

Recursion

There is an easy way to generate row n from row $n-1$ in Pascal's triangle. The formula that indicates how this is done can be written as follows.

Formula 1.6 $\quad \binom{n}{k} = \binom{n-1}{k-1} + \binom{n-1}{k}$

Let's prove this formula, which we shall soon use to establish the claim that the entries of the triangle are indeed the binomial coefficients. Replacing each term in Formula 1.6 by using Formula 1.4, we obtain

$$\frac{n!}{k!(n-k)!} = \frac{(n-1)!}{(k-1)![(n-1)-(k-1)]!} + \frac{(n-1)!}{k![(n-1)-k]!}$$

After a little simplification of the right-hand side, we can rewrite this equation as

$$\frac{n!}{k!(n-k)!} = \frac{(n-1)!}{(k-1)!(n-k)!} + \frac{(n-1)!}{k!(n-k-1)!}$$

To verify the truth of this equation, we show that the expression on the right side of the equation equals that on the left. Begin by finding the least common denominator, which is $k!(n-k)!$. Now multiply the first fraction on the right side of the equation by k/k and the second fraction on that side by $(n-k)/(n-k)$.

$$\frac{(n-1)!}{(k-1)!(n-k)!} + \frac{(n-1)!}{k!(n-k-1)!}$$
$$= \frac{k}{k} \cdot \frac{(n-1)!}{(k-1)!(n-k)!} + \frac{n-k}{n-k} \cdot \frac{(n-1)!}{k!(n-k-1)!}$$
$$= \frac{k(n-1)!}{k!(n-k)!} + \frac{(n-k)(n-1)!}{k!(n-k)!}$$

Now combine the fractions and factor out the common factor $(n-1)!$ from the numerators to get

$$\frac{[k+(n-k)](n-1)!}{k!(n-k)!} = \frac{n(n-1)!}{k!(n-k)!} = \frac{n!}{k!(n-k)!}$$

which proves Formula 1.6. ∎

That was somewhat tedious and not very illuminating. So let's prove Formula 1.6 in a different way. The technique we shall now use is called a **combinatorial argument**.

Suppose we have n people in a room and we wish to count the number of possible committees of size k, that is, we wish to determine $\binom{n}{k}$. Just one of the people in the room is an accountant, C.P.A. Smith. Now let's separate the committees of size k into two types. The first type will consist of possible committees containing Smith, and the other will consist of possible

1.4 Combinatorial Identities

committees without Smith. The total number of committees will be the sum of the number of committees in each of the two types.

To form a committee in the first type (with Smith), we must select $k-1$ additional members from the $n-1$ available candidates. Yes, we know there are n people, but Smith is already on the committee, leaving a pool of $n-1$ other people. So there are $\binom{n-1}{k-1}$ such committees.

On the other hand, to form a committee of the second type (without C.P.A. Smith), we must select all k committee members from a pool of $n-1$ candidates (everyone other than Smith), yielding $\binom{n-1}{k}$ such committees. Since the total number of committees is the sum of these two quantities, we have verified Formula 1.6. ∎

The Rows of Pascal's Triangle

If you compute the row sums of the first several rows of Pascal's triangle, you obtain 1, 2, 4, 8, 16, 32, 64, and 128. These numbers are powers of 2 (recall that $2^0 = 1$) and strongly suggests that the sum of the entries of row n is 2^n. Let's state this formally.

Theorem 1.1 The sum of the entries of row n of Pascal's triangle is 2^n.

Proof. We shall use mathematical induction. It is certainly true for the first few cases. To show that the truth of the proposition for row $k-1$ carries over to row k, observe that each entry of row $k-1$ is counted twice in the next row. It is part of the sum for the lower entry to the left and the lower entry on the right according to Formula 1.6. Hence the sum of the entries of row k is twice the sum of the entries of row $k-1$. Since the latter is assumed to be 2^{k-1}, the sum for row k is indeed $2 \cdot 2^{k-1}$, which equals 2^k, thereby completing the proof. ∎

An alternate proof of this will teach us an important lesson. How many subsets are there for a set of cardinality n? For each k satisfying the condition $0 \le k \le n$, there are $\binom{n}{k}$ subsets, right? So we have to add these binomial coefficients; that is, the answer is the sum of the entries in row n of Pascal's triangle. But as we saw in Example 1.20 of Section 1.2, the power set of a set with n elements has cardinality 2^n. Thus, $\sum_{k=0}^{n} \binom{n}{k} = 2^n$.

Notice that the power set includes the empty set, obtained by voting "exclude" for each member of S. It also includes the set S, obtained by voting "include" for each member of S. Between these two extremes lie all the other subsets of various sizes, ranging from size 1 to $n-1$.

We will soon give a third proof that the sum of the entries in row n of Pascal's triangle is 2^n. The method can be used to prove other things as well. First, we must prove the following important fact, mentioned earlier.

Formula 1.7 (Binomial Theorem) $(a+b)^n = \sum_{k=0}^{n} \binom{n}{k} a^{n-k} b^k$

We prove this result using mathematical induction. It is true for small values of n, for example, $(a+b)^0 = 1$ and $\sum_{k=0}^{0} \binom{0}{k} a^{0-k} b^k = \binom{0}{0} a^0 b^0 = 1$. Similarly, for $n = 1$, we have $(a+b)^1 = a+b$ and $\sum_{k=0}^{1} \binom{1}{k} a^{1-k} b^k = \binom{1}{0} a^1 b^0 + \binom{1}{1} a^0 b^1 = a+b$. Thus assume when $n = t$ that $(a+b)^t = \sum_{k=0}^{t} \binom{t}{k} a^{t-k} b^k$. That is our inductive hypothesis. Now we show for $n = t+1$ that $(a+b)^{t+1} = \sum_{k=0}^{t+1} \binom{t+1}{k} a^{t+1-k} b^k$. Beginning with the left side $(a+b)^{t+1} = (a+b) \cdot (a+b)^t$, we use the inductive hypothesis to replace the second factor in the last quantity. Thus we get $(a+b) \cdot \sum_{k=0}^{t} \binom{t}{k} a^{t-k} b^k$. By the distributive property, a will multiply each term in the summation as will b. Thus we obtain $a \sum_{k=0}^{t} \binom{t}{k} a^{t-k} b^k + b \sum_{k=0}^{t} \binom{t}{k} a^{t-k} b^k$. This boosts up the exponent on a in the first sum and the exponent on b in the second sum. Thus we get $\sum_{k=0}^{t} \binom{t}{k} a^{t-k+1} b^k + b \sum_{k=0}^{t} \binom{t}{k} a^{t-k} b^{k+1}$. Analyzing the two sums, we find that the coefficient for a given term $a^x b^y$ is $\binom{t}{y}$ in the first sum and $\binom{t}{y-1}$ in the second sum. Thus, the coefficient of $a^{t+1-k} b^k$ is $\binom{t}{k} + \binom{t}{k-1}$. Using Formula 1.6, this gives $\binom{t+1}{k}$, which is what we had to prove. Thus Formula 1.7 is true for all $n \in N \cup \{0\}$. ∎

By proving Formula 1.7, we have confirmed that the entries in Pascal's triangle are indeed the binomial coefficients. Note that the exponent on b, namely k, indicates the number of times that b was selected as a factor from the n parentheses in $(a+b)^n$, hence the coefficient $\binom{n}{k}$. If we set a and b equal to 1 in Formula 1.7, we obtain

$$2^n = \sum_{k=0}^{n} \binom{n}{k} \tag{1.11}$$

Several Nice Identities

As a bonus, let's set $a = 1$ and $b = -1$ in Formula 1.7, yielding

$$0 = \sum_{k=0}^{n} \binom{n}{k}(-1)^k = \binom{n}{0} - \binom{n}{1} + \binom{n}{2} - \binom{n}{3} + \cdots + (-1)^n \binom{n}{k} \tag{1.12}$$

The last term is positive if n is even and negative if n is odd. Note also that if we group all of the positive terms of (1.12) and group all of the negative terms of (1.12), those sums must be additive inverses of one another since their total is zero according to (1.12). If we call the sum of the positive terms A and the sum of the negative terms B, then $B = -A$ and $A + (-B) = 2^n$. This implies that $\sum_{k \text{ odd}} \binom{n}{k} = \sum_{k \text{ even}} \binom{n}{k} = 2^{n-1}$.

1.4 Combinatorial Identities

If we set $a = 1$ and $b = 2$, we get the beautiful identity
$$3^n = \sum_{k=0}^{n} \binom{n}{k} 1^{n-k} 2^k = \sum_{k=0}^{n} \binom{n}{k} 2^k$$
or, equivalently,
$$\binom{n}{0} + \binom{n}{1} 2^1 + \binom{n}{2} 2^2 + \binom{n}{3} 2^3 + \cdots + \binom{n}{n} 2^n = 3^n \quad (1.13)$$

In fact, if we let $a = 1$ and $b = x$, this generalizes to the following formula:
$$(1+x)^n = \sum_{k=0}^{n} \binom{n}{k} x^k$$
which is equivalent to
$$(1+x)^n = \binom{n}{0} + \binom{n}{1} x^1 + \binom{n}{2} x^2 + \binom{n}{3} x^3 + \cdots + \binom{n}{n} x^n \quad (1.14)$$

If we let x be a continuous variable, we can apply calculus to (1.14). Integration, differentiation and distribution of powers of x yield more amazing identities.

Example 1.29 (Some basic knowledge of calculus is needed for this example.) Differentiate both sides of equation (1.14) to obtain another combinatorial identity.

Solution
$$[(1+x)^n]' = \left[\binom{n}{0} + \binom{n}{1} x^1 + \binom{n}{2} x^2 + \binom{n}{3} x^3 + \cdots + \binom{n}{n} x^n\right]'$$
$$n(1+x)^{n-1} = \binom{n}{1} + \binom{n}{2}(2x) + \binom{n}{3}(3x^2) + \cdots + \binom{n}{n}(nx^{n-1})$$
$$= \binom{n}{1} + 2\binom{n}{2} x + 3\binom{n}{3} x^2 + \cdots + n\binom{n}{n} x^{n-1}$$

So we get
$$n(1+x)^{n-1} = \sum_{k=1}^{n} k \binom{n}{k} x^{k-1} \quad (1.15) \diamond$$

Note that if we multiply both sides of (1.15) by x, we get an additional identity, namely,
$$nx(1+x)^{n-1} = \sum_{k=1}^{n} k \binom{n}{k} x^k \quad (1.16)$$

Here is another powerful identity. We start with the last entry in row k of Pascal's triangle and proceed down and toward the left until row $k+r$.

The sum of these $r+1$ entries is the entry in the next row to the right. In symbols, we have the following theorem.

Theorem 1.2 The binomial coefficients satisfy

$$\binom{k}{k} + \binom{k+1}{k} + \binom{k+2}{k} + \cdots + \binom{k+r}{k} = \binom{k+r+1}{k+1} \quad (1.17)$$

Proof. We shall prove (1.17) by repeatedly using Formula 1.6. Begin by replacing $\binom{k}{k}$ in (1.17) by the equivalent number $\binom{k+1}{k+1}$ and then adding this to the next term, $\binom{k+1}{k}$, obtaining $\binom{k+2}{k+1}$, using Formula 1.6. Now add that quantity to the next term, $\binom{k+2}{k}$, obtaining $\binom{k+3}{k+1}$, employing Formula 1.6 once again. Notice that the number of terms decreases by one each time we repeat this process. At the last stage, we add $\binom{k+r}{k+1}$ to $\binom{k+r}{k}$, obtaining the right side of equation (1.17), thereby completing the proof. ∎

Theorem 1.2 is sometimes called **the hockey stick theorem** because if you encircle the entries involved as you move down from the last entry in row k toward the left to entry $k+1$ in row $k+r$ together with the entry in the next row and to the right, namely, entry $k+2$ in row $k+r+1$, the resulting figure has the shape of a hockey stick. Try it using Figure 1.3.

Note that when $k = 1$, equation (1.17) becomes $1+2+3+\cdots+(r+1) = \binom{r+2}{2} = \frac{(r+2)(r+1)}{2}$, which is the same as equation (1.3) in Section 1.2, if we let $n = r+1$. If we let $k = 2$ in equation (1.17), we get

$$\binom{2}{2} + \binom{3}{2} + \binom{4}{2} + \cdots + \binom{r+2}{2} = \binom{r+3}{3} \quad (1.18)$$

This formula is usually presented in a slightly modified form obtained as follows. First note that Formula 1.6 can be rewritten as

$$\binom{n}{k} - \binom{n-1}{k-1} = \binom{n-1}{k} \quad (1.19)$$

Now begin with (1.18) and subtract $\binom{r+2}{2}$ and $\binom{r+1}{2}$ from both sides to get

$$\binom{2}{2} + \binom{3}{2} + \binom{4}{2} + \cdots + \binom{r}{2} = \binom{r+3}{3} - \left[\binom{r+2}{2} + \binom{r+1}{2}\right]$$

$$= \binom{r+3}{3} - \binom{r+2}{2} - \binom{r+1}{2}$$

$$= \left[\binom{r+3}{3} - \binom{r+2}{2}\right] - \binom{r+1}{2}$$

Now use (1.19) with $n = r+3$ and $k = 3$ to replace the difference in the brackets and get $\binom{r+2}{3}$. Thus the right side of the last equation becomes

1.4 Combinatorial Identities

$\binom{r+2}{3} - \binom{r+1}{2}$. Now use (1.19) again, this time with $n = r+2$ and $k = 3$ to get $\binom{r+1}{3}$. Thus we get the following modified version of equation (1.17):

Theorem 1.3 $\binom{2}{2} + \binom{3}{2} + \binom{4}{2} + \cdots + \binom{r}{2} = \binom{r+1}{3}$ ∎

The rth **tetrahedral number** is the total number of points in a pyramid with a triangular base, where the number of points in layer i equals the ith triangular number. Thus the formula in Theorem 1.3 says that the rth tetrahedral number is equal to $\binom{r+1}{3}$. For example, letting $r = 6$, we get $1 + 3 + 6 + 10 + 15 = 35$. Thus a triangular pyramid of cannon balls with five levels contains 35 cannon balls. The base of the pyramid is a triangular array of 15 balls (similar to a triangular array of 15 billiard balls). The successive layers of the pyramid are shown in Figure 1.5.

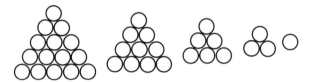

Figure 1.5 The successive layers of a five-layer pyramid of cannon balls.

Here, finally, is the last identity we shall look at.

Example 1.30 Prove that $\sum_{k=0}^{n} \binom{n}{k}^2 = \binom{2n}{n}$.

Solution
Note that this is equivalent to $\binom{n}{0}\binom{n}{0} + \binom{n}{1}\binom{n}{1} + \binom{n}{2}\binom{n}{2} + \cdots + \binom{n}{n-1}\binom{n}{n-1} + \binom{n}{n}\binom{n}{n} = \binom{2n}{n}$. By applying Formula 1.5, namely, $\binom{n}{k} = \binom{n}{n-k}$, to one factor in each term on the left side of the equation, we change the preceding equation into $\binom{n}{0}\binom{n}{n} + \binom{n}{1}\binom{n}{n-1} + \binom{n}{2}\binom{n}{n-2} + \cdots + \binom{n}{n-1}\binom{n}{1} + \binom{n}{n}\binom{n}{0} = \binom{2n}{n}$, which is easier to prove.

Imagine a room containing n men and n women. We claim that the left and right sides of this last equation count the number of possible committees of exactly n people, regardless of gender, which may be selected from the $2n$ people in the room. The two sides of the equation must, therefore, be equal.

The right side counts the number of committees of size n, which can be selected from the $2n$ people. Why does the left side count the same quantity? We partition the committees into $n+1$ mutually disjoint groups on the basis of their gender makeup. A committee of n people must consist of exactly one of the following: no men and n women, 1 man and $n-1$ women, 2 men and $n-2$ women, ..., or, finally, n men and no women. Using the multiplication principle, each term on the left counts how many of each of these committees can be formed. For each k from 0 to n, a committee of k men and $n-k$ women involves two independent tasks that can be completed in $\binom{n}{k}$ and $\binom{n}{n-k}$ ways, respectively. So we add the terms, and the proof is complete. ∎

Figure 1.6 illustrates the selection process.

$$\boxed{\begin{array}{c|c} n \text{ Men} & n \text{ Women} \\ \hline \text{choose } k & \text{choose } n-k \end{array}} \rightarrow \binom{2n \text{ people}}{n}$$

$$0 \le k \le n$$

Figure 1.6

The proof in Example 1.30 is another example of a *combinatorial argument*.

Exercises 1.4

1. Prove the following identity: $1 \times 3 \times 5 \times \cdots \times (2n-1) = \frac{(2n)!}{n!2^n}$.

2. Show that the entries of row n in Pascal's triangle, excluding the initial and final 1's, are all even if and only if n is a power of 2. The entries of row 4, for example, are 1, 4, 6, 4, 1, which, except for the initial and final 1's, are even.

3. Prove Theorem 1.2 by mathematical induction.

4. Prove by mathematical induction: $\sum_{i=1}^{n} i(i!) = (n+1)! - 1$.

5. Prove by mathematical induction: $\sum_{i=1}^{n} \frac{i}{(i+1)!} = \frac{(n+1)!-1}{(n+1)!}$.

6. Find row 9 of Pascal's triangle.

7. Prove by a combinatorial argument that $\binom{n}{r} = \binom{n-2}{r} + 2\binom{n-2}{r-1} + \binom{n-2}{r-2}$.

8. Use Formula 1.6 to prove that $r\binom{n}{r} = n\binom{n-1}{r-1}$.

9. Use the identity of Exercise 8 and a change of variables to produce the formula $\sum_{r=1}^{n} r\binom{n}{r} = n \cdot 2^{n-1}$.

1.4 Combinatorial Identities

In Exercises 10–13, find appropriate choices for a and b in the binomial theorem (Formula 1.7) in order to evaluate the given quantity as a function of n.

10. $\sum_{k=0}^{n} \binom{n}{k}(-1)^k$

11. $\sum_{k=0}^{n} \binom{n}{k} 5^k$

12. $\sum_{k=0}^{n} \binom{n}{k} 10^{n-k}$

13. $\sum_{k=0}^{n} \binom{n}{k} 3^k$

14. Using Formula 1.6, prove that $\binom{n}{r}\binom{r}{k} = \binom{n}{k}\binom{n-k}{r-k}$.

15. Redo Exercise 14, but this time prove the identity using a combinatorial argument. *Hint*: It is probably easiest to do this by thinking of each side of the equation as counting subcommittees of a committee.

16. Prove by induction on r that $\sum_{k=0}^{r} \binom{n+k}{k} = \binom{n+r+1}{r}$.

17. Determine the coefficient of $x^3 y^{12}$ in $(x + y^2)^9$.

18. Determine the coefficient of $a^4 b^3$ in $(a^2 - 2b)^5$.

19. Since $\binom{n}{r}$ counts the number of subsets of size r in an n-element set, $\binom{n}{r}$ is always an integer. Use that fact to prove that if p is a prime, then p divides $\binom{p}{r}$ for $1 \leq r \leq p-1$.

Problems and Projects for Chapter 1

P1-1. Determine the total number of integers that have all digits distinct.

P1-2. Write a program that prints the first 15 factorials.

P1-3. Write a program that reads in an integer k and finds the first value of n for which $n! > 90$.

P1-4. Write a program that inputs integers m and q and prints all values $P(n,r)$ for $1 \leq n \leq m$ and $1 \leq r \leq q$.

P1-5. Find all values of $n \in N$ for which $P(n,4) = (n-3)!$.

P1-6. Write a program that will read a positive integer n and print all values of $\binom{n}{r}$ for $0 \leq r \leq n$ ($n \leq 10$).

P1-7. Write a program that calculates and prints the first fifteen rows of Pascal's triangle.

P1-8. As an extra challenge, redo P1-7 to make the computer print the output in a triangular pattern.

References for Chapter 1

1. Loomis, E. S., *The Pythagorean Proposition*. National Council of Teachers of Mathematics, Washington, D.C. (1968).

2. Molluzzo, J. C., and F. Buckley, *A First Course in Discrete Mathematics*, Waveland Press, Prospect Heights, Illinois (1986).

Additional Readings

3. Calinger, R., Leonhard Euler: The First St. Petersburg Years (1727–1741). *Historia Mathematica* 23:2 (May 1996) 121–166.

4. Gadbais, S., Poker with wild cards—a paradox? *Mathematics Magazine* 69 (1996) 283–285.

5. Gindikin, S., Carl Friedrich Gauss. *Quantum* 10:2 (November/December 1999) 14-19, and 10:3 (January/February 2000) 10–15.

6. Graver, J. E., and L. J. Lardy, Linear functions and rounding. *College Mathematics Journal* 31 (2000) 132–136.

7. Hill, T. P., Mathematical devices for getting a fair share. *American Scientist* 88:4 (2000) 325–331.

8. Hoffman, P. Man of numbers. *Discover* 19 (1998) 118–123.

9. Kalajdzievski, S., Some evident summation formulas. *Mathematical Intelligencer* 22:3 (Summer 2000) 47–49.

10. Klastmeyer, W. F., M. E. Mays, L. Soltes, and G. Trapp, A Pascal rhombus. *Fibonacci Quarterly* 35 (1997) 318–328.

11. Leigh, R. B., and T. T. Ng, Minimizing aroma loss. *College Mathematics Journal* 30 (1999) 356–358.

12. McCutcheon, R., Analyzing games of information. *College Mathematics Journal* 32 (2001) 82–90.

13. Stewart, I., Your half's bigger than my half!, *Scientific American* 279:6 (December 1998) 112,114.

14. Stewart, I., Division without envy. *Scientific American* 280:1 (January 1999) 110–111.

15. Walser, H. The Pascal pyramid. *College Mathematics Journal* 31 (2000) 383–392.

Chapter 2

Introduction to Graphs and Their Uses

The structure displayed in Figure 2.1 is an example of a graph. You will observe from the diagram in Figure 2.1 that the graphs that we are studying here are not the ones you might first think of—that is, functions plotted on an (x, y)-coordinate system. Instead, they are, loosely speaking, composed of points, some of which are joined by line segments. Graphs have a wide variety of applications. In this chapter, we introduce the most basic concepts of graph theory and illustrate some of the areas where graphs are used. We will see many other applications throughout the book.

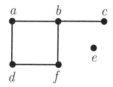

Figure 2.1

2.1 Graphs as Models

Graph theory is a delightful subject with a host of applications in such diverse fields as computer science, chemistry, business, and sociology, to name just a few. The mathematics of graph theory, which is combinatorial and geometric in nature, requires cleverness at every step. Each theorem is a work of art to be savored. Remarkably, graph theory has important pragmatic applications as well.

Graphs

A **graph** is a mathematical representation of a relationship. Specifically, a graph G consists of two sets—a nonempty finite set V of **vertices** and a finite set E of **edges** consisting of unordered pairs of distinct vertices from V. If $V = \{v_1, v_2, \ldots, v_n\}$ is a set of n vertices, the edge set $E = \{e_1, e_2, \ldots, e_m\}$ consists of m two-element subsets of V; that is, each edge is of the form $\{v_i, v_j\}$. However, we generally write $v_i v_j$ or $v_j v_i$ rather than $\{v_i, v_j\}$ to denote the edge. The cardinality n of $V(G)$ is the **order** of G and the cardinality m of $E(G)$ is the **size** of G. A pair of vertices v_i and v_j in V are **adjacent** if $v_i v_j \in E(G)$—that is, if $v_i v_j$ is an edge of G. Otherwise, v_i and v_j are **nonadjacent**. The **degree deg(v)** [or $\deg_G(v)$ if we wish to identify the graph in question] of v is the number of vertices adjacent to v. Edge $v_i v_j$ is **incident with** v_i and v_j.

Example 2.1 Let G be the graph in Figure 2.1.
a. List the vertex set and edge set of G.
b. Let the vertices of G represent the students in your graph theory class, and suppose that two vertices are adjacent if the corresponding people have worked on homework problems together. Who has worked on homework problems with the most classmates? Who has never worked on homework problems with any of the classmates?
c. List the degrees of the vertices of G.
d. Suppose that the vertices of G represent the computers in the mathematics department computer lab and an edge indicates that the corresponding computers can communicate directly. Assuming that the computers all have the same speed and memory capacity, which computer would be most appropriate as a network server? (Such a computer should be able to communicate directly with as many other computers as possible.) Which computer is completely isolated from the network server?

Solution
a. $V(G) = \{a, b, c, d, e, f\}$ and $E(G) = \{ab, bc, ed, da, bc\}$.
b. Student b has worked on homework problems with three other students, the most in the class. Student e has never worked on homework problems with any of the classmates.
c. The degrees are as follows: $\deg(a) = 2$, $\deg(b) = 3$, $\deg(c) = 1$, $\deg(d) = 2$, $\deg(e) = 0$, and $\deg(f) = 2$.
d. Computer b would be best as the server because vertex b has the largest degree. Computer e is completely isolated from the network server. ◊

A vertex of degree zero is called an **isolated vertex**. In Figure 2.1, vertex e is isolated. A vertex of degree one is called an **end vertex** (or sometimes a **leaf** when the graph has a "treelike" structure; see Figure 2.2). Vertex c in Figure 2.1 is an end vertex. The degrees of the vertices in a graph play an important role in many problems. The **minimum degree**

2.1 Graphs as Models

among vertices in graph G is denoted $\delta(G)$ and the **maximum degree** is $\mathbf{\Delta(G)}$. If all vertices have the same degree, then $\delta(G) = \Delta(G)$ (little delta equals big delta) and the graph is called **regular**. An **r-regular** graph has all vertices of degree r. In any graph, the degrees satisfy a very simple relationship stated in the following theorem due to the Swiss mathematician Leonhard Euler (1707–1783), one of the most prolific mathematicians of all time with 886 papers and books that fill about ninety volumes. He was totally blind when he produced most of those papers during the last twenty years of his life. There are many theorems and formulas in various branches of mathematics with the name Euler (pronounced *"oiler"*) attached to them. Graph theorists often refer to Euler as the father of graph theory due to the following result, which is often called the first theorem in graph theory.

Theorem 2.1 (Euler) In any graph G, the sum of the degrees equals twice the number of edges.

Proof. Each edge contributes two toward the degree sum, one at one end of the edge and one at the other end. ∎

Corollary 2.1a In any graph G, the sum of the degrees of the vertices is a nonnegative even integer. ∎

The vertices of a graph can represent people, computers, cities, or just about anything. An edge shows which vertices are related—for example, two people who know each other, two computers that can communicate, or two cities that are linked by a direct airline flight. Graphs are used in many different ways. Chemists use graphs to study molecules, environmentalists use graphs to ensure that two neighboring toxic dump sites don't have chemicals that are dangerous when mixed, businesses use graphs to keep track of warehouses and retail outlets, city planners use graphs to analyze traffic flow, and so on.

Mathematical Models

Mathematical models are used to abstract and simplify real-world situations. For example, in business, the function $A = P(1 + r)^n$ represents the accumulated amount A in a bank account when the principal P is deposited and kept in an account paying a rate r for n years. Graph theory also abstracts and simplifies real-world problems so that we can better understand relationships between things. Using "dots" (that is, vertices) to represent people, computers, cities, and so on and "line segments" (that is, edges) to demonstrate which vertices are related, graphs are a simple yet powerful tool in business, science, social science, and many other areas.

We should mention at the outset that there are occasions in graph theory modeling when we need structures that are slightly different from graphs. For example, in modeling a problem where there are multiple flights available between two cities, we might represent the cities as vertices and join two cities that have, say, four available flights between them on a given day by four edges rather than just one. The resulting structure, which permits multiple edges between given pairs of vertices, is called a **multigraph**. We will encounter multigraphs in Chapter 5 as well as in Chapter 9. Note that Theorem 2.1 happens to hold true for multigraphs as well. When we use the term "graph," multiple edges are *not* permitted.

Chemistry Application

The nineteenth-century mathematician Arthur Cayley used graph theory to study simple hydrocarbons—that is, molecules that are composed of hydrogen and carbon atoms. Specifically, he modeled such hydrocarbons using a special type of graph called a tree. Let $V(T) = \{c_1, c_2, c_3, c_4, h_1, h_2, h_3, h_4\}$ be the vertices of our tree and let $E(T) = \{c_1h_1, c_1c_2, c_1h_9, c_1h_{10}, c_2h_2, c_2h_8,$ $c_2c_3, c_3h_3, c_3h_7, c_3c_4, c_4h_4, c_4h_5, c_4h_6\}$. The graph T is shown in Figure 2.2.

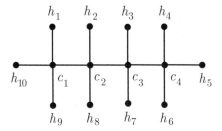

Figure 2.2

This graph has order 14 (that is, there are 14 vertices) and size 13 (that is, 13 edges). Using models like this, Cayley abstracted hydrogen and carbon atoms as vertices, and bonds between atoms as edges. The graph of Figure 2.3 depicts a different hydrocarbon. Although it has the same vertex set, it is a different tree (and, chemically speaking, has different physical properties). Both molecules have chemical formula C_4H_{10}.

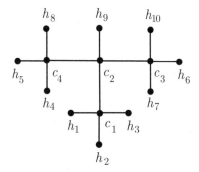

Figure 2.3

2.1 Graphs as Models

More abstractly, we could study the graph shown in Figure 2.4. You can begin to see why they're called trees.

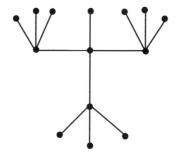

Figure 2.4

Trees are also studied in computer science as a type of data structure and in social science to represent hierarchies. Trees will be studied in Chapter 3.

Business Application: Warehouses/Retail Stores

Businesses use graphs to model which of several warehouses service retail stores. Specifically, suppose that $V = \{w_1, w_2, w_3, r_1, r_2, r_3, r_4, r_5\}$ represents three warehouses and five retail stores, and $E = \{w_1r_1, w_1r_2, w_2r_2, w_2r_3, w_2r_4, w_3r_3, w_3r_5\}$ represents the relation of each warehouse to each retail store. That is, w_ir_j is an edge in E if and only if warehouse w_i services retail store r_j ($1 \leq i \leq 3$, $1 \leq j \leq 5$). The geometric representation of this model is shown in Figure 2.5.

Example 2.2 Consider the graph in Figure 2.5.
a. Which warehouse services the most retail stores?
b. If store r_2 needed merchandise but w_2 was out of stock, which warehouse might be able to service r_2?

Solution
a. Warehouse w_2 services the most retail stores, namely, r_2, r_3, and r_4.
b. Figure 2.5 indicates that w_1 (in addition to w_2) services r_2 and should, therefore, be contacted. ◇

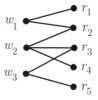

Figure 2.5

Note that we can split V into two disjoint (that is, nonoverlapping) subsets, $\{w_1, w_2, w_3\}$ and $\{r_1, r_2, r_3, r_4, r_5\}$, such that *every* edge contains exactly one vertex from each set. Such graphs are called **bipartite graphs** since they have two parts—namely, the two disjoint subsets that partition V. Bipartite graphs will be studied in Chapter 3.

Application: Distance between Two Cities

Consider the graph G shown in Figure 2.6, with $V(G) = \{a, b, c, d, e\}$ representing five cities and $E(G) = \{ab, ad, bc, be, de\}$ representing direct airline flights. The numbers on each edge represent distances traveled by each flight. This is an example of a **weighted graph**, which we will encounter several times throughout the text.

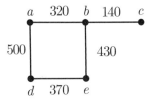

Figure 2.6

Example 2.3 Find the shortest way to get from d to c in Figure 2.6.
Solution
We can get from d to c by following the route d, a, b, c or the route d, e, b, c. Adding the distances along each of the routes, we get $500 + 320 + 140 = 960$ for the first one and $370 + 430 + 140 = 940$ for the second one. So d, e, b, c would be the best route as 940 is the shortest distance. Distances in graphs will be studied in Chapter 4. ◇

Application: Ice Cream Truck Routing

In the model depicted in Figure 2.7, graph G represents streets in a town as edges and street intersections as vertices. Here we have 16 streets with 10 intersections.

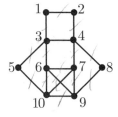

Figure 2.7 The streets of a town.

2.1 Graphs as Models

Example 2.4 If an ice cream truck begins at vertex 5 in Figure 2.7, can the ice cream salesperson travel along each street exactly once and return to vertex 5?

Solution
Here is one solution: 5, 3, 1, 2, 4, 3, 6, 10, 9, 8, 4, 7, 9, 6, 7, 10, 5. An alternative solution is 5, 10, 9, 8, 4, 2, 1, 3, 6, 10, 7, 9, 6, 7, 4, 3, 5. Can you find other solutions? ◇

Note that if certain streets (edges) were not yet constructed between particular intersections (vertices) in Figure 2.7, the desired truck route would not be possible. For example, if the street joining 6 to 7 were missing, then the ice cream salesperson would need to travel along some streets more than once to complete the route. This problem is related to **eulerian graphs**, which will be studied in Chapter 5.

Application: Traveling Salesman Problem

Example 2.5 Suppose a computer representative has to travel by plane to five cities and return to her home city. The direct routes between cities (vertices) are indicated by edges, where h is her hometown. This situation is illustrated in Figure 2.8. How can she accomplish this assignment if she is to visit each city only once?

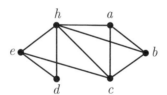

Figure 2.8 Flight routes.

Solution
We can see in this example that the computer representative can only get to city d by way of e or h. Thus both ed and dh must be flight routes that she travels. That forces her to also use flight route ce, because if she uses eh, she would end her trip too early with the cycle h, d, e, h. By the same reasoning, she can't use flight route ch. With a little analysis, we find that there are two ways she can complete the desired route, namely with cb, ba, and ah or with ca, ab, and bh. Thus she has two possible routes h, d, e, c, a, b, h or h, d, e, c, b, a, h. She could also fly those routes in reverse order, giving a total of four different travel itineraries. ◇

This problem is related to **hamiltonian graphs**, which will be studied in Chapter 5. The traveling salesman problem is discussed in greater detail there as well. In the actual traveling salesman problem, we are generally concerned with total distance traveled, so a *weighted* graph, like Figure 2.6, is used to model the problem. We also want to minimize total distance. Depending on the flight routes that are available, it is not always possible to complete an itinerary without returning to a city already visited.

Application: Scheduling Exams

A professor needs to schedule final exams so that no student has a conflict (that is, has two exams scheduled at overlapping times). We model this situation with a graph G by letting the vertices represent the courses giving final exams. We join two vertices with an edge if the two courses they represent have at least one student in common. Different colors can represent different time slots. For example, red can represent Thursday morning, blue can represent Thursday afternoon, and white can represent Friday morning.

Example 2.6 Let $V = \{a, b, c, d, e, f\}$ be the six courses offering finals and let G be the graph of Figure 2.9. Scheduling exams is equivalent to assigning each vertex (that is, course) a color (that is, a time slot) while assuring that courses connected by an edge have different colors (time slots).

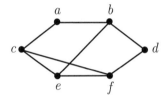

Figure 2.9

Solution
One possibility is to assign a, d, and e the Thursday morning (red) slot, assign b and f the Friday morning (white) slot, and assign c the Thursday afternoon (blue) slot. An alternative is to assign a and e the Thursday morning (red) slot, assign c and d the Friday morning (white) slot, and assign b and f the Thursday afternoon (blue) slot. ◇

Graph coloring problems will be studied in Chapter 6.

Application: An Assignment Model

Example 2.7 Consider a graph whose vertices represent people, where an edge connects two people who are friends. These people are planning to go on a long bus ride, and seats are assigned in pairs. Therefore, the travel director would like to pair friends so that they will have an enjoyable trip. Let G be the graph shown in Figure 2.10 with vertex set $V = \{$Ann, Bob, Cal, Don, Ed, Fran, Grace, Heidi$\}$.

2.1 Graphs as Models

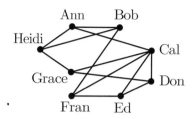

Figure 2.10

Solution
One possible solution is {Ann, Cal}, {Bob, Heidi}, {Don, Grace}, and {Ed, Fran}. An alternative solution is {Ann, Bob}, {Cal, Don}, {Ed, Fran}, and {Grace, Heidi}. Can you find others? ◊

Notice that if there were an odd number of vertices, then no satisfactory pairing can be made since someone would have to sit alone. When the number of vertices is even, in such problems there may still be adjacency conditions that make pairings impossible. This problem is related to matchings in graphs, which will be studied in Chapter 3.

Exercises 2.1

1. Let G be the graph with vertex set $V(G) = \{a, b, c, d, e, f, g\}$ and edge set $E(G) = \{ab, ac, ae, af, bc, ce, ef, fg\}$.

   (a) Draw a geometric representation of the graph G.

   (b) Suppose that the vertices of G represent people, and two vertices are adjacent if the corresponding people are friends. Find two groups of three mutual friends.

   (c) List the degrees of the vertices of G.

   (d) Verify Theorem 2.1 for the graph you have drawn.

2. Explain why it is impossible for the list of degrees of the vertices of a graph to be

   (a) 5, 4, 2, 2, 2, 1, 1   (b) 10, 6, 3, 2, 2, 1, 1, 1

3. Each of the following can be modeled by a graph. Explain what the vertices would represent and what the edges would correspond to.

   (a) A road map   (b) A molecule   (c) A subway system
   (d) A family tree   (e) Jobs and applicants for those jobs

4. Construct (that is, draw a geometric representation of) graphs with the following vertex and edge sets.

(a) $V(G) = \{a, b, c, d\}$, $E(G) = \{ab, cd, bd, bc\}$

(b) $V(H) = \{e, f, g, h, i, j\}$, $E(H) = \{eh, ej, hi, ij, fh, hj\}$

(c) $V(M) = \{k, m, n, p, q, r\}$, $E(M) = \{km, mp, kp, qr\}$

5. A **digraph** is a structure similar to a graph except that there are directions on the edges. Digraphs can be used to describe the structural hierarchy in a corporation (that is, the chain of command). Each employee corresponds to a vertex of the digraph. If u is the direct superior of v, there is an arrow directed from u toward v. Draw the digraph for the following corporate structure. The chairman c of the board is the boss of the president p, who has three vice presidents under his direct control: VP for finance f, VP for administration a, and VP for sales s. The VP for finance is in charge of the controller t and the manager for research r. The VP for administration is in charge of each division manager, which includes the data processing manager md, manager for research mr, sales manager ms, and advertising manager ma. The VP for sales is in charge of the sales manager, the advertising manager, and the purchasing agent pa.

6. Draw three different 2-regular graphs.

7. Draw a 3-regular graph on six vertices.

8. Explain why there is no 3-regular graph on seven vertices.

9. Consider the street map (Figure 2.7) for the ice cream truck routing problem. Suppose that there is road construction on the street joining 3 to 4, so that street has been closed to traffic. Explain why it is impossible to begin and end a route at 5 while passing along each open street (that is, delete edge 34 from the graph) exactly once. Find a route in the modified graph that begins at vertex 3 and ends at vertex 4 while passing along each street exactly once.

10. Suppose that the graph in Figure 2.11 represents the available flights between various cities, with the number on each edge being the distance.

    (a) Find a shortest route from city h to f.

    (b) If a salesman must travel from location a to each city and return to a, explain why it would be impossible to do so if each city is to be visited just once.

    (c) Determine a best possible route if we are allowed to revisit cities.

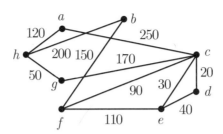

Figure 2.11

2.2 Subgraphs and Types of Graphs

When studying graphs, certain classes of graphs arise rather frequently. In this section, we introduce many of those graph classes. Additional graph classes will be encountered when we study graph operations in Section 2.4.

Take a Long Walk

Given two vertices, u and v, in a graph G, we define a **u–v walk** as an alternating sequence of vertices and edges beginning with u and ending with v such that consecutive vertices and edges are incident. In practice, it is not necessary to list the edges because they are implied by the sequence of vertices that are used.

The number of edges in a walk is its **length**. Note that vertices and edges may be repeated in a walk. Imagine a leisurely stroll in your favorite city in which you suddenly notice you've just walked by a particular store or along a favorite street for the third time.

If no edge in a walk is repeated, we call it a **trail**. Note that vertices may be repeated in a trail. If the trail begins and ends at the same vertex—that is, if the trail is **closed**—we call it a **circuit**.

A walk in which no vertex is repeated is called a **path**. A path that joins vertices u and v is called a **u–v path**. Obviously, an edge cannot be repeated if the two vertices of that edge aren't repeated, so a path is also a trail. A closed path is a **cycle**.

Example 2.8 In the graph of Figure 2.12,
a. Find a walk that is not a trail.
b. Find a trail that is not a path.
c. Find five b–d paths.
d. Determine the length of each path of part c.
e. Find a circuit that is not a cycle.
f. Find all distinct cycles that are present.

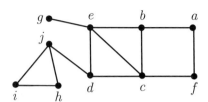

Figure 2.12

Solution

a. g, e, d, c, b, e, d is an example of a walk that is not a trail. It repeats the edge ed.

b. g, e, d, c, b, e, c is a trail that is not a path. It repeats the vertices c and e, which is not allowed for a path.

c. b, c, d; b, e, d; b, c, e, d; b, e, c, d; and b, a, f, c, d. There is just one other b–d path. Find it.

d. Remember that the length is the number of edges in the path, not the number of vertices. The lengths of the given paths are 2, 2, 3, 3, and 4.

e. c, b, a, f, c, e, d, c is a circuit in the graph. Note that it repeats vertices but does not repeat any edges.

f. This is done best by organizing the cycles by length. There are three cycles of length 3: i, j, h, i; c, d, e, c; and b, c, e, b. There are two cycles of length 4: b, c, d, e, b and a, b, c, f, a. There is one cycle of length 5, a, b, e, c, f, a, and one of length 6, a, b, e, d, c, f, a. ◇

A graph is **connected** if for every two vertices u and v, there is a u–v path joining them. Otherwise, it is **disconnected**. We can easily see that the graph in Figure 2.12 is connected. A **geodesic path** (or **geodesic** for short) between vertices u and v of a graph G is a u–v path of minimum length. It is an excellent path to traverse when you are in a hurry to get from u to v. Note, however, that there may be several geodesic u–v paths. Thus paths b, c, d and b, e, d are both b–d geodesics in the graph of Figure 2.12. The path b, c, e, d is not. Its length is 3, while both b–d geodesics have length 2. Geodesic paths are also called **shortest paths**.

Subgraphs

For a graph G with vertex set $V(G)$ and edge set $E(G)$, we call a graph H a **subgraph** of G if the vertex set $V(H)$ and the edge set $E(H)$ are subsets of $V(G)$ and $E(G)$, respectively, and for each edge $e = uv \in E(H)$, both u and v are in $V(H)$. In other words, we can obtain H from G by deleting edges and/or vertices from G. Note, however, that for each vertex v that is removed, all edges incident with v must also be removed. We call G a **supergraph** of H. In Figure 2.13, graph H is a subgraph of G.

2.2 Subgraphs and Types of Graphs

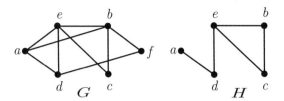

Figure 2.13 A graph and one of its subgraphs.

Example 2.9 Consider graph G of Figure 2.13. Draw all subgraphs of G that use vertex set $\{a, b, e, f\}$.

Solution

The subgraphs are drawn in Figure 2.14. Note that in each case, we must use all of the specified vertices in $\{a, b, e, f\}$. We then may use any of the edges that join those vertices in G. Thus we must select our edges from the set $\{ab, ae, be, bf\}$. ◊

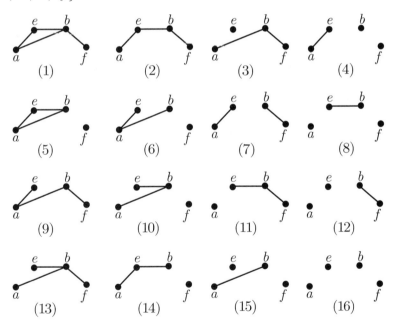

Figure 2.14 The subgraphs of graph G having vertex set $\{a, b, e, f\}$.

At first it is probably surprising that there are so many subgraphs with the given vertex set $\{a, b, e, f\}$ in Example 2.9. We shall now examine why there are so many. First, we note that it is not enough to specify the vertices of a subgraph. We must also list the edges. There is one special exception. If S

is a subset of the vertices of G—that is, if $S \subseteq V(G)$—then the **subgraph induced by S**, denoted $\langle S \rangle$, is the subgraph consisting of S and all edges of the form uv of G, where $u \in S$ and $v \in S$. Among all subgraphs of G that use S as the vertex set, $\langle S \rangle$ has the maximum number of edges possible. A **labeled graph** is a graph whose vertices have labels attached to them. If there are no labels attached to the vertices, the graph is **unlabeled**. So, for example, in Section 2.1, all of the graphs displayed in the figures were labeled graphs except the graph in Figure 2.4. To see how many subgraphs there are for a given labeled graph, note that for each edge we must make a decision worthy of Shakespeare: to include or not to include—that is the question. Thus, the next observation follows directly from the multiplication principle.

Note 2.1
If $S \subseteq V(G)$ and there are a total of k edges that join the vertices of S within a labeled graph G, then there are 2^k subgraphs of G that use S as the vertex set.

Of the sixteen subgraphs in Figure 2.14, graph (1) is the *induced* subgraph $\langle \{a, b, e, f\} \rangle$. Four of the subgraphs are connected—namely, graphs (1), (2), (9), and (13). The other twelve subgraphs are disconnected. In a disconnected graph, the various *"pieces"* of the graph are called components. Formally, a **component** of a graph is a maximal connected subgraph. That is, it is a largest connected subgraph that is not contained in a larger connected subgraph. *Maximal* and *maximum* are related but not equivalent concepts. When we are looking for an object that is maximum, we want the object whose size or cardinality is as large as possible. On the other hand, when we are considering maximality, we are looking for an object that is as large as possible *subject to some condition*. Thus, when we speak of the components as maximal connected subgraphs of a graph G, we mean that we are looking for the largest possible subgraphs H that preserve the property of being connected. For example, graph (4) of Figure 2.14 has three components—namely the component consisting of the edge ae, the isolated vertex b, and the isolated vertex f. Note that a by itself is not a component of graph (4) because it is not maximal—we can enlarge a to ae and still preserve connectedness. Note that if we try to include either vertex b or f along with the edge ae for graph (4), the resulting subgraph would no longer be connected. Thus ae is indeed a maximal connected subgraph and therefore a component of graph (4). A connected graph has just one component—namely, the graph itself. The graphs (13), (14), (15), and (16) of Figure 2.14 have 1, 2, 3, and 4 components, respectively.

A subgraph H of graph G is called a **spanning subgraph** if $V(H) = V(G)$. In that case, we say that "H **spans** G." We may form a spanning

2.2 Subgraphs and Types of Graphs

subgraph of a given graph by simply deleting edges. It follows from Note 2.1 that a labeled graph with k edges has 2^k spanning subgraphs.

Important Graph Types

Certain types of graphs occur frequently in graph theory. Sometimes they occur naturally because of a particular applied problem that is being studied. At other times, when studying a new graph concept, we begin by examining particular classes of graphs to gain insight on that concept. We now look at several standard classes of graphs. As we have already seen (refer to Figure 1.4), the **complete graph** on n vertices, denoted $\boldsymbol{K}_n$, is the graph where all pairs of vertices are adjacent. Since there are n vertices, this implies that

$$|E(K_n)| = \binom{n}{2} = \frac{n(n-1)}{2}. \tag{2.1}$$

It also follows that this number is an upper bound for the number of edges of any graph on n vertices. This can be written

$$|V(G)| = n \implies |E(G)| \leq \frac{n(n-1)}{2}. \tag{2.2}$$

Thus, for example, a graph on five vertices can have at most ten edges. Graph K_n is the unique connected $(n-1)$-regular graph on n vertices. The graph K_1 is often called the **trivial graph**. The complete graph K_5 is displayed in Figure 2.15.

For $n \geq 3$, the **cycle** on n vertices denoted $\boldsymbol{C}_n$ is exactly what you think it is—namely, a cycle on n vertices! You may note that C_3 and K_3 are the same graph. C_n is the unique connected 2-regular graph on n vertices. The cycle C_6 is shown in Figure 2.15.

The **path** on n vertices is denoted $\boldsymbol{P}_n$. Here you may note that $P_1 = K_1$ and $P_2 = K_2$. Note also that P_n is a spanning subgraph of C_n, obtained by deleting a single edge. (In graph theory, the deletion of an edge e does not delete either of the two vertices incident with e, while the deletion of a vertex v always entails the deletion of all edges incident with v. Similarly, the deletion of several vertices results in the deletion of all edges incident with any of those vertices.) The path P_4 is shown in Figure 2.15.

The **complete bipartite graph** $\boldsymbol{K}_{m,n}$ is the graph whose vertex set can be partitioned into nonempty sets A and B of order m and n, respectively, such that each vertex of set A is adjacent to each vertex of B and there are no other adjacencies. When $m = n$, the complete bipartite graph becomes $K_{n,n}$, which is n-regular. Graphs $K_{2,3}$ and $K_{1,4}$ are shown in Figure 2.15. We shall discuss more general bipartite graphs in Chapter 3. These graphs are important in job assignment problems where there are, say, m applicants (set A) and n different jobs (set B). An edge is placed between the vertex representing a given applicant to the vertices representing each job for which he or she is qualified. If each applicant is qualified for every job, the resulting graph is $K_{m,n}$. In reality, this usually does not happen. Instead, we usually have some subgraph of $K_{m,n}$ in the job qualification graph. We then try to find a maximum assignment of applicants to the jobs (one applicant per job) for which the applicants are qualified.

The **star**, $\boldsymbol{K_{1,n}}$, is the complete bipartite graph that consists of one vertex of degree n, while the remaining n vertices are **end vertices**, that is, they have degree one. The star $K_{1,n}$ would be a graph used to model a computer network that has one network server (the vertex of degree n) linked to the n other computers in the network.

In closing this section, we display in Figure 2.15 examples of each type of graph that we discussed.

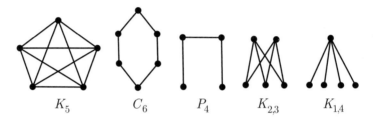

Figure 2.15 A complete graph, a cycle, a path, a complete bipartite graph, and a star.

Exercises 2.2

1. Describe all 2-regular graphs (connected and disconnected).

2. Why is K_n $(n-1)$-regular?

3. Draw each of the following graphs.

   a. K_6 b. $K_{3,5}$ c. P_7 d. C_4 e. $K_{1,6}$

4. Find a formula in terms of m and n for the number of edges in the complete bipartite graph $K_{m,n}$.

5. Determine the number of edges in each of the following graphs without actually drawing the graphs.

   a. K_{14} b. $K_{9,6}$ c. P_{25} d. C_{80} e. $K_{1,19}$

6. Draw a 3-regular disconnected graph on eight vertices.

7. Determine which of the graphs of Figure 2.16 are subgraphs of graph G. Explain.

2.2 Subgraphs and Types of Graphs

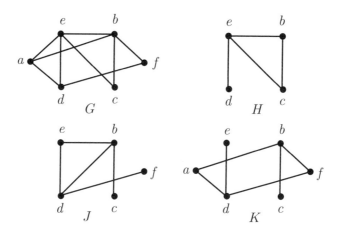

Figure 2.16

8. Which of the graphs in Figure 2.16 are induced subgraphs of graph G? Explain.

9. Find all a–f walks of length four in graph G of Figure 2.16.

10. Which of the walks that were found in Exercise 9 are paths?

11. Find an a–f trail of length six in graph G of Figure 2.16.

12. Why can there be no a–f path of length six in graph G of Figure 2.16?

13. Find all geodesics between b and d in graph G of Figure 2.16.

14. How many spanning subgraphs does graph K of Figure 2.16 have?

15. Find a spanning path for graph J of Figure 2.16.

16. Prove that there is no spanning path for graph K of Figure 2.16.

17. Find a 3- regular graph M such that graph H of Figure 2.16 is an *induced* subgraph of M. Prove that it is impossible to find such an M that has just two vertices more than H has.

18. Show that a graph with n vertices and k edges such that $k < n - 1$ is disconnected. In other words, show that a graph on n vertices requires at least $n - 1$ edges to be connected. To see that this is merely a necessary condition—that is, a prerequisite—produce a graph with n vertices, having $n - 1$ edges, which is disconnected. By the way, a connected graph with n vertices and $n - 1$ edges is called **minimally connected**. P_n and $K_{1,n}$ are examples of minimally connected graphs.

19. Consider the disconnected graph on n vertices, consisting of two components: K_{n-1} and K_1. How many edges does it have? Show that this is the maximum number of edges a disconnected graph on n vertices can have. *Hint*: To maximize the number of edges in a disconnected graph on n vertices, we would presumably create two components, each a complete graph. Let their orders be k and $n-k$, where $1 \leq k \leq n-1$. Now let $f(k)$ be the total number of edges. Show that the maximum value for $f(k)$ is obtained when $k = 1$ or $k = n - 1$.

20. Show that if G, H, and L are graphs such that H spans G and L spans H, then L spans G.

2.3 Isomorphic Graphs

Although the vertex set and edge set of a graph completely determine the graph, two people may draw that graph quite differently. Suppose that graph G has vertex set $V(G) = \{a, b, c, d, e\}$ and edge set $E(G) = \{ab, ac, ad, cd, ce, de\}$. One person might draw G as in Figure 2.17(a), while another might draw it as in Figure 2.17(b). The two graphs are the same, even though they appear different.

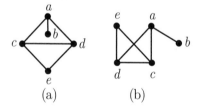

Figure 2.17 Two drawings of the same graph.

Now consider the two graphs in Figure 2.18. They have different vertex sets and hence are not equal, yet they look the same. In most problems, we are trying to determine whether two graphs are *structurally* the same. We shall see that the graphs in Figure 2.18 are. In some problems, the vertices have not yet been labeled to help us in our decision. We will be able to handle those cases as well.

Graphs G and H are **isomorphic**, denoted $G \cong H$, if they can be labeled so that u and v are adjacent in G if and only if the corresponding vertices are adjacent in H. Otherwise, G and H are **nonisomorphic**. Described in terms of function concepts, we can say that graphs G and H are **isomorphic** if there is a one-to-one onto function (see Section 1.1, if necessary) $f : V(G) \to V(H)$ such that any pair of vertices u and v are adjacent

2.3 Isomorphic Graphs

in G if and only if $f(u)$ and $f(v)$ are adjacent in H. Such a function f is called an **isomorphism** from G onto H. For example, for the graphs in Figure 2.18, we have $f(a) = 2$, $f(b) = 3$, $f(c) = 1$, $f(d) = 5$, and $f(e) = 4$. It is easy to see that f is a one-to-one onto function mapping $\{a, b, c, d, e\}$ onto $\{1, 2, 3, 4, 5\}$, so the two graphs in Figure 2.18 are isomorphic.

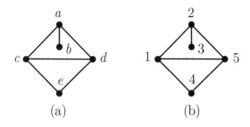

Figure 2.18 Structurally the same graphs with different vertex sets.

We sometimes say that f *preserves adjacency*. Any statement about one of the graphs applies to the other (when we change the names of the vertices and edges to correspond to their names in the other graph). Note that a prerequisite for an isomorphism between two graphs is that they have the same order and the same size. These conditions are necessary but not sufficient. Two graphs may have the same order and size and still be nonisomorphic. For example, graph (2) and graph (5) of Figure 2.14 each have four vertices and three edges. However, those graphs are not isomorphic. Note that an isomorphism must preserve the degrees of the vertices. Graph (2) has a vertex of degree 1, but graph (5) does not, so the two graphs can't possibly be isomorphic. Isomorphism or nonisomorphism for large graphs is usually very difficult to determine.

One basic technique we may use when trying to show that two graphs are isomorphic is the following.

> **Note 2.2**
> Let $f : V(G) \to V(H)$ be an isomorphism between graphs G and H. Then for any vertex $u \in V(G)$, we have $\deg(u) = \deg(f(u))$. In other words, if two graphs are isomorphic, corresponding vertices have the same degree.

Example 2.10 For the pair of graphs G_1 and G_2 in Figure 2.19, describe a function that shows that the graphs are isomorphic. Then show that H_1 and H_2 are isomorphic.

Solution
a. To show that G_1 and G_2 are isomorphic, we must give a correspondence between their vertices that preserves adjacency. We help narrow down the possible functions by using Note 2.2 (corresponding vertices must have the

same degree) and focusing on vertices of unique degree. In G_1 there is just one vertex of degree 3, and one vertex of degree 1. Thus, vertex a (having degree 3) must correspond to vertex 4, and vertex e (having degree 1) must correspond to vertex 3. All remaining vertices have degree 2, so now we focus on adjacencies. Vertex d is the unique vertex adjacent to vertex e. Since e corresponds to vertex 3, d must correspond to the unique vertex adjacent to vertex 3 in G_2—namely, vertex 2. The last two vertices, b and c, are adjacent to one another and mutually adjacent to vertex a. Vertices 1 and 5 satisfy the same condition in G_2. It does not matter which of 1 or 5 gets assigned to vertex b, but whichever one does, we assign the remaining vertex to vertex c (this means that there are, in fact, two possible isomorphism functions). Here is one of them: $f(a) = 4$, $f(b) = 1$, $f(c) = 5$, $f(d) = 2$, and $f(e) = 3$. The function f *induces* a correspondence from the edges of G_1 to the edges of G_2 as follows: $ab \rightarrow 41$; $ac \rightarrow 45$; $ad \rightarrow 42$; $bc \rightarrow 15$; and $de \rightarrow 23$. Thus function f preserves adjacency, so we have shown that G_1 is isomorphic to G_2.

b. For graphs H_1 and H_2, there are two vertices of degree 2 and two vertices of degree 3 in each graph. By the various symmetries involved, it does not matter how we pair up the vertices as long as the vertices of degree 2 of H_1 are paired with the vertices of degree 2 in H_2. Thus, one possible isomorphism is $g(h) = 6$, $g(i) = 9$, $g(j) = 7$, and $g(k) = 8$. It is easy to verify that g preserves adjacencies. Note that as a direct consequence of the multiplication principle, we know that there are four different possible isomorphisms between H_1 and H_2, depending on how we make the assignments. $\diamond$

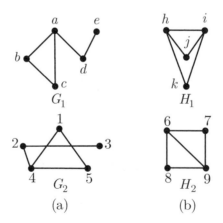

Figure 2.19

2.3 Isomorphic Graphs

Since an isomorphism must preserve adjacencies, all adjacencies along a given geodesic are also necessarily preserved. Thus we have the following observation.

Note 2.3
Let G and H be isomorphic with isomorphism $f : V(G) \to V(H)$. If $v_1, v_2, v_3, \ldots, v_k$ is a geodesic between vertices v_1 and v_k in G, then $f(v_1), f(v_2), f(v_3), \ldots, f(v_k)$ is a geodesic between vertices $f(v_1)$ and $f(v_k)$ in H.

Degree Sequences

If G has n vertices, its **degree sequence** is the ordered n-tuple of the degrees in nonincreasing order. Note that we list repeated degrees as many times as they occur. The degree sequence of P_5, for example, is $2, 2, 2, 1, 1$, while that of C_5 is $2, 2, 2, 2, 2$. We may even write the degree sequence of C_n for any n as $2, 2, 2, \ldots, 2$, where it is clear that there are n 2's. The degree sequence of P_n is easily determined to be $2, 2, 2, \ldots, 2, 1, 1$ where there are $n-2$ twos in the sequence.

If two graphs have the same degree sequence, they need not be isomorphic, as the two graphs of Figure 2.20 show. Of course, isomorphic graphs have the same degree sequence.

Figure 2.20 Two nonisomorphic graphs with degree sequence $3, 2, 2, 1, 1, 1$.

A nonincreasing sequence of nonnegative integers, S, is called **graphical** if there is a graph whose degree sequence is S. It is often easy to show that a given sequence is *not* graphical. Consider the sequence $5, 1$. How can a graph with two vertices have a vertex of degree 5? It can't! What about the sequence $1, 1, 1, 1, 1$? We know that a vertex of degree 1 must be a part of a K_2. Five vertices may form two K_2's and an isolated vertex of degree 0, not 1. Thus $1, 1, 1, 1, 1$ also cannot be the degree sequence of a graph. More subtly, the sequence $3, 3, 3, 2, 2, 2, 2, 1, 1$ violates the fact that the degree sum of a graph is even, while the sum of these numbers is 19. The degree sum condition is the first thing we should check. Note that the sequence $1, 1, 1, 1, 1$ also violated that condition.

We have a prerequisite for a nonincreasing sequence to be graphical. Its sum must be even, as was indicated in Corollary 2.1a. This condition, however, is not sufficient, as the nongraphical sequence $4, 4, 4, 4$ shows. A graph with this degree sequence would have eight edges since the degree sum would be 16. But a graph with four vertices has at most $\binom{4}{2} = 6$ edges.

To determine if a given sequence is graphical, we need the following construction. Given a nonincreasing sequence S of n nonnegative integers, we attempt (see step 2) to form a new sequence S' with $n-1$ integers, as in Algorithm 2.1. (An **algorithm** is a sequence of steps used to solve a problem.)

Algorithm 2.1 (Graphical Degree Sequence)
1. Delete the first number, say k, from S.
2. Subtract 1 from each of the next k terms of S if this is possible—that is, if the next k integers are each at least 1. Call the resulting sequence S'. If S' cannot be formed, stop; the original sequence is not graphical. If all terms of the current sequence are zero, stop; the sequence is graphical.
3. Rearrange the sequence obtained so that it is a sequence $S*$ in nonincreasing order.
4. Let $S = S*$, and return to step 1.

If S' cannot be formed in Algorithm 2.1, then S is obviously not graphical (because some vertex is claiming to be adjacent to more vertices than there are available). It is amazing that the converse is also true. This was discovered independently by Havel [3] and Hakimi [2]. A proof can be found in Buckley and Harary [1, p. 174].

Theorem 2.2 If S' is obtained from S by the aforementioned procedure, then S is graphical if and only if S' is graphical. ∎

Of course, it may be just as difficult to decide whether S' is graphical as it is to decide whether S is. Notice, however, that S' is shorter than S. Step 4 of Algorithm 2.1 tells us to repeat the construction, producing an even shorter sequence. This algorithm yields a succession of sequences, which get progressively shorter until we obtain a sequence short enough to analyze. This last sequence is graphical if and only if S is graphical.

Example 2.11 Use Algorithm 2.1 to determine whether $3, 2, 2, 1, 1, 1$ is graphical.

Solution
Step 1 of Algorithm 2.1 yields $2, 2, 1, 1, 1$. Step 2 yields $1, 1, 0, 1, 1$. (We subtracted 1 from the first three numbers.) We rearrange this sequence according to step 3 and obtain $1, 1, 1, 1, 0$. We return to step 1 and apply the procedure to our new sequence. Step 1 now produces $1, 1, 1, 0$. After step 2, we have $0, 1, 1, 0$. In step 3, we rearrange to get $1, 1, 0, 0$. Then we return to step 1 once again. On this round, we get $1, 0, 0$ after step 1, and then get $0, 0, 0$. Step 2 then tells us that since all terms of our current sequence are zero, the original sequence is graphical, and we stop. ◇

2.3 Isomorphic Graphs

Note that to test a sequence by computer, we would generally write a computer program that would accept a sequence as input and run through the steps of the algorithm until it terminates. If we, as humans, are testing a sequence, we can sometimes stop a little earlier—namely, when we reach a sequence that we are sure is either graphical or not graphical. For example, after step 3 on the first pass through the algorithm, we obtained $1, 1, 1, 1, 0$, which is obviously graphical. It is the degree sequence of a graph that has three components: two K_2's and a K_1. We may therefore conclude that $3, 2, 2, 1, 1, 1$ is graphical. In fact, this is the degree sequence of each of the graphs of Figure 2.20. Thus, when doing this problem by hand, we require only one application of the procedure before we can draw our conclusion. We aren't always this fortunate, as our next example shows.

Example 2.12 Determine whether $5, 4, 4, 3, 3, 3, 3, 2, 2, 1$ is graphical. Stop as soon a sequence that is definitely graphical of nongraphical is recognized.

Solution
By using step 1 of Algorithm 2.1, we get $4, 4, 3, 3, 3, 3, 2, 2, 1$; step 2 then gives $3, 3, 2, 2, 2, 3, 2, 2, 1$, which becomes $3, 3, 3, 2, 2, 2, 2, 1$ at step 3. Since it is not clear whether $3, 3, 3, 2, 2, 2, 2, 1$ is graphical, we repeat the process on $3, 3, 3, 2, 2, 2, 2, 1$. At step 1, we get $3, 3, 2, 2, 2, 2, 1$; step 2 then gives $2, 2, 1, 2, 2, 2, 1$, which becomes $2, 2, 2, 2, 2, 1, 1$ at step 3. From our earlier observations, we recognize that $2, 2, 2, 2, 2, 1, 1$ is the degree sequence of the path P_7 (there are three other graphs with the degree sequence $2, 2, 2, 2, 2, 1, 1$; see Exercise 4). Since $2, 2, 2, 2, 2, 1, 1$ is graphical, so is $3, 3, 3, 2, 2, 2, 2, 1$, and therefore $5, 4, 4, 3, 3, 3, 3, 2, 2, 1$ is graphical as well. ◊

We shall now look at the problem of generating graphs with a given degree sequence. We begin with a case where there is just one graph with the given sequence.

Example 2.13 Show that there is precisely one graph that has degree sequence $5, 3, 2, 2, 1, 1$.

Solution
To generate the graph, first note that it must have six vertices because there are six numbers in the degree sequence, one for each vertex. Now begin with the maximum degree, which is five. Some vertex, call it a, is adjacent to five other vertices, call them b, c, d, e, f. We have now accounted for all six vertices. Any additional adjacencies can only occur between the new vertices b, c, d, e, f. Currently our graph is the star $K_{1,5}$ [see Figure 2.21(a)] and, except for their labels, the vertices b, c, d, e, f are equivalent to one another. Since the next highest degree is 3, one of the vertices b through f, say b, must have two additional adjacencies, say c and d [see Figure 2.21(b)]. But now the remaining vertices have degrees 2, 2, 1, and 1. Thus no additional edges may be added. We remove the labels and get the unique graph having degree sequence $5, 3, 2, 2, 1, 1$ [see Figure 2.21(c)]. ◊

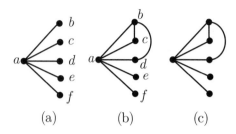

(a)   (b)   (c)

Figure 2.21 Construction of the graph with degree sequence $5, 3, 2, 2, 1, 1$.

Example 2.14 There are two nonisomorphic graphs having degree sequence $4, 4, 3, 2, 2, 1$. Draw them and explain why they are not isomorphic to one another.

Solution
We begin with a similar procedure to Example 2.13. First, note that there are six vertices and eight edges (the sum of the degrees is sixteen). Begin with a vertex of degree 4, call it a, and join it to four other vertices, say b, c, d, e. The graph so far is displayed in Figure 2.22(a). At this point, there are five vertices. Next we consider where the other vertex of degree 4 is. It cannot be the missing vertex, say f. To see why, observe that if it were, then f would have to be adjacent to each of b, c, d, e. But then there could be no vertex of degree 1 as required. Thus, one of b, c, d, e must be the other vertex of degree 4. All four vertices are equivalent, except for their labels, so let b have degree four. Thus, vertex b has three adjacencies in addition to vertex a. Now there are two possibilities. Either b is adjacent to c, d, and e (case 1) or b is adjacent to f and two of c, d, and e, say c and d, without loss of generality (case 2).

In case 1, our graph currently looks as in Figure 2.22(b). Note that c, d, and e are equivalent (they have the same degree and the same neighbors). The missing vertex f is adjacent to one of them, so without loss of generality, let it be e. This completes that graph, so remove the labels to get the graph in Figure 2.22(d).

In case 2, our graph currently looks as in Figure 2.22(c). Note that c and d are equivalent. By the symmetry in the graph, so are e and f. We need to preserve a vertex of degree 1, so let it be f. Then e must be adjacent to one of c or d (which are equivalent), so say d. Finally, remove the labels to get the graph in Figure 2.22(e). To see that the graphs in Figure 2.22(d) and (e) are nonisomorphic, observe, that in (d) the vertex of degree 1 is adjacent to a vertex of degree 3, whereas in (e), the end vertex is adjacent to a vertex of degree 4. ◇

2.3 Isomorphic Graphs

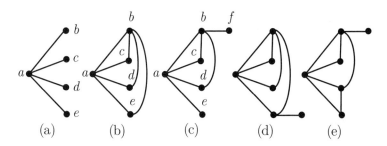

Figure 2.22 Construction of the graphs with degree sequence 4, 4, 3, 2, 2, 1.

There is no efficient algorithm for distinguishing nonisomorphic graphs. Indeed, finding better isomorphism testing algorithms is an active area of graph theory research. There are, however, several techniques that work well for graphs of moderate size. They are based for the most part on graph invariants. A **graph invariant** is a function defined on graphs with the property that isomorphic graphs take on the same function value. You already know several graph invariants: the order, the size, the degree sequence, and the number of components. To explore this further, note that if $G \cong H$, then it is necessarily true that G and H have the same order. Thus the order is a graph invariant. Similarly, if $G \cong H$, then G and H must have the same degree sequence. Therefore, the degree sequence is a graph invariant. We will learn about numerous additional invariants throughout the text. The important thing to note at this point is that if f is a graph invariant and $f(G) \neq f(H)$, then $G \not\cong H$. So showing that graphs G and H have a different value for a given graph invariant proves that G and H are *nonisomorphic*.

> **Note 2.4** Some items to check when trying to show that a pair of graphs are not isomorphic are as follows:
> (1) the number of vertices
> (2) the number of components
> (3) the number of edges
> (4) the degree sequence
> (5) the length of a geodesic between a pair of vertices with a given unique degree
> (6) the length of the longest path in the graph
> (7) degree of neighbors for a vertex of unique degree

We used part (7) of Note 2.4 to distinguish the graphs in Figures 2.22(d) and (e) in Example 2.14. We could have used part (5) of Note 2.4 instead. A geodesic joining the vertex of degree 2 to the vertex of degree 3 in graph (d) has length 1, whereas in graph (e) it has length 2. Thus, graphs (d) and (e) of Figure 2.22 are not isomorphic.

Exercises 2.3

1. Three nonisomorphic graphs have degree sequences $3, 2, 2, 1, 1, 1$. Construct them.

2. Draw the graph having degree sequence 2,1,1,0.

3. Three nonisomorphic graphs have degree sequence $5, 3, 2, 2, 1, 1, 1, 1$. Construct them. *Hint*: One of them is disconnected.

4. There are three nonisomorphic graphs (all disconnected) besides P_7 that have degree sequence $2, 2, 2, 2, 2, 1, 1$. Draw them.

5. Show that the two graphs of Figure 2.20 are nonisomorphic.

6. Formally prove that isomorphic graphs have the same degree sequence.

7. Prove Note 2.2.

8. Show that the sequence $3, 3, 3, 3, 3, 3, 3, 3$ is graphical. Then draw two graphs with this degree sequence, one connected and the other disconnected.

9. Show that the sequence $n - 1, 3, 3, 3, \ldots, 3$ of length $n \geq 4$ is graphical.

10. If the first integer in a nonincreasing sequence of nonnegative integers is m, at least how long must the sequence be in order for it to (possibly) be graphical?

11. Given a graph G with degree sequence $d_1, d_2, d_3, \ldots, d_k, d_{k+1}, \ldots, d_n$, show that there exists a graph H with degree sequence (out of order, perhaps) $k, d_1 + 1, d_2 + 1, d_3 + 1, \ldots, d_k + 1, d_{k+1}, \ldots, d_n$ by showing how to construct H from G. Of course, Theorem 2.2 guarantees that this new sequence is graphical.

12. Show that the sequence $k, k, k, \ldots, k$ of length $k + 1$ is graphical using Algorithm 2.1, thereby proving that there exist k-regular graphs of order $k + 1$. Identify this class of graphs.

13. How many vertices and edges does the graph G have if its degree sequence is $d_1, d_2, d_3, \ldots, d_n$? Using your answer and equation (2.2), explain why $5, 4, 4, 4, 4$ is not graphical. Now find an even easier way to do this.

14. Prove that C_4 and $K_{2,2}$ are isomorphic.

15. Prove that $K_{1,2}$ and P_3 are isomorphic.

2.4 Graph Operations

16. Draw two different 3-regular graphs on six vertices.

17. Draw two different 4-regular graphs on seven vertices. Then prove that they are nonisomorphic.

18. Draw two nonisomorphic disconnected subgraphs of C_5 that have four vertices.

19. Consider the complete graph K_7 with its vertices having distinct labels. How many of its labeled subgraphs on five vertices are isomorphic to K_5?

20. Prove that the two graphs in Figure 2.23 are isomorphic by finding an appropriate isomorphism. Show all vertex and edge correspondences.

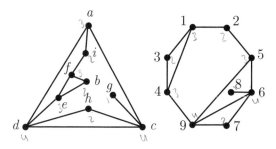

Figure 2.23

21. Explain why the graphs G and H in Figure 2.24 are *not* isomorphic by finding some *characteristic* that distinguishes them from one another. (Simply quoting the definition of isomorphism is not sufficient.)

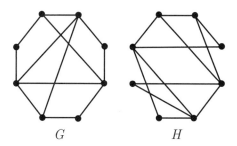

Figure 2.24

2.4 Graph Operations

Many interesting graphs are obtained by combining two or more graphs or by operating on a single graph in a particular way. In this section, we describe numerous operations that are used to obtain new graphs from old ones.

Unions and Joins

Perhaps the simplest graph operation is the union of two graphs. If G and H are disjoint graphs, their **union** $G \cup H$ is the graph with $V(G \cup H) = V(G) \cup V(H)$ and $E(G \cup H) = E(G) \cup E(H)$. Thus, $G \cup H$ consists of a copy of G together with a copy of H. Figure 2.25(a) shows $P_3 \cup P_4$. The **join** of disjoint graphs G and H, denoted $\boldsymbol{G + H}$, is formed by appending to $G \cup H$ edges that have one end vertex in G and the other in H. If G and H have m and n vertices, respectively, we must add mn edges to $G \cup H$. The graph of Figure 2.25(b) shows $P_3 + P_4$.

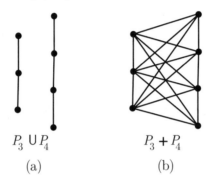

Figure 2.25 The union and the join of the paths P_3 and P_4.

The wheels are a well-studied class of graphs constructed using the join operation. For $n \geq 3$, the **wheel** $\boldsymbol{W_{1,n}}$ is the join of K_1 with C_n—that is, $\boldsymbol{W_{1,n} = K_1 + C_n}$.

Example 2.15 Draw the wheels $W_{1,n}$ for $3 \leq n \leq 5$.

Solution

For each n, we place the K_1 inside the cycle C_n and then join the K_1 to each vertex of C_n. The resulting graphs are displayed in Figure 2.26. You may notice that $W_{1,3} \cong K_4$. ◇

Figure 2.26 The wheels $W_{1,3}$, $W_{1,4}$, and $W_{1,5}$.

2.4 Graph Operations

Another graph class that uses joins has been useful in recent research of the authors. The **sequential join** $G_1 + G_2 + \cdots + G_k$ of graphs $G_1, G_2, \ldots, G_k$ is the graph formed by taking one copy of each of the graphs $G_1, G_2, \ldots, G_k$ and adding in additional edges from each vertex of G_i to each vertex of G_{i+1}, for $1 \leq i \leq k-1$. The sequential join $K_1 + K_2 + K_2 + P_3 + K_1$ is shown in Figure 2.27.

Figure 2.27 The sequential join $K_1 + K_2 + K_2 + P_3 + K_1$.

Edge or Vertex Deletion

If v is a vertex of G, the graph $\boldsymbol{G-v}$ is the graph formed from G by removing v and all edges incident with v. See Figures 2.28(a) and (b). When we remove an edge from a graph, we do not remove th vertices incident with that edge. If e is an edge of G, the graph $G-e$ is the graph formed from G by removing e from G. See Figure 2.28(a) and (c).

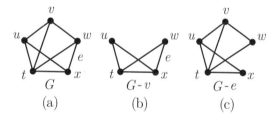

Figure 2.28

Complements

The complement $\bar{\boldsymbol{G}}$ of a graph G has $V(\bar{G}) = V(G)$ and $uv \in E(\bar{G})$ if and only if $uv \notin E(G)$. In plain English, G and its complement have the same vertices, while the complement of G has precisely the edges that G lacks. Of course, we could say that G has exactly the edges that its complement lacks. We could say that $E(G)$ and $E(\bar{G})$ *complement* each other; hence the terminology. It is clear then that the complement of $\bar{G}$ is G itself—that is, performing the complement operation twice in succession gets one back to the original graph. You may notice a connection between this operation and the complements of sets that we examined in Section 1.1. In finding the complement of a graph, we know that $V(\bar{G}) = V(G)$. The only item in question is, What is $E(\bar{G})$? In the context of sets, if our graph has n vertices, then $E(K_n)$ plays the role of the universal set, and then $E(\bar{G}) = (E(G))'$.

Figure 2.29 shows an (unlabeled) graph G and its complement. Notice that G is connected, while its complement is not. This need not happen in general. This might leave you wondering whether a graph and its complement might both be disconnected. Our next theorem guarantees that this cannot happen.

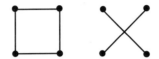

Figure 2.29 A graph G and its complement $\bar{G}$.

Theorem 2.3 *The complement of a disconnected graph is connected.*

Proof. For a disconnected graph G, the order must be at least 2. Thus let x and y belong to different components of a disconnected graph G. Then no vertex of G is adjacent to both x and y. Now consider $\bar{G}$. First, x and y are adjacent in $\bar{G}$. Second, any other vertex t is adjacent to at least one of x and y. To see this, note that t cannot be in the same component as both x and y in G. So, for example, if t is in a different component than x in G, then t is adjacent to x in $\bar{G}$. Now let w and z be any two vertices of $\bar{G}$. If w and z were nonadjacent in G, then they are adjacent in $\bar{G}$ and so are connected by a w–z path of length one in $\bar{G}$.

Now suppose that w and z were adjacent in G. Then they were in the same component in G. If they were both in the same component as x in G, then w, y, z is a w–z path in $\bar{G}$. Similarly, if they were both in the same component as y in G, then w, y, z is a w–z path in $\bar{G}$. Finally, if neither was in a component with x or y in G, then w, y, z is a w–z path in $\bar{G}$. Thus it follows that given any two vertices w and z of $\bar{G}$, there exists a w–z path. In fact, we can deduce even more. If G is disconnected, then any two vertices of $\bar{G}$ are at most three edges apart! ∎

Note that for any two graphs G and H, we have $\overline{(G + H)} = \bar{G} \cup \bar{H}$. This is because the construction of the complement of $G + H$ requires that we remove all the "join" edges and take the complements of G and H.

There is an important class of graphs related to complements. A graph G is **self-complementary** if G is isomorphic to $\bar{G}$. When we consider graphs with a small number of vertices, it is easy to determine which ones are self-complementary. When the number of vertices gets large, the problem is much more difficult. However, the following theorem helps.

2.4 Graph Operations

Theorem 2.4 For any self-complementary graph, the order n is of the form $4k$ or $4k+1$, where k is a nonnegative integer. Moreover, G has $n(n-1)/4$ edges.

Proof. Let G be a self-complementary graph of order n. Since G is isomorphic to $\bar{G}$, half of the possible $\binom{n}{2}$ edges joining the n vertices are in G and the other half are in $\bar{G}$. Therefore G and $\bar{G}$ each contain $\frac{1}{2}\binom{n}{2} = \frac{1}{2}\frac{n(n-1)}{2} = \frac{n(n-1)}{4}$ edges. Since $|E(G)|$ must be an integer, 4 must divide evenly into $n(n-1)$. Since just one of n and $n-1$ is even (n and $n-1$ are consecutive integers, so one is odd while the other is even), we must conclude that either 4 divides n or 4 divides $n-1$. If 4 divides n, then $n = 4k$; if 4 divides $n-1$, then $n = 4k+1$, where k is a nonnegative [to ensure that $|E(G)|$ is nonnegative] integer. ∎

Example 2.16 Find all self-complementary graphs of order at most 5.

Solution
By Theorem 2.4, the order of G must either be 1, 4 or 5. The only graph of order 1 is K_1, which is self-complementary. For order 4, Theorem 2.4 implies that the number of edges in G is three. There are three graphs that have four vertices and three edges—namely, $K_3 \cup K_1$, P_4, and $K_{1,3}$. Of those, just P_4 is self-complementary. For order 5, Theorem 2.4 implies that G must have five edges. There are six graph with five vertices and five edges (tables of such graphs may be found in Buckley and Harary [1] or Read and Wilson [4]). Of those, C_5 is self-complementary, as is the graph consisting of a triangle with an edge attached at each of two distinct vertices. All self-complementary graphs of order at most 5 are displayed in Figure 2.30. ◊

Figure 2.30 The four self-complementary graphs of order at most 5.

Cartesian Product

Given graphs G and H with vertex sets $\{u_1, u_2, \ldots, u_m\}$ and $\{v_1, v_2, \ldots, v_n\}$, respectively, their **cartesian product**, $\boldsymbol{G \times H}$, (read "G cross H") is the graph with vertex set consisting of mn vertices labeled (i, j), where $1 \leq i \leq m$ and $1 \leq j \leq n$. Two vertices (i, j) and (h, k) are adjacent in $G \times H$ if either

1. $i = h$ and v_j is adjacent to v_k in graph H, or
2. $j = k$ and u_i is adjacent to u_h in G.

We shall call an edge of the first type an $\boldsymbol{H}$ **edge**, and an edge of the second type a $\boldsymbol{G}$ **edge**. Each vertex (i, j) of $G \times H$ may be thought to have two "parents"—u_i in G and v_j in H. For each i, the induced subgraph on the vertices (i,j) where $j = 1, 2, \ldots, n$ is a copy of H, called the **ith H copy**. Likewise, for each j, the induced subgraph on the vertices (i,j) where $i = 1, 2, \ldots, m$ is called the **jth G copy**. Figure 2.31 exhibits graphs $G = K_1 + K_1 + K_2$, $H = P_3$ and their cartesian product $G \times H$.

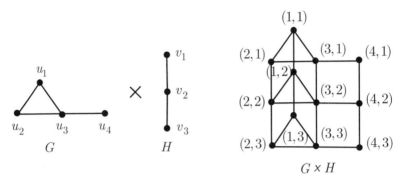

Figure 2.31 G, H, and $G \times H$.

The graph $G \times H$ in Figure 2.31 may be viewed as three copies of G corresponding to the three vertices of H. Think of replacing the vertices of H by copies of G. Notice that each G copy has a fixed second coordinate (see Figure 2.31). Furthermore, the vertices of the first G copy are rendered adjacent to the corresponding vertices of the second G copy, because v_1 is adjacent to v_2 in H. Similarly, we render the vertices of the second G copy adjacent to the corresponding vertices of the third G copy, because v_2 is adjacent to v_3 in H. On the other hand, corresponding vertices of the first and third G copies are not adjacent since v_1 and v_3 are not adjacent in H.

The preceding paragraph can be rewritten with minor adjustments if we switch our point of view and start with four H copies. These are the four "vertical" P_3's, each of which is characterized by a fixed first coordinate. In fact, $G \times H$ and $H \times G$ are isomorphic for any two graphs G and H, differing only in their labeling. It can also be shown that the cartesian product is associative—that is, $(G \times H) \times K = G \times (H \times K)$ for any three graphs G, H, and K.

Hypercubes

An important class of graphs that can be defined in terms of the cartesian product operation are the hypercubes. These graphs are important because of their usefulness in computer architecture of massively parallel computing. The **hypercube Q_n** is defined *recursively*—that is, each hypercube is constructed from the previous one (after we define the first one!). The hypercubes are defined by $Q_1 = K_2$ and $Q_n = K_2 \times Q_{n-1}$. The second

2.4 Graph Operations

equation says that we get Q_n by taking two copies of Q_{n-1} and rendering corresponding vertices adjacent. The name "hypercube" comes from the fact that Q_3 may be drawn as a cube in three dimensions and the concept is then extended to higher dimensions.

The two vertices of Q_1 are labeled 0 and 1. As Q_2 contains two copies of Q_1, we label the vertices in the first copy 00 and 01, while the vertices in the second copy are labeled 10 and 11. Let's do this one more time before we generalize. Since Q_3 consists of two copies of Q_2, we attach a 0 to the labels of the vertices of the first copy of Q_2 and a 1 to the labels of the vertices of the second copy. Thus, $V(Q_3) = \{000, 001, 010, 011, 100, 101, 110, 111\}$. In Figure 2.32, the third graph is Q_3 which is composed of labeled "inner" and "outer" copies of Q_2. The outer labels start with 0's, while the inner ones start with 1's.

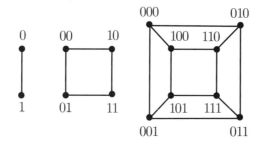

Figure 2.32 The first three hypercubes.

More generally, the vertex set of Q_n consists of all of the binary strings of length n. Since each digit is a 0 or a 1, there are 2^n such strings, implying that $|V(Q_n)| = 2^n$. This can also be seen by induction. It is obviously true when $n = 1$, since Q_1 has two vertices. Since Q_n consists of two copies of Q_{n-1}, the order of Q_n is twice that of Q_{n-1} and the inductive leap follows.

By the recursive definition, we see that the degree of each vertex of Q_n is n—that is, Q_n is n-regular. This is certainly true when $n = 1$. Each time we increment n, we attach another edge to each vertex, causing the regularity to persist for all n.

Two vertices in Q_n are adjacent if and only if their labels differ in exactly one place. Thus 11100 and 11000 are adjacent in Q_5, as they differ precisely in the third place. By the same reasoning, 11100 and 11001 are not adjacent. They differ in two places—the third and the fifth.

Meshes

Another graph that can be formed using the cartesian product is a **mesh**, also called a **grid** or **lattice**. The **2-mesh $M(m,n)$** consists of the cartesian product of P_m with P_n—that is, $M(m,n) = P_m \times P_n$. The **3-mesh**

$M(a, b, c)$ consists of the cartesian product of P_a with P_b and P_c. Thus $M(a, b, c) = (P_a \times P_b) \times P_c = P_a \times P_b \times P_c$. Since the cartesian product is associative, we can omit the parentheses. This can be extended to n-meshes. The **n-mesh** $\boldsymbol{M(a_1, a_2, \ldots, a_n)}$ is the cartesian product of paths of orders $a_1, a_2, \ldots, a_n$; $M(a_1, a_2, \ldots, a_n) = P_{a_1} \times P_{a_2} \times \cdots \times P_{a_n}$. The wonders of mathematics never cease. The mesh $M(4, 3)$ is shown in Figure 2.33, in which the four vertices of degree 2 are labeled.

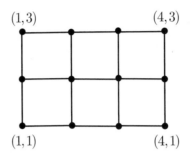

Figure 2.33 The mesh $M(4, 3)$.

The 3-mesh $M(3, 2, 3)$ is shown in Figure 2.34 with several vertices labeled. The general 3-mesh $M(a, b, c)$ may be viewed as a parking garage with several identical levels, c to be exact. The third coordinate identifies the level. Each level is a 2-mesh with b rows and a columns. If a, b, and c are large, it is best to memorize them well; otherwise, you may not find your car in the parking lot after you have finished shopping.

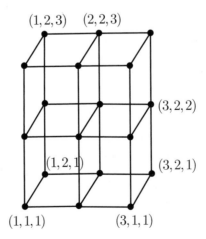

Figure 2.34 The 3-mesh $M(3, 2, 3)$.

2.4 Graph Operations

Line Graphs

For any graph G, its **line graph** $\boldsymbol{L(G)}$ has vertex set consisting of the edges of G. Two vertices of $L(G)$ are adjacent if the corresponding edges of G have a vertex in common. It helps to label the edges when forming the line graph. This is done for graph G in Figure 2.35, but not graph H. In graph G edge ab shares a vertex with ac as well as bc. Edges ac and bc each share an end vertex with all other edges. Edge cd just shares a vertex with ac and bc. Although the edges of H are not labeled, can you see why the line graph must be the wheel $W_{1,4}$?

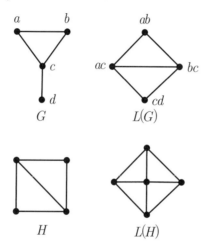

Figure 2.35 Two graphs and their line graphs.

Edge Contraction

As a final graph operation, we consider edge contraction. Let uv be an edge in graph G. Then the graph G/uv is the graph obtained from G by removing both u and v as well as any edges incident with them and then inserting a new vertex, which we will call uv^*. Vertex uv^* is then rendered adjacent to each vertex that had been adjacent to u or v (or both). In Figure 2.36, we display a graph G with two different edges contracted.

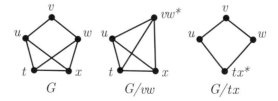

Figure 2.36 A graph and two graphs obtain by edge contraction.

Exercises 2.4

1. Draw the union $G \cup H$ and the join $G + H$ for graphs G and H from Figure 2.31.

2. Give a formula as a function of n for the number of edges in the hypercube Q_n for all n and explain why your formula is correct.

3. Draw the wheel $W_{1,6}$.

4. How many edges does $W_{1,n}$ have?

5. Prove that Q_3 is isomorphic to the mesh $M(2,2,2)$.

6. If G is a k-regular graph on n vertices, show that $\bar{G}$ is $(n-1-k)$-regular.

7. Show that the wheel has a spanning subgraph that is a cycle—that is, show that C_{n+1} is a spanning subgraph of $W_{1,n}$.

8. Consider K_3 with vertices a, b, and c. Now obtain a new graph, A, by adding two more vertices, r and s, and edges ra and sb. Show that A is self-complementary.

9. Let G be the graph in Figure 2.17(a). Construct each of the following graphs: $G - a$, $G - d$, G/ab, G/cd, and G/ad.

10. Show that every 3-mesh has a spanning 2-mesh.

11. Given the n-mesh $M = M(a_1, a_2, \ldots, a_n)$, where $a_i \geq 3$, for each i, show that $\delta(M) = n$ and $\Delta(M) = 2n$. If k satisfies $n < k < 2n$, show that there is a vertex $v \in M$, such that $\deg(v) = k$. Is this still true without the condition $a_i \geq 3$?

12. Assume that G and H are graphs where $V(G) = \{u_1, u_2, \ldots, u_m\}$ and $V(H) = \{v_1, v_2, \ldots, v_n\}$. Let (i,j) be a vertex in $G \times H$. Prove that $\deg(i,j) = \deg(u_i) + \deg(v_j)$.

13. Show that given an positive integer n, there exists a self-complementary graph G with $|V(G)| = 4n$.

14. Show that the graph obtained by the following elaborate construction is self-complementary: Start with any self-complementary graph A and any graph B. In the third graph displayed in Figure 2.30, replace the vertex of degree 2 by a copy of A, replace the two vertices of degree 3 by copies of B, and replace the two end vertices by copies of $\bar{B}$. Finally, replace the edges of G by joins.

15. Show that $K_{m,n} \cong \bar{K}_m + \bar{K}_n$.

2.4 Graph Operations

16. What familiar graph is $K_m + K_n$ isomorphic to?

17. Explain why the complement of $G + H$ is disconnected for all pairs of graphs G and H.

18. If $G \cong H$, show that $\bar{G} \cong \bar{H}$.

19. Given a graph G on six vertices, show that G or $\bar{G}$ must contain K_3 as a subgraph. Hint: Select a vertex $u \in V(G)$. Show that u must have degree at least 3 in exactly one of G and $\bar{G}$. Without loss of generality, assume $\deg(u) \geq 3$ in G. Let x, y, and z be *neighbors* of u in G—that is, x, y, and z are each adjacent to u in G. Now consider the possible adjacencies among x, y, and z in G and in $\bar{G}$. After you finish the proof, explain why this fact is not true if a G has only five vertices. Why is the fact always true if G has more than six vertices? (*Note*: We will study more problems of this sort when we study Ramsey theory in Section 11.1.)

References for Chapter 2

1. Buckley, F., and F. Harary, *Distance in Graphs*, Addison-Wesley, Redwood City, CA (1990).

2. Hakimi, S. L., On the realization of a set of integers as the degrees of the vertices of a graph. *Journal SIAM Applied Mathematics* 10 (1962) 496–506.

3. Havel, V., A remark on the existence of finite graphs, [in Czech]. *Časopis Pěst. Mat.* 80 (1955) 477–480.

4. Read, R. C., and R. J. Wilson, *An Atlas of Graphs*, Oxford University Press, New York (1998).

Additional Readings

5. Bain, V., An algorithm for drawing the n-cube. *College Mathematics Journal* 29 (1998) 320–322.

6. Bertram, E., and P. Horak, Some applications of graph theory to other parts of mathematics. *The Mathematical Intelligencer* 21:3 (Summer 1999) 6–10.

7. Bivens, I., and S. L. Davis, Some graphs whose vertices pair off by degree: Part 1. *College Mathematics Journal* 27 (1996) 127–135.

8. Chartrand, G., P. Erdős, and O. R. Oellermann, How to define an irregular graph. *College Mathematics Journal* 19 (1988) 36–42.

9. Fan, C. K., B. Poonen, and G. Poonen, How to spread rumors fast. *Mathematics Magazine* 70 (1997) 40–42.

10. Hayes, B., How to avoid yourself. *American Scientist* 86 (1998) 314–319.

Chapter 3

Trees and Bipartite Graphs

One class of graphs is so important that it deserves treatment in its own chapter. These graphs are trees. A graph with the property that no subgraph is a cycle is called **acyclic**. A connected acyclic graph is called a **tree**. Thus a tree is connected but has no cycle of either parity. Trees arise in many applications, such as analyzing business hierarchies and determining minimum cost transportation networks, and are the basis of important data structures in computer science. A tree is a special type of bipartite graph. In this chapter, we shall examine trees and bipartite graphs and their uses.

3.1 Properties of Trees

You have already met two important classes of trees—namely, stars and paths. The path P_n may be defined as a tree T on n vertices, such that $\Delta(T) \leq 2$. (Recall that Δ means maximum degree.) The star $K_{1,n}$ may be defined as the tree or order $n+1$, such that one vertex, v, satisfies $\deg(v) = n$. Since $K_{1,n}$ is a tree, the other vertices must be end vertices (sometimes called **leaves** of the tree), or we would have a cycle. Refer to the tree, T, of Figure 3.1 to help verify the following observations.

Some Properties of Trees

P3.1. Every nontrivial tree (the trivial graph K_1 is also a tree) has at least two end vertices.

P3.2. The deletion of any edge of a tree disconnects it. (An edge of a connected graph whose deletion disconnects the graph is called a **bridge**. Thus all edges of a tree are bridges. Incidentally, even if G is disconnected, an edge of G is a bridge if its deletion increases the number of components of G.)

P3.3. Given two vertices, x and y, of a tree, there is a *unique* x–y path.

Hence, that path is a geodesic. We might say that there are no travel alternatives in a tree. Contrast this with the two geodesics between a first and fourth consecutive vertex on the cycle C_6.

P3.4. A tree with n vertices and q edges satisfies $q = n - 1$. Thus, a tree is minimally connected. (See property P3.2.)

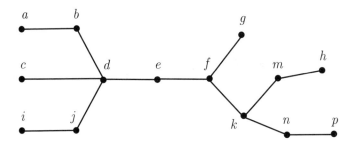

Figure 3.1 A tree, T, with $n = 14$ and $q = 13$.

You will be asked to prove several of these properties in the Exercises. We will encounter additional properties of trees in Chapter 4 when we study distance concepts for graphs. Trees are so important in graph theory that a whole book was written just on the topic of counting labeled trees (see Moon [7]). One reason that trees are so important is that their simple structure makes them ideal for testing conjectures. That is, sometimes when we are trying to prove that a statement is true for all graphs, we first prove the statement true for trees. Another reason for the importance of trees is that they form an important class of data structures in computer science.

Let's prove property P3.4. We use induction on n. When $n = 1$, the tree is K_1, which has $q = n - 1 = 1 - 1 = 0$ edges. So P3.4 is true when $n = 1$. Now assume that P3.4 is true when $n = k$—that is, assume that any tree with k vertices has $k - 1$ edges (this is the inductive hypothesis). Let T be a tree with $n = k + 1 \geq 2$ vertices. We must show that T has $q = n - 1 = (k + 1) - 1 = k$ edges (this is the *goal*). Consider an end vertex $v \in T$ incident with edge vu. There must be such a vertex (in fact, there must be at least two; see property P3.1) or T would have a cycle. If we remove v and its incident edge vu, we obtain a tree T' with k vertices. By the inductive hypothesis, T' has $k - 1$ edges. Now reattach edge vu at vertex u of T'. By doing so, we increase the number of vertices by 1 (with vertex v added) to $k + 1$ and the number of edges by 1 (with vu added) to k. Thus T has $n = k + 1$ vertices and $q = k$ edges. We have reached the goal. Thus property P3.4 is true for all $n \in N$. ∎

Note that a graph satisfying the relation $q = n - 1$ need not be a tree. If the graph is disconnected and has cycles, the relation $q = n - 1$ could still hold but will rarely do so. However, if we also insist that the graph be connected in addition to satisfying $q = n - 1$, then the graph will definitely be a tree. (See Exercise 8).

Graphical Sequences of Trees

In Section 2.3, we presented a procedure for deciding when a nonincreasing sequence of numbers is graphical—that is, whether a graph exists with degrees given by the sequence. We shall now present a simple way to decide whether such a sequence is the degree sequence of a tree.

Theorem 3.1 Given a sequence S of n positive integers $d_1, d_2, \ldots, d_n$ such that $d_1 \geq d_2 \geq \cdots \geq d_n$ and $d_1 + d_2 + \cdots + d_n = 2(n-1)$, then there exists a tree whose degree sequence is S.

Proof. We shall use induction. The theorem is obvious when $n = 1$ or 2. Let us assume that it is true when $n = k$ and show that the theorem is also true when $n = k + 1$. We must, therefore, show that given a nonincreasing sequence $S = d_1, d_2, \ldots, d_{k+1}$ such that $d_1 + d_2 + \cdots + d_{k+1} = 2(k+1-1) = 2k$, there exists a tree T on $k + 1$ vertices whose degree sequence is S.

For starters, at least one of the numbers in S must be 1, or their sum would be at least $2(k+1) > 2k$. Then $d_{k+1} = 1$. Let's form a new sequence S' by deleting d_1 and d_{k+1} from S and adding $d^* = d_1 - 1$ in its correct position. Then S' has k entries and has sum $2k - 2 = 2(k-1)$, thereby satisfying the inductive hypothesis. We can, therefore, construct a tree T' with degree sequence S'. Now add a single vertex v to S' and add an edge from v to the vertex of degree d^* (raising its degree to d_1 and giving v degree 1), thereby producing a tree T with degree sequence S. ∎

Nonisomorphic Trees

In Section 2.3, we studied techniques to determine when two graphs are isomorphic. Graphs G and H are isomorphic if there is a one-to-one onto function f that maps $V(G)$ into $V(H)$ and preserves adjacency of the vertices. That is, $f(u)$ and $f(v)$ are adjacent in H precisely when u and v are adjacent in G. For trees, the most useful of these techniques is comparing degree sequences, looking at longest paths, and looking at shortest paths between vertices of a given unique degree.

Example 3.1 Find a structural property of the trees T_1 and T_2 in Figure 3.2 that show they are not isomorphic.

Solution

Trees T_1 and T_2 both have nine vertices, eight edges, and degree sequence $4, 3, 3, 1, 1, 1, 1, 1, 1$, so those items do not help distinguish T_1 from T_2. Also, they each have a longest path of length 4, so that doesn't help. But note that in T_1, the vertices of degree 3 are joined by a geodesic of length 2, whereas in T_2, the vertices of degree 3 are adjacent. Another distinguishing feature is that in T_1, the vertex of degree 4 has two neighbors of degree 1, but in T_2, the vertex of degree 4 has three neighbors of degree 1.   ◇

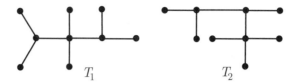

Figure 3.2

Example 3.2 There are six trees that have six vertices. Draw them.
Solution
The trees are displayed in Figure 3.3. Note that in doing a problem such as this, we find it easiest to organize our graphs according to the length of a longest path in the tree.   ◇

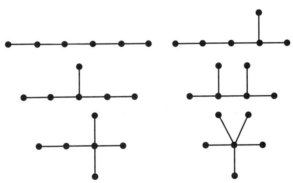

Figure 3.3

The Number of Leaves of a Tree

Given a nontrivial tree T, define n_i as the number of vertices of degree i. It is an interesting fact that the number of leaves (or end vertices) of a tree (that is, n_1) can be determined from the numbers $n_3, n_4, n_5, \ldots$.

Theorem 3.2 The number of leaves of a nontrivial tree is given by the formula

$$n_1 = 2 + n_3 + 2n_4 + 3n_5 + \cdots = 2 + \sum_{i=3}^{\infty} (i-2)n_i \qquad (3.1)$$

(Before we get to the proof, first note that the term n_2 is not present.)
Proof. This formula clearly works for paths of any length greater than zero, since n_2 is conspicuously absent and $n_i = 0$ for $i \geq 3$, for any path. So we

3.1 Properties of Trees

get $n_1 = 2$ for any nontrivial path. This will help us prove equation (3.1) for arbitrary nontrivial trees. Assume that T is a nontrivial tree that is not a path. To do this, we will draw the given tree as a **rooted tree**. We first arbitrarily choose a vertex of degree 1 as the root. (Note that a root need not have degree 1 in general, but it is convenient for us to choose such a vertex as the root and does not affect the result). The location of the root in the drawing is called row (or level) zero. The neighbors of the root are drawn in one horizontal row below the root. These vertices are called the "children of the root." The children of all of the vertices in the first row are drawn in a row beneath the first row. This process continues until all vertices have been listed. The number of rows below the root is called the **height** of the tree rooted at r. This is also the length of a longest path emanating from r. See Figure 3.4.

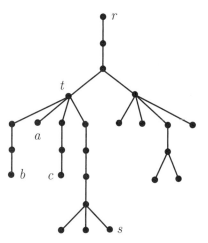

Figure 3.4 A rooted tree, T, of height 8.

Note that we may generate the tree of Figure 3.4 (or any other rooted tree) in stages. Start with a path from the root to some leaf, yielding the two of the leaves of equation (3.1)—namely, the root and the leaf we ended up at. (In Figure 3.4, this might be the r–c path, for example.) Once this is done, for each vertex v on this path whose degree in T is greater than 2, generate a new path (starting at v) that terminates at a new leaf for each *new* edge incident with v. These paths must terminate at new leaves since the tree is finite and has no cycles. Iterate this process for vertices of degree greater than 2 in T that are on any of the subsequent paths that are generated. A vertex u of degree i will be incident with $i - 2$ new edges, since two edges incident with u are on the original paths. This explains the $i - 2$ coefficient of n_i in equation (3.1). (In Figure 3.4, vertex t has degree 5 and generates three new paths with leaves a, b and s, respectively; vertex c had already been accounted for from the original r–c path.) This procedure will eventually generate the entire tree. This is so because if a vertex v in an intermediate tree does not have its degree fully accounted for, we span out from v via additional paths. The tree is connected, so every leaf must have a path to the root. The tree is finite, so the process must eventually terminate. When it does, all degrees will have been accounted for and we will find that equation (3.1) adds up the number of leaves in all the paths (and therefore the tree) that were generated along the way. ∎

Saturated Hydrocarbons

A **hydrocarbon** is a chemical compound whose molecules contain only carbon and hydrogen atoms, denoted by C and H, respectively. Some hydrocarbons whose names may be familiar to you are methane, ethylene, acetylene, propane, and benzene. Hydrocarbons are generally flammable.

The star $K_{1,4}$ models the methane molecule, which has one carbon and four hydrogen atoms. The chemical formula for methane is CH_4, which is displayed in Figure 3.5. When chemists who work with hydrocarbons draw this molecule, they leave out the hydrogen atoms, considering the hydrogens to be "understood." Thus, they would draw the molecule for methane as K_1 with the vertex labeled C.

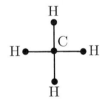

Figure 3.5 $K_{1,4}$ models the methane molecule.

It is helpful to note that in hydrocarbons, the carbon atoms always have degree (chemists say *valence*) 4 and hydrogen atoms have degree 1.

A **saturated hydrocarbon** has the maximum possible number of hydrogen atoms for the number of carbon atoms in the molecule. It can be shown by using Theorem 2.1 that the number of hydrogen atoms is $2n + 2$ when there are n carbon atoms (see Exercise 23). Thus the general formula for a saturated hydrocarbon is C_nH_{2n+2}. Why are we considering saturated hydrocarbons here? Because the graph that models them is always a tree. Methane is the simplest saturated hydrocarbon. If two distinct molecules have the same chemical formula (that is, the same number of each type of atom composing them), those molecules are called **isomers**. Example 3.3 deals with isomers.

Example 3.3 There is a pair of isomers with chemical formula C_4H_{10}. Each has a treelike structure. Find the isomers.

3.1 Properties of Trees

Solution
Since each carbon atom has degree 4 and each hydrogen atom has degree 1, we are looking for two nonisomorphic trees with degree sequence $4, 4, 4, 4, 1, 1, 1, 1, 1, 1, 1, 1, 1, 1$. Taking a hint from chemists, we focus on the four carbon atoms and how they can be linked to one another in the tree. There are only two ways—namely, as a path P_4 or as a star $K_{1,3}$. We begin with these two structures and attach hydrogen atoms to fill out the degree of each carbon to 4. The resulting molecules are displayed in Figure 3.6 and are called butane and isobutene, respectively. ◇

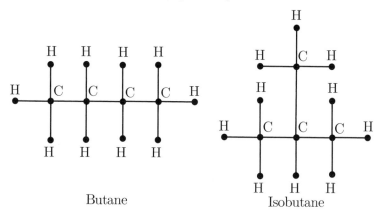

Figure 3.6 The chemical isomers with formula C_4H_{10}.

Exercises 3.1

1. A **forest** is a graph whose components are trees. There are six nonisomorphic forests that have four vertices. Find them.

2. Draw all nonisomorphic trees having five or fewer vertices.

3. There are eleven nonisomorphic trees that have seven vertices. Draw them.

4. There are ten nonisomorphic forests that have five vertices. Draw them.

5. Suppose that a tree has 50 vertices. How many edges does it have?

6. Show that if a forest F contains c trees and a total of n vertices, then the number of edges in F is $n - c$. (Hint: Let tree i have n_i vertices, where $\sum_{i=1}^{c} n_i = n$, and apply property P3.4 to each tree.)

7. Determine all trees that are regular graphs and explain why there are no more than the ones you have found.

8. Suppose that G is a graph having n vertices. Prove that G is a tree if and only if G is connected and has $n-1$ edges. (You may not use properties P3.1–P3.3 to prove this.)

9. Suppose that G is a graph having n vertices. Prove that G is a tree if and only if G has $n-1$ edges and contains no cycles. (You may not use properties P3.1–P3.3 to prove this.)

10. Prove property P3.3—that is, show that any two vertices x and y of a tree are joined by precisely one x–y path.

11. Prove that if T is a tree of order at least 3, then the addition of any edge to T (without adding any new vertices) will produce exactly one cycle in the resulting graph.

12. Prove that if T_1 and T_2 are trees with n_1 and n_2 vertices, respectively, then the join $T_1 + T_2$ has $n_1 + n_2$ vertices and $(n_1+1)(n_2+1)-3$ edges.

13. A rooted tree is called **binary** if each vertex has at most two children. A finite binary tree is called **complete** if each vertex, except each leaf, has exactly two children. How many vertices are there in a complete binary tree of height k? How many leaves are there?

14. A graph is **unicyclic** if it is connected and contains exactly one cycle. Prove that a connected graph G is unicyclic if for some edge e of G, $G - e$ is a tree.

15. If G has n vertices and $n-1$ edges, must G be a tree? Explain.

16. Suppose that a rooted tree T with root r (at level 0) has the property that each vertex at level i has $i+1$ children. Determine the number of vertices at level k of tree T.

17. Verify, using Theorem 3.1, that $3, 3, 2, 2, 1, 1, 1, 1$ is the degree sequence of a tree. Then construct three nonisomorphic trees with this degree sequence.

18. Draw a disconnected graph with the degree sequence of the previous exercise. *Hint*: One of the components is not a tree.

19. If a saturated hydrocarbon has 16 hydrogen atoms, how many carbon atoms must it have?

20. Propane is a saturated hydrocarbon with chemical formula C_3H_8. Draw the tree that models the propane molecule. Explain why there is no chemical isomer for propane.

21. Draw all possible chemical isomers having formula C_5H_{12}.

22. Prove property P3.1. That is, prove that every nontrivial tree has at least two end vertices. *Hint*: Begin a path at any vertex v of tree T and extend the path as far as possible. If v has degree 1, show that the path ends in another vertex of degree 1. If $\deg(v) > 1$, extend the path in two directions away from v as far as possible. Show that path ends in two vertices of degree 1. Use the fact that T is finite and has no cycles.

23. In a saturated hydrocarbon, every carbon atom has degree 4 while each hydrogen atom has degree 1 and the graph that models the molecule is a tree. Prove that the number of hydrogen atoms is $2n + 2$ when there are n carbon atoms in a saturated hydrocarbon.

24. Prove that the deletion of any edge from a nontrivial tree separates the tree into exactly two components. (This proves property P3.2.)

3.2 Minimum Spanning Trees

Given a connected graph G, a spanning subgraph of G that is a tree is called **spanning tree**. In applications involving spanning trees, there are sometimes numbers called **weights**, associated with each edge of a graph. We then must find a spanning tree for the graph for which the total of the weights in the tree is minimum. For example, suppose that a county is to build a passenger railroad to serve its eight major towns. Joining all $\binom{8}{2}$ pairs of town directly would be too expensive and wasteful; railroad tracks would clutter the county. Since the county is on a tight budget, the county supervisors want to build a rail system that would allow a person to travel from each of the eight towns to any of the others by using the system. Furthermore, they want to do so by spending the least amount of money. To achieve both of those goals, it turns out that the railroad system should have a tree structure. We study such problems in this section.

Spanning Trees

Since a tree with n vertices has $n - 1$ edges, to generate a spanning tree of a connected graph G having n vertices and q edges, we must delete all but $n - 1$ edges from G. We cannot do so randomly, because a tree is connected and has no cycles. So we must delete $q - (n - 1) = q - n + 1$ edges, none of which is a bridge, to form a spanning tree of G.

Even a rather simple connected graph can have many spanning trees, as is illustrated in our next example.

Example 3.4 Find all spanning trees for labeled graph G in Figure 3.7.

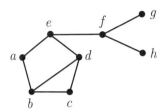

Figure 3.7

Solution
First we shall determine how many spanning trees we will get. Since G has eight vertices and nine edges, we must delete two edges. Note that ef, fg, and fh must remain. Also, we can delete at most one of the edges ab, ae, de, de, or else G would become disconnected. Thus we will either (1) delete none of ab, ae, de, in which case we must delete bd and either cd or bc; or (2) delete one of ab, ae, de, in which case we must delete one edge from the triangle bcd. In case (1), there are two choices, so two spanning trees will result. In case (2), there are three choices among ab, ae, de and three choices from the triangle. Thus, by the multiplication principle (see Note 1.6 of Section 1.3) there are $3 \cdot 3 = 9$ spanning trees produced in case (2). Hence we get a total of 11 spanning trees for graph G. They are displayed in Figure 3.8. ◇

k-Deficient Vertices of Spanning Trees

A vertex v in a spanning tree T of a graph G is called **k-deficient** if its degree satisfies the equation $\deg_G(v) - \deg_T(v) = k$. The integer, k, is called the **deficiency** of vertex v. If $k = 0$, v is a **degree-preserving vertex**—that is, $\deg_G(v) = \deg_T(v)$. Observe that for any k-deficient vertex of a nontrivial tree T, we have $k \leq \Delta(G) - 1$, since $\deg_T(v) > 0$.

The following theorem yields the sum of the deficiencies of the vertices of a spanning tree of a given graph.

Theorem 3.3 Let G be a connected graph on n vertices and q edges. Then the sum of the deficiencies of the vertices of any spanning tree of G is $2(q - n + 1)$.

Proof. Let T be a spanning tree of G. Since the deficiency of a vertex $v \in V(T)$ is $\deg_G(v) - \deg_T(v)$, the sum of the deficiencies of the vertices of T may be obtained by adding the degrees of the vertices in G and then subtracting the sum of the degrees in T. We know that the sum of the degrees of the vertices in G is $2q$. Since T has $n - 1$ edges, the sum of the degrees of the vertices in T is $2(n - 1)$. The difference is $2q - 2(n - 1) = 2(q - n + 1)$. ∎

3.2 Minimum Spanning Trees

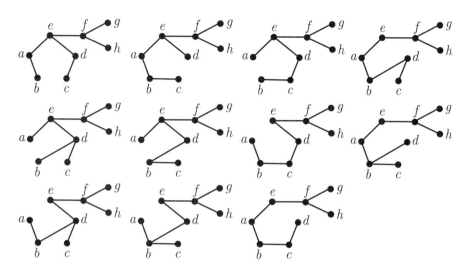

Figure 3.8 The spanning trees of graph G in Figure 3.7.

Minimum-Cost Spanning Trees

We now consider an important problem—the creation of a minimum-cost spanning tree, often called the **minimal-connector problem**. Given a graph G such that each edge has a positive weight, which represents its cost, the **cost** of a spanning tree T is the sum of the weights of the edges of T. We present an algorithm due to Kruskal [6] for the construction of a spanning tree such that no other spanning tree has a smaller cost. Such a tree is called a **minimum spanning tree**.

The graph G of Figure 3.9 will be used to illustrate the algorithm. To add drama to the problem, let's assume that these weights are the projected costs in millions of dollars for building railroad lines between 14 towns represented by the vertices. The edges in the graph represent possible rail lines that may be built to join pairs of towns. The idea is to build the cheapest railroad network so that we have a connected community. Obviously, this calls for a spanning tree of minimum cost, which means we must select exactly 13 edges because there are 14 towns.

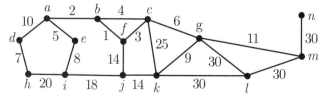

Figure 3.9 A graph with weighted edges.

The algorithm is called **greedy**, since it tells us to proceed at every stage by selecting the best (that is, cheapest) available edge that does not form a cycle. The algorithm is self-terminating in the sense that we stop as soon as we obtain a spanning tree, because the addition of any remaining edge will create a cycle. We depict the result after six stages in Figure 3.10. Note that edge bc of G, with cost 4, was skipped to avoid forming a cycle. Also, note that at various stages of the algorithm, the current graph may be disconnected, as in Figure 3.10.

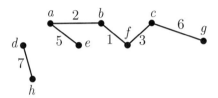

Figure 3.10 A minimum spanning tree in the making.

Figure 3.11 shows the tree after five more stages. Notice that we had a choice of two edges with weight 14—namely, fj and jk. We arbitrarily chose fj. When we continue the algorithm, we will not be able to choose jk, lest we create a cycle.

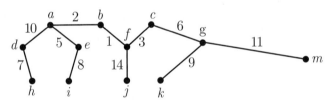

Figure 3.11 Just a few edges short of a minimum spanning tree.

Figure 3.12 displays the final product—a minimum spanning tree.

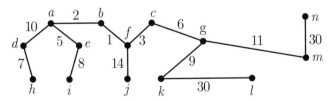

Figure 3.12 A minimum spanning tree for the graph of Figure 3.9.

It should be noted that there are several minimum spanning trees for the weighted graph of Figure 3.9. This is because there are choices to be made between edges of equal cost. The total cost for the project—that is, building

3.2 Minimum Spanning Trees

the least expensive railroad network connecting the 14 towns, is the sum of the weights (in millions of dollars) on the edges of the minimum spanning tree. Adding, we get a total cost of $136 million.

We now give an algorithmic description of the process we just discussed. We will assume that a weighted connected graph G and its order n have been stored, its number of edges q have been counted, and the edges have been presorted by weight from smallest to largest in a list L. Edges whose weights are the same are stored in their proper place alphabetically. So, for the graph of Figure 3.9, edge fj of weight 14 would be stored just before edge jk, also of weight 14. The four edges of weight 30 would be stored in the order gl, kl, lm, mn. A set S will be used to collect the edges of the spanning tree as the edges are selected. At step 1, S is initialized to be the empty set.

The algorithm uses a special type of subgraph. For a set A of edges from a graph G, the **edge-induced subgraph** $\langle A \rangle$ is the subgraph H with $E(H) = A$ and $V(H) = \{v : v$ is incident with an edge in $A\}$.

Algorithm 3.1 (Kruskal's Algorithm)
1. $S = \emptyset$.
2. Let e be the next edge on sorted list L for which $e \notin S$ and the edge-induced subgraph $\langle S \cup \{e\} \rangle$ is acyclic. Let $S = S \cup \{e\}$.
3. If $|S| = n - 1$, stop and output set S. Otherwise, return to Step 2 and continue along list L.

Algorithm 3.1 produces a minimum-weight spanning tree. A more formal pseudocode description of Kruskal's algorithm appears in Buckley and Harary [2, p. 262].

Let's examine why Algorithm 3.1 works. First, step 2 always guarantees that we have an acyclic graph, and step 3 tells us to stop when we have achieve $n - 1$ edges. Exercise 9 of Section 3.1 then implies that $\langle S \rangle$ is a tree on n vertices when the algorithm terminates. Thus we know we have a spanning tree. Why is it minimum? Suppose that $\langle S \rangle$ is not minimum. Then there must be some other set for which the total weight is smaller than that of S. Let $R \neq S$ be a set of edges producing a minimum spanning tree such that R has as many edges in common with S as possible. Label the edges of S as they entered the (eventual) tree by $e_1, e_2, \ldots, e_{n-1}$. So $w(e_1) \leq w(e_2) \leq \cdots \leq w(e_{n-1})$. Let e_i be the edge that entered set S earliest during Algorithm 3.1 that is not in R. (There must be such an edge since $S \neq R$.) Consider $R \cup e_i$. This set of edges induces a graph that has precisely one cycle C, since it consists of a tree of order at least 3 with one edge added (see Exercise 11 of Section 3.1). Since $\langle S \rangle$ is a tree, it has no cycles, so some edge $f \neq e_i$ of C is not in S. Also, $(R \cup e_i) - f$ induces a spanning tree T^* of G, since it breaks the cycle C by removing edge f. The weight of T^* equals $w(R) + w(e_i) - w(f)$. Since R induces a minimum spanning tree, we must have $w(R) \leq w(T^*)$. But this implies that $w(f) \leq w(e_i)$. By Algorithm 3.1, the set $\{e_1, e_2, \ldots, e_{i-1}, e_i\}$ has minimum weight such that the graph induced by those edges has no cycles. Since $w(f) \leq w(e_i)$, the set $\{e_1, e_2, \ldots, e_{i-1}, f\}$ also has that property. But this implies that $w(e) = w(e_i)$, which means that $w(R) = w(T^*)$. So T^* is also a minimum spanning tree, but it has more edges in common with S than R does, which is a contradiction. Thus the tree $\langle S \rangle$ produced by Algorithm 3.1 is indeed a minimum spanning tree. ∎

Example 3.5 Apply Algorithm 3.1 to the weighted graph G of Figure 3.13 and show the order in which the edges enter set S. Then draw the resulting minimum spanning tree for G.

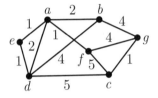

Figure 3.13

Solution
Assume that the edges are presorted according to weight: $ae, af, cg, de, ab, ad, bd, bg, fg, cd, cf$. G has eleven edges and seven vertices, so a spanning tree for G has six edges. The algorithm begins with $S = \emptyset$. The successive edges selected are then ae, af, cg, de, ab (ad, and bd are skipped; each would form a cycle) bg. We stop because S now has six edges. The resulting minimum spanning tree is displayed in Figure 3.14, and the total weight is 10. ◇

There are other algorithms to generate a minimum spanning tree. We now present an algorithm due to Prim [8] that is also greedy, but unlike Kruskal's algorithm, the edge-induced subgraph is connected at every step. The way the algorithm works is to start at some vertex, which we could think of as the root of the tree. It then successively adds new vertices not yet in the tree by selecting a vertex that can be attached to the tree using a minimum-weight edge.

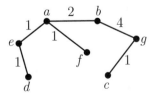

Figure 3.14

3.2 Minimum Spanning Trees

Algorithm 3.2 (Prim's Algorithm)
1. Select a vertex v, and let $V(T) = \{v\}$, $E(T) = \emptyset$.
2. Among all $u \notin V(T)$, let $e = uw$ be a minimum-weight edge joining u to w, where $w \in V(T)$. Let $V(T) = V(T) \cup \{u\}$ and $E(T) = E(T) \cup \{uw\}$.
3. If $|E(T)| = n - 1$, stop and output $E(T)$. Otherwise, return to step 2 and add another vertex to the tree.

The proof that Prim's algorithm does indeed produce a minimum spanning tree is quite similar to that for Kruskal's algorithm. First establish that a spanning tree T is produced. Then use a proof by contradiction by supposing that T is not minimum. Use a minimum spanning tree T' that has the maximum number of edges in common with T. Find the earliest edge e_i that entered T that is not in T'. Consider the cycle formed if e_i is added to T', and proceed in a similar manner as in the argument for Kruska's algorithm. The main difference here is to focus on the order in which *vertices* were attached to the tree. With these hints given, the proof is left as an exercise.

Example 3.6 Find a minimum spanning tree for graph G in Figure 3.13, but this time using Prim's algorithm. Show the tree as it grows during the algorithm when a is used as the root.

Solution
The edges would enter the tree in the order ae, af, de, ab, bg, cg. At each stage we have a tree. Note that the edges do not enter in weight order as in Kruskal's algorithm. For example, the weight of cg is 1 but cg enters after both ab and bg, which have weights 2 and 4, respectively. Prim's algorithm attaches a new vertex to the tree in the cheapest possible way. To attach c to the tree earlier would have cost 5 units (by using either cd or cf) rather than only 1 unit. The successive trees are shown in Figure 3.15. ◇

We mention here that Prim's algorithm applied to a connected graph where all weights are equal is equivalent to an important search technique called **breadth-first search** when ties are suitably broken (usually by order in adjacency lists). Breadth-first search will be discussed in Chapter 8 along with a variety of other algorithms.

Exercises 3.2

1. Find all nonisomorphic spanning trees for the following graphs:

   a. The wheel $W_{1,5}$ b. $K_{3,3}$

2. Determine the deficiency for each vertex of the spanning tree (see Figure 3.12) for the graph in Figure 3.9. Then verify that your result is consistent with Theorem 3.3.

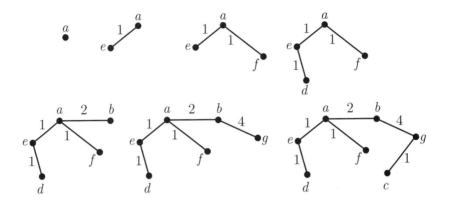

Figure 3.15

3. Using Theorem 3.3, determine the sum of the deficiencies of the vertices in any spanning tree of

   a. $W_{1,n}$ b. $K_{m,n}$ c. K_n

4. Draw all spanning trees for the labeled graph in Figure 3.16.

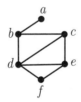

Figure 3.16

5. Draw all nonisomorphic spanning trees for the mesh $M(2,4)$.

6. Produce spanning trees of $M(3,3)$ with 2, 3, and 4 degree-preserving vertices.

7. Produce spanning trees of $M(3,3)$ with 2, 3, 4, 5 and 6 end vertices.

8. Prove that a connected weighted graph where all weights are distinct has a unique minimum weight spanning tree.

9. Describe an algorithm for finding a *maximum*-weight spanning tree of a connected graph.

3.2 Minimum Spanning Trees

10. Describe an algorithm for finding a minimum-weight spanning *forest* in a weighted graph that is not necessarily connected. How can you simplify the algorithm if it is known that the graph has k components?

11. (a) For the weighted graph in Figure 3.17, list the edges of the spanning tree in the order in which they would be selected if Kruskal's algorithm were used. Then draw the resulting minimum spanning tree.

    (b) List the edges of the spanning tree in the order in which they would be selected if Prim's algorithm were used *beginning at vertex c*. Then draw the resulting spanning tree.

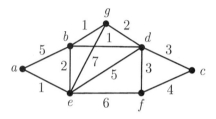

Figure 3.17

12. (a) For the weighted graph in Figure 3.18, list the edges of the spanning tree in the order in which they would be selected if Kruskal's algorithm were used. Then draw the resulting minimum spanning tree.

    (b) List the edges of the spanning tree in the order in which they would be selected if Prim's algorithm were used *beginning at vertex g*. Then draw the resulting spanning tree.

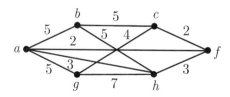

Figure 3.18

13. Find a minimum spanning tree for the weighted graph in Figure 3.19 using your favorite method.

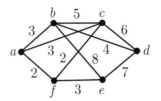

Figure 3.19

14. Prove that if T_1 and T_2 are spanning trees of a connected graph G such that if e_1 is in T_1 but not T_2, then there exists and edge e_2 in T_2 but not T_1 such that $T_2 + e_1 - e_2$ is a spanning tree of G. *Hint*: Adding an edge to a tree creates exactly one cycle.

15. Suppose that for the graph in Figure 3.9, the county commissioner lives in town k. He expects to make a lot of trips to towns c and l and has used his political clout to force the construction of the direct routes ck and kl even though those rail lines will be rather expensive. Find a minimum spanning tree for the graph in Figure 3.9 subject to the restriction that edges ck and kl are in the tree. Explain why there can be no spanning tree containing ck and kl that has smaller cost than the tree that you have found.

16. Using the hints following Algorithm 3.2, prove that Prim's algorithm does indeed produce a minimum spanning tree for any connected graph G.

3.3 A Characterization of Bipartite Graphs

A graph G is called **bipartite** if $V(G)$ can be partitioned into two nonempty subsets V_1 and V_2, such that if xy is an edge of G, then x and y belong to different subsets. Sets V_1 and V_2 are the **parts** or **bipartition sets** of G. It is common practice to assign one color, say black, to the vertices of V_1 and another color, say white, to the vertices of V_2. Thus a bipartite graph has the property that each of its vertices may be colored black or white so that adjacent vertices are oppositely colored. An immediate example is P_6 whose vertices may be colored white, black, white, and so on. This will partition the vertices into three whites and three blacks.

Now we see that our old friend the complete bipartite graph $K_{m,n}$ is indeed bipartite. The vertices of the two parts can be colored black and white, respectively, so that each edge of $K_{m,n}$ is incident with one black vertex and one white vertex.

3.3 A Characterization of Bipartite Graphs

It should be clear that, in fact, all trees are bipartite. To see this, select a vertex as the root and color it white. Color all of its neighbors black—that is, all of its children are black. Now color the next row white, and continue this process. The labels of the tree of Figure 3.20 indicate color. It isn't drawn as a rooted tree.

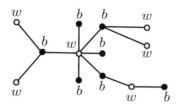

Figure 3.20 Trees are bipartite.

Notice that the tree in Figure 3.20 has color sets of cardinalities 6 and 7. The cardinality of the color sets is an invariant of a bipartite graph. If we partition P_6, its subsets have equal cardinalities, while the star, $K_{1,10}$, for example, produces very unequal subsets of cardinalities 1 and 10. We classify bipartite graphs, therefore, as follows. Let G be bipartite and let its vertex set be partitioned into the subsets V_1 and V_2, such that $|V_1| \geq |V_2|$. This entails no loss in generality, since we are permitting equality. If there is inequality, let the larger subset be called V_1.

1. G is called **equitable** if $|V_1| = |V_2|$.
2. G is **nearly equitable** if $|V_1| - |V_2| = 1$.
3. G is **skewed** if $|V_1| - |V_2| \geq 2$.

The bipartite graph of Figure 3.20 is nearly equitable. Notice that a path is either equitable or nearly equitable. Path P_n is equitable when n is even and is nearly equitable when n is odd.

An analysis of C_n is most illuminating. When n is even, C_n is bipartite—in fact, equitably so. On the other hand, when n is odd, C_n is not bipartite at all! Imagine attempting to color the vertices of C_5, adhering to the rule that adjacent vertices must be oppositely colored, which for a cycle means that we must alternate colors. Proceeding around the cycle (optimistically) with the sequence white, black, white, black, we find that as we arrive at the fifth vertex—say v, which happens to be a neighbor of the initial (white) vertex—v must be black. Impossible! It is also a neighbor of vertex 4 in our sequence, which happens to be black. Convince yourself that this scenario will occur whenever we attempt to color any odd cycle with only two colors. Odd cycles are not bipartite!

We can extend the process just described to graphs in general. That is, the simple labeling procedure described in Algorithm 3.3 will determine whether G is bipartite.

> **Algorithm 3.3 (Determining If G Is Bipartite)**
> In this procedure, we refer to the labels a and b as opposite labels.
> 1. Label any vertex a.
> 2. Label all vertices adjacent to a with the label b.
> 3. If there are unlabeled vertices adjacent to a labeled vertex v, then label all vertices adjacent to v by using the opposite label that v was assigned.
> 4. Repeat step 3 until there are no unlabeled vertices adjacent to any labeled vertex.
> 5. If there is an unlabeled vertex, select one such vertex (which is necessarily in a new component), label it a, and return to step 3.
> 6. If adjacent vertices get distinct labels, then graph G is bipartite. The set V_1 in the definition of a bipartite graph is then the set of all vertices with label a; the set V_2 is the set of all vertices with label b. If some pair of adjacent vertices were assigned the same label, then G is not bipartite.

The proof that Algorithm 3.3 works is left as an exercise.

Example 3.7 Label the vertices in graph G and H in Figure 3.21 to determine whether they are bipartite. If bipartite, redraw the graph to display the vertex sets V_1 and V_2 in the definition of bipartite.

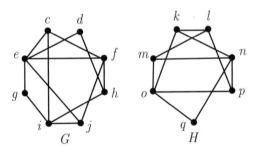

Figure 3.21

Solution
We show the labelings in Figure 3.22. The subscript on each label indicates at which round of labeling during Algorithm 3.3 a particular vertex was labeled. Graph H is bipartite and is redrawn in Figure 3.22(b). Graph G is not bipartite because there are pairs of adjacent vertices labeled a at step 3. The vertex e labeled a_3 is adjacent to the two other vertices labeled a_3—namely, d and f. ◇

Can we tell whether a graph is bipartite without using the labeling procedure of Algorithm 3.3? The answer is yes. If G contains an odd cycle—that is, a cycle containing an odd number of vertices—then G is not bipartite.

3.3 A Characterization of Bipartite Graphs

The interesting thing is that the argument works in the other direction as well, as the following theorem asserts. The proof is left as an exercise.

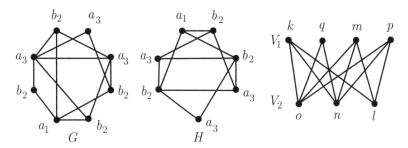

Figure 3.22

Theorem 3.4 A graph G is bipartite if and only if G does not contain an odd cycle. ∎

Corollary 3.4a Trees are bipartite.

Proof. A tree has no cycles, so it has no odd cycles. ∎

A corollary is a fact whose truth is an immediate consequence of a theorem. That is certainly the case here.

Example 3.8 Draw all connected bipartite graphs of order 5.

Solution
We organize our graphs according to the number of edges. When there are four edges, we get $K_{1,4}$, P_5, and the sequential join $K_1 + K_1 + K_1 + \bar{K}_2$. When there are five edges, we get the sequential join $K_1 + K_1 + \bar{K}_2 + K_1$, and for six edges, we get $K_{2,3}$. The graphs are displayed in Figure 3.23. ◇

Figure 3.23 Connected bipartite graphs with five vertices.

Exercises 3.3

1. Draw the seven bipartite graphs (both connected and disconnected) that have four vertices.

2. Draw all connected bipartite graphs with six vertices.

3. Suppose that G and H are graphs, at least one of which has an edge. Show that the join $G + H$ is not bipartite.

4. Prove Theorem 3.4: G is bipartite if and only if G has no odd cycles.

5. Prove that any subgraph of a bipartite graph is bipartite.

6. Graph G is **k-partite** if its vertex set can be partitioned into nonempty sets $V_1, V_2, \ldots, V_k$, so that if uv is an edge of G, then vertices u and v are in different subsets V_i and V_j. Show that graph G of Figure 3.21 is 3-partite by drawing it with vertices of the three different subsets on different levels.

7. Prove that if a bipartite graph with parts V_1 and V_2 is regular, then $|V_1| = |V_2|$.

8. A graph is **semiregular bipartite** if vertices in part V_1 all have degree s and vertices in part V_2 all have degree t. Prove that if G is semiregular bipartite, then the line graph $L(G)$ is regular of degree $s + t - 2$.

9. Prove that a bipartite graph of order 8 has at most 16 edges.

10. Prove that both G and $\bar{G}$ are connected if and only if no complete bipartite graph spans G or $\bar{G}$.

11. Prove that if G_1 and G_2 are bipartite, then so is the cartesian product $G_1 \times G_2$.

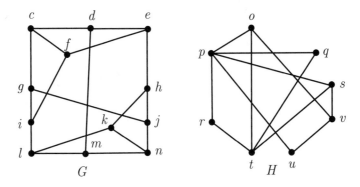

Figure 3.24

12. Apply the labeling procedure of Algorithm 3.3 to the graphs G and H of Figure 3.24 to determine whether each is bipartite.

13. If G is semiregular bipartite with n vertices of degree s and m vertices of degree t, determine the number of edges in G.

14. Let G be a semiregular bipartite graph with parts X and Y such that $|X| = m$ and $|Y| = n$, $\deg(v) = r$ if $v \in X$, and $\deg(u) = s$ if $u \in Y$. Show that $mr = ns$. Then show that $r \leq n$ and $s \leq m$.

15. Construct a semiregular bipartite graph with parts V_1 and V_2 of orders 6 and 8, respectively, and degrees 4 and 3, respectively. Are there any other values of possible pairs of degrees r and s that can be used here?

16. Suppose that G is bipartite with parts A and B and consider the two copies of G in the cartesian product $G \times K_2$. Call them G_1 and G_2 and their related bipartition sets A_1, B_1, and A_2, B_2, respectively. Show that $G \times K_2$ is bipartite with parts $A_1 \cup B_2$ and $A_2 \cup B_1$.

17. Prove that Algorithm 3.3 works. That is, prove that if G is bipartite, then the algorithm will label the vertices so that those labeled a are mutually nonadjacent, and those labeled b are also mutually nonadjacent. Also, show that if H is not biparite, then some pair of adjacent vertices of H receive the same label as one another.

3.4 Matchings and Job Assignments

A **matching** M in a graph is a set of edges, no two of which are incident with one another. Bipartite graphs are important, particularly in applications involving matching problems. For example, a company may have a number of employees qualified to perform various jobs. If there are several tasks to be done, how should the company assign its employees to the tasks to maximize efficiency? This job assignment problem is of utmost importance when there is a strike and management personnel must try to cover the jobs of the striking employees. In this section, we discuss matchings in bipartite graphs, with particular attention to the job assignment problem. We then discuss matchings in more general graphs and how to find matchings of maximum size in bipartite graphs.

Matchings in a Bipartite Graph

Let G be a bipartite graph with parts X and Y. Clearly, every edge in a matching has one vertex in X and the other in Y. (Actually, this is true for every edge in G.) A matching M such that every vertex in X is incident with an edge of M is called a **complete matching from X to Y**. Needless to say, M might not also be a complete matching from Y to X. However, if it is, then M is called a **perfect matching**. In particular, this requires that $|X| = |Y|$—that is, that G is equitable. A complete matching from X to Y requires merely that $|X| \leq |Y|$.

It is important in applications to determine whether a bipartite graph with parts X and Y has a complete matching from X to Y. Here's an example of such an application. Suppose we have several applicants for several jobs. Each job requires a certain skill, and the resume of each applicant lists his or her skills.

The situation can be described by a bipartite graph G as follows. Let the vertices of X represent the applicants and let the vertices of Y represent the job openings. We draw an edge from vertex v in X to vertex w in Y whenever the applicant represented by v, say Virgil, has the skill required by the job represented by w, say writer. Of course, $\deg(v)$ indicates how many jobs Virgil is qualified to do, and $\deg(w)$ indicates how many applicants are qualified to hold the job of writer.

A complete matching from X to Y in our bipartite graph G means that there is a job for each applicant. Note that some jobs will be unfilled if $|X| < |Y|$. The bipartite graph of Figure 3.25 depicts that situation for applicants A, B, C, and D and available jobs a, b, c, d, and e.

One complete matching from X to Y is $\{Aa, Bd, Cc, De\}$. Another one is $\{Ab, Bd, Cc, De\}$. A third complete matching is $\{Ab, Bd, Cc, Da\}$. Can you find any more? Notice that in all of the preceding examples, applicant C, having only one skill, must be assigned to job c. Perhaps he should have studied more mathematics and computer science in college.

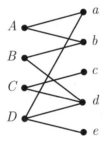

Figure 3.25 Jobs for everyone!

We shall soon be able to describe precisely when a bipartite graph has a complete matching. But in order to establish that result, we first consider a technique for obtaining a matching of largest possible size.

Finding a Maximum Matching

The darkened edges ah, bd, and ef of Figure 3.26(a) constitute a matching. M is a **maximum matching** for G if M has largest cardinality among all possible matchings—that is, if M' is any other matching for G, then $|M'| \leq |M|$. A **maximal matching** is a matching M such that there is no larger matching M' *containing* M. Thus a maximal matching is one that

3.4 Matchings and Job Assignments

cannot be enlarged by adding additional edges. It is important to recognize the difference between a *maximal* matching and a *maximum* matching. The matching in Figure 3.26(a) is maximal, but not a maximum matching. The only vertices not incident with the matching edges are c and g, but they are not adjacent, so the matching cannot be enlarged by adding edges. The matching in Figure 3.26(b) is a maximum matching. It has the largest possible cardinality.

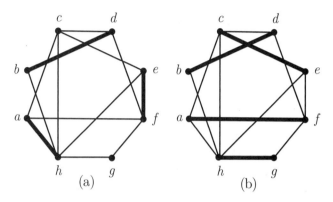

Figure 3.26 A maximal matching and a maximum matching.

The most useful technique for finding a maximum matching in a graph is based on the concept of an augmenting path. Consider a matching M in a graph G. An **M-alternating path** in G is a path whose edges are alternately in M and not in M. Vertex v is **M-matched** by the matching M if v is incident with one of the edges of M. Otherwise, v is **M-unmatched**. An **M-augmenting path** is an alternating path joining two M-unmatched vertices. Note that an M-augmenting path need not use all the edges of M. In the matching M of Figure 3.26(a), c, e, f, g is an M-augmenting path, because $ce \notin M$, $ef \in M$, and $fg \notin M$. Path c, e, f, a, h, g is also an M-augmenting path since $ce \notin M$, $ef \in M$, $fa \notin M$, $ah \in M$, and finally $hg \notin M$. Thus a graph may have several M-augmenting paths. Notice, however, that an M-augmenting path always begins and ends with an edge that is not in M. The following theorem of Berge [1] characterizes precisely when a matching is maximum.

Theorem 3.5 (Berge's Matching Theorem) A matching M in a graph G is maximum if and only if there is no M-augmenting path in G.

Proof. Suppose that matching M is maximum. Then there can be no M-augmenting path, or otherwise we can find a larger matching M' as follows. Let P be an M-augmenting path joining vertices u and v in G. Then P necessarily has odd length with one more edge not in M than in M. Let M_P be the edges of P that are in M, let M'_p be the edges of P that are not in M, and let $M' = (M - M_p) \cup M'_p$. Thus we remove the edges of M_P and include the edges of M'_p. Then $|M'| > |M|$, a contradiction. Thus there is no M-augmenting path.

Conversely, suppose that there is no M-augmenting path in G. We prove that M is indeed a maximum matching. Suppose M^* is a maximum matching. Then we know that there is no M^*-augmenting path. Consider each component of the subgraph H containing all vertices of G, where $E(H) = (M - M^*) \cup (M^* - M)$. [$E(H)$ is the symmetric difference of the two matchings; that is, $E(H)$ is the set of edges that appear in one of the matchings but not in the other.] Note that the maximum degree of each vertex in H is at most 2 since otherwise one of the matchings M or M^* has two edges incident with the same vertex. Thus each component of H consists of either a isolated vertex, a path whose edges are alternately in M and M^*, or an even cycle whose edges are alternately in M and M^*. Since neither M nor M^* has an augmenting path, any path component of H has one endpoint covered by M and the other covered by M^*. Thus each path component has even length with edges alternately in M and M^* and each cycle component has even length with edges alternately in M and M^*. This implies that $|M| = |M^*|$, so M is a maximum matching. ∎

Note that the M-augmenting path c, e, f, a, h, g is the one that is used to augment the matching in the graph of Figure 3.26(a) to become the maximum matching of Figure 3.26(b).

Matchings can be found in any nontrivial graph; however, most matching problems involve a bipartite graph. As described earlier, the vertices of A could represent people, and the vertices of B could represent jobs. There is an edge ab if person a is qualified for job b. For maximum employment, we want a maximum matching. The edges in the matching tell us whom to hire for each job.

Example 3.9 TT&A Corporation has just been hit by a strike. Management has decided that four jobs (showroom sales, repairs, security, and account clearance) are essential if business is to continue during the strike. Two people are needed for showroom sales. Find a maximum matching to determine whether all key jobs can be covered by the various managers, one manager per key position, if their capabilities are as described in Table 3.1.

Manager	Can Do Job
Jay	accounts, sales
Kay	sales, security
May	sales, security, repairs
Ray	repairs
Wai	security, repairs

Table 3.1 Employees and their qualifications.

3.4 Matchings and Job Assignments

Solution
We draw the bipartite graph where each vertex in part X corresponds to a manager and each vertex in part Y corresponds to a job. Since there are two people needed for showroom sales, we have two separate vertices for those two positions (labeled S_1 and S_2). An edge is present between person a and job b if person a is qualified for job b. The resulting graph is shown in Figure 3.27(a). In Figure 3.27(b), we show a maximum matching. Note that all the key jobs are covered. ◇

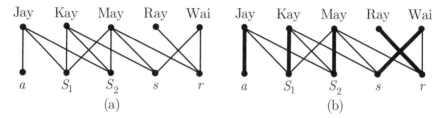

Figure 3.27 Matching people with appropriate jobs.

For small problems such as that in Example 3.9, a maximum matching is rather easy to find. In larger problems, algorithms using labeling procedures are usually used. See, for example, Chartrand and Oellermann [3, pp. 171–172]. In Section 10.2, we show how to find a maximum matching in a bipartite graph by using networks.

A **perfect matching** M is a matching for which every vertex is M-matched.

The matchings in Figure 3.26(b) and Figure 3.27(b) are perfect matchings. It is clear that a graph must have an even number of vertices to have a perfect matching. Furthermore, in a bipartite graph G with partite sets X and Y, a perfect matching can exist only if there are the same number of vertices in parts X and Y; that is, G is equitable. This condition is not sufficient. An example of an equitable bipartite graph having no perfect matching is given in Figure 3.28.

Figure 3.28 An equitable bipartite graph with no perfect matching.

Now we are ready to consider the problem of when a bipartite graph has a complete matching or when it has a perfect matching.

Complete Matchings in Bipartite Graphs

The following theorem due to Philip Hall [5] characterizes precisely when a bipartite graph has a complete matching. For a vertex v, its neighborhood $n(v)$ is the set of all vertices adjacent to v. For any set S of vertices in a graph G, the neighborhood of S is the union of the neighborhoods of vertices in S. That is, $N(S) = \bigcup_{v \in S} n(v)$.

Theorem 3.6 (Hall's Matching Theorem) A bipartite graph G with parts X and Y has a complete matching from X to Y if and only if $|N(S)| \geq |S|$ for every subset $S \subseteq X$.

Proof. If G has a complete matching M from X to Y, then in M, each vertex of X is matched by M to a distinct vertex of Y. This means that for every subset $S \subseteq X$, $|N(S)| \geq |S|$.

Conversely, let G be bipartite where $|N(S)| \geq |S|$ for every subset $S \subseteq X$. Suppose that G does not have a complete matching from X to Y. Let M be a maximum matching in G and let v be an M-unmatched vertex of X. Let X^* be the set of M-matched vertices of X, let Y^* be the set of M-matched vertices of Y, and let $X' = X^* \cup \{v\}$. Since $\{v\} \subseteq X$ and $|N(\{v\})| \geq |\{v\}| = 1$, v has at least one neighbor in Y. If one such neighbor $v*$ is not in Y^*, then $M' = M \cup \{vv^*\}$ is a matching whose size exceeds M, a contradiction, which would imply that G has a complete matching from X to Y.

If every neighbor v^* of v is in Y^*, then we can begin an M-alternating path at v that will eventually become an M-augmenting path because for every subset $S \subseteq X$, we have $|N(S)| \geq |S|$. But then M can be augmented into a matching whose size exceeds M, a contradiction that would imply that G has a complete matching. ∎

When $|X| = |Y|$, Theorem 3.6 is sometimes referred to as Hall's marriage theorem because it can be interpreted in the following way. Suppose that set X is a set of men and set Y is a set of women such that for every subset of n men there are at least n suitable women. Then it is possible to find a perfect matching of the men with women and marry them all off in pairs. Thus we have the following corollary.

Corollary 3.6a (Hall's Marriage Theorem) A bipartite graph G with parts X and Y, where $|X| = |Y|$, has a perfect matching if and only if $|N(S)| \geq |S|$ for every subset $S \subseteq X$. ∎

Let us now use Theorem 3.6 to show that a regular bipartite graph G has a perfect matching. To begin with, we will need a little lemma.

Lemma 3.7 A regular bipartite graph must be equitable—that is, parts X and Y have the same number of vertices.

Proof. Assume that $|X| = k$ and that G is r-regular. Then $|E(G)| = rk$. These rk edges must each be incident with a vertex in Y. Since graph G is

3.4 Matchings and Job Assignments

r-regular, each vertex in Y must be incident with r of these edges, yielding the equation $r|Y| = rk$, implying that $|Y| = k = |X|$. ∎

Theorem 3.7 If a bipartite graph is regular, then it has a perfect matching.

Proof. If we can show that G has a complete matching from X to Y, it will follow immediately that G has a perfect matching. To apply Theorem 3.6, we must show that given any set S of m vertices in X, there are at least m vertices in Y that are adjacent to at least one member of S. Another way of saying this is that the cardinality of the union of the neighborhoods of the vertices in S is at least m. Even shorter—we must show that $|N(S)| \geq m$.

Since each vertex in S has degree r, it follows that S is incident with a total of rm edges, each of which is incident also to a vertex in Y. Since each vertex in Y has degree r, those vertices are each incident with r edges, each incident with a vertex of X. Thus there must be at least m vertices in $N(S)$, completing the proof that $|N(S)| \geq |S|$. Theorem 3.6 now guarantees the existence of a perfect matching. ∎

The converse is, of course, false, as is shown by the bipartite graph G with a perfect matching shown in Figure 3.29. Note that G is not regular. (The **converse** of the statement $p \Rightarrow q$, that is, "p implies q" is $q \Rightarrow p$. The converse attempts to reverse an implication. Unfortunately, the converse of a statement may or may not be true. The correct reverse implication, $\sim q \Rightarrow \sim p$, "not q implies not p," is called the **contrapositive**. The contrapositive $\sim q \Rightarrow \sim p$ is true whenever $p \Rightarrow q$ is true.)

Systems of Distinct Representatives

Suppose that a small department at a certain college has four professors A, B, C, and D who serve on four committees a, b, c and d as shown in Table 3.2. Suppose, furthermore, that they agree to form a new "umbrella committee" with one representative from each of the committees a, b, c and d. The catch is that no professor who serves on the umbrella committee should represent more than one of the individual committees. The professor must not wear two different hats, so to speak, but must faithfully represent the committee that chose him as its representative. In the language of combinatorics, we seek a system of distinct representatives (SDR) for the committees a, b, c, and d.

Professor	Committees
A	a, d
B	a, b, d
C	b, c, d
D	c, d

Table 3.2

This situation is depicted in Figure 3.29. The professors are represented by the vertices A, B, C, D, and the committees are represented by the vertices a, b, c, d. The search for an SDR is exactly the problem of finding a matching from Y to X. (Y to X rather than X to Y because each committee must be represented, but not all professors would need to serve if there were a large enough pool of professors.) Since the graph has a perfect matching, this can be accomplished.

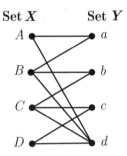

Figure 3.29 G has a perfect matching but is not regular.

More General Matchings

It is important to stress that the M-augmenting path technique applies to all graphs, not just bipartite graphs. Indeed, the graph from Figure 3.26 on which we illustrated that technique was not bipartite. There are other techniques for finding maximum matchings, most notably, the network flow technique used for the case of a bipartite graph. We will discuss that technique in Chapter 10.

Matchings are also found in weighted graphs. In a weighted job assignment problem, we have a bipartite graph where the parts consist of the applicants and the jobs, respectively. The weight on a particular edge uj gives a comparative measure of how qualified applicant u is for job j. A weight $w(uj) > w(vj)$ indicates that between applicants u and v, applicant u is more qualified for job j. We may think of the weights as scores on some job-related exam. What we usually want to do in such problems is maximize the sum of the weights on the edges of the matching. Try to find a maximum-weight matching in the graph of Figure 3.30. You will see that even for this relatively small weighted graph, the problem is rather challenging.

For the interested reader, we mention that Gould [4] gives an excellent discussion of weighted matchings and other general matching problems, including algorithms for their solution.

3.4 Matchings and Job Assignments

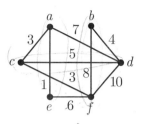

Figure 3.30

Exercises 3.4

1. Find all perfect matchings of a labeled copy of Q_3.

2. Prove that $G \times K_2$ always has a perfect matching for all graphs G.

3. Given a positive integer n, construct a graph of order n such that a maximum matching has exactly one edge.

4. Show that the wheel $W_{1,n}$ has a perfect matching if and only if n is odd.

5. Show that every even mesh has a perfect matching.

6. Let T be a spanning tree of G. Show that a perfect matching for T is also a perfect matching for G. Find an example to show that the converse is not true.

7. Construct four connected graphs of order n, where n is even and $n \geq 8$, that have no perfect matchings.

8. Find all perfect matchings of the mesh $M(2,5)$.

9. Show that the number of perfect matchings of K_{2n} is $1 \cdot 3 \cdot 5 \cdots (2n-1)$. *Hint*: Select a vertex v. There are $2n-1$ vertices left to match with v. Once we select a vertex w, delete v and w from K_{2n}, yielding K_{2n-2}. Repeat the process by selecting an arbitrary vertex u. Then there are $2n-3$ vertices to match with u. Continue this process until you obtain K_2, which has one perfect matching.

10. How many maximum matchings does K_{2n+1} have? *Hint*: Begin by deleting a vertex, and then use the previous exercise.

11. Find two maximum matchings for each of the graphs in Figure 3.31.

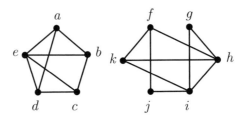

Figure 3.31

12. (a) Let G be the cycle C_{2n} with vertices labeled $1, 2, 3, \ldots, 2n$. How many different maximum matchings does G have?
    (b) Let H be the cycle C_{2n+1} with vertices labeled $1, 2, 3, \ldots, 2n+1$. How many different maximum matchings does H have?

13. Prove by induction that the hypercube Q_n has a perfect matching.

14. Ann, Dan, Fan, Nan, and Van are applying for jobs $a, b, c, d,$ and e. Ann is qualified for $a, b,$ and c; Dan for $b, d,$ and e; Fan for b and c; Nan for c and d; and Van for b.
    (a) Draw the associated bipartite graph.
    (b) Find a maximum matching to create maximum employment.

15. Applicant A is qualified for jobs $a, b, d,$ and e. Applicant B is qualified for $b, c,$ and e; Applicant C is qualified for $b, d,$ and e; Applicant D is qualified for $a, c,$ and e; and Applicant E is qualified for a and b.
    (a) Draw the associated bipartite graph.
    (b) Find a maximum matching to create maximum employment.

16. Applicant A is qualified for jobs $a, d,$ and e. Applicant B is qualified for a and e; Applicant C is qualified for $a, c,$ and e; Applicant D is qualified for d and e; and Applicant E is qualified for a and d.
    (a) Draw the associated bipartite graph.
    (b) Find a maximum matching to create maximum employment.
    (c) Use Corollary 3.6a to explain using a specific subset of vertices why there is no perfect matching.

17. Let H be the complete bipartite graph $K_{m,n}$ with vertices labeled $a_1, a_2, \ldots, a_m; b_1, b_2, \ldots, b_n$, where $m \leq n$. How many different maximum matchings does H have?

18. Prove that if a tree has a perfect matching, it has only one.

3.4 Matchings and Job Assignments

For a collection S of sets $A_1, A_2, \ldots, A_n$, a **system of distinct representatives** is a set of distinct elements $a_1 \in A_1$, $a_2 \in A_2$, ..., $a_n \in A_n$. Philip Hall originally proved Theorem 3.6 in this context as follows: S has a system of distinct representatives if and only if for any subcollection of k of the sets $A_{t_1}, A_{t_2}, \ldots, A_{t_k}$, we have $|A_{t_1} \cup A_{t_2} \cup \cdots \cup A_{t_k}| \geq k$.

If possible, find a system of distinct representatives for each of the collections of sets in Exercises 19–22.

19. $A_1 = \{2, 3, 5, 7\}$, $A_2 = \{1, 2, 5\}$, $A_3 = \{1, 3, 7\}$, $A_4 = \{2, 5, 7\}$, $A_5 = \{1, 3, 5\}$

20. $A_1 = \{1, 3, 4, 6\}$, $A_2 = \{2, 4, 5\}$, $A_3 = \{1, 2, 6\}$, $A_4 = \{1, 5, 7\}$, $A_5 = \{1, 3, 4, 5\}$

21. $A_1 = \{4, 5, 6\}$, $A_2 = \{1, 2, 3, 5\}$, $A_3 = \{2, 4, 6, 8\}$, $A_4 = \{1, 2, 8\}$, $A_5 = \{3, 6, 8\}$, $A_6 = \{1, 4, 6\}$

22. $A_1 = \{1, 6\}$, $A_2 = \{2, 6\}$, $A_3 = \{2, 5, 7\}$, $A_4 = \{1, 7\}$, $A_5 = \{1, 5, 7\}$, $A_6 = \{1, 2, 6\}$

23. The board of directors of *XYZ* Corporation has 13 members identified by letter as $a, b, c, \ldots, m$. They have six important committees composed as follows: Acquisitions = $\{d, g, h, m\}$, Sales = $\{a, g, h\}$, Budget = $\{d, h, m\}$, Planning = $\{b, g, m\}$, Finance = $\{a, b, d, h\}$, and Legal = $\{a, d, h\}$. They would like to form an additional committee, Oversight, containing six members, one from each of the other six committees. Is such a committee possible? If so, find an appropriate committee. If not, explain why not.

References for Chapter 3

1. Berge, C., Two theorems in graph theory. *Proceedings of the National Academy of Sciences, U.S.A.* 43 (1957) 842–844.

2. Buckley, F., and F. Harary, *Distance in Graphs*, Addison-Wesley, Redwood City, CA (1990).

3. Chartrand, G., and O. R. Oellermann, *Applied and Algorithmic Graph Theory*, McGraw-Hill, New York (1993).

4. Gould, R., *Graph Theory*, Benjamin Cummings, Menlo Park, CA (1988).

5. Hall, P., On representations of subsets. *Journal of the London Mathematical Society* 10 (1935) 26–30.

6. Kruskal, Jr., J. B., On the shortest spanning subtree of a graph and the Traveling Salesman Problem. *Proceedings of the American Mathematical Society* 7 (1956) 48–50.

7. Moon, J. W., Counting labeled trees. *Canadian Mathematical Congress*, Montreal (1970).

8. Prim, R. C., Shortest connection networks and some generalizations. *Bell Systems Technical Journal* 36 (1957) 1389–1401.

Additional Readings

9. Crilly, T., Arthur Cayley as Sadlerian Professor: a glimpse of mathematics teaching at 19th century Cambridge. *Historia Mathematica* 26:2 (1999) 125–160.

10. DePalma, R., and M. Lewinter, Which starlike trees span n-meshes? *Graph Theory Notes of New York* 17 (1989) 12–15.

11. Kalman, D., Marriages made in the heavens: a practical application of existence. *Mathematics Magazine* 72 (1999) 94–103.

12. Lewinter, M., and W. Widulski, Which trees span ternary cubes? *Graph Theory Notes of New York* 17 (1989) 16–19.

Chapter 4

Distance and Connectivity

The concept of distance is widely used throughout graph theory and its applications. Distance is used in various graph operations, in isomorphism testing, and in convexity problems and is the basis of several graph symmetry concepts. Distance is used to define many graph centrality concepts, which in turn are useful in facility location problems. Numerous graph algorithms are distance related in that they search for paths of various lengths within the graph. Distance is an important factor in extremal problems in graph connectivity. Graph connectivity is important in its own right because of its strong relation to the reliability and the vulnerability of computer networks. In this chapter, we discuss some of the many important concepts and results concerning distance and connectivity in graphs.

4.1 Distance in Graphs

Given two vertices, u and v, in a graph G, the **distance** between them, denoted $d(u,v)$, is defined as the number of edges in any u–v geodesic in G. Note that we don't care how long each edge is or whether the edges are drawn as straight lines or as curved lines. We are merely interested in the *number* of edges in the geodesic. In fact, the relevant information supplied by the edges of a graph consists of the adjacencies.

The distance function in graph theory must behave like distance functions (technically **metrics**) in all fields of mathematics. Here is a list of the axioms that all distance functions obey:

1. $d(u,v) \geq 0$, and $d(u,v) = 0$ if and only if $u = v$.
2. $d(u,v) = d(v,u)$ for all u,v. (4.1)
3. $d(u,v) \leq d(u,w) + d(w,v)$ for all u,v,w.

The third axiom is called the **triangle inequality** because of its obvious interpretation in plane geometry—the shortest distance between two points is a straight line. Of course, this is graph theory, not plane geometry, but the idea is the same. Think of the left side of the inequality as the result of a request to a taxi driver to make a stop at w while taking you from u to v. The new stop may or may not be on a direct route from u to v, but it can't possibly shorten the trip.

Eccentricity, Center, Radius, and Diameter

Given a vertex v of a graph G, the **eccentricity** of v, denoted $e(v)$, is the distance to a vertex farthest from v in G. Mathematically, this is written

$$e(v) = \max_{u \in V(G)} d(u, v)$$

where the information under the word *max* tells us that we compute the *maximum* of $d(u, v)$ as u varies over the entire set of vertices of G. Let H be the graph of Figure 4.1. Then we see that $e(j) = 5$ since a farthest vertex (either i or q) is five edges away from j. That is, $d(j, i) = d(j, q) = 5$, and $d(j, x) < 5$ for $x \neq i, q$. The number listed at each vertex in graph H is the eccentricity of that vertex.

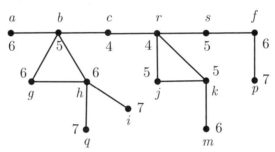

Figure 4.1 A graph called H.

Note 4.1 If $e(v) = t$ for some vertex v in a graph G, then
1. The distance from v to any other vertex of G is no more than t, and
2. There is at least one vertex whose distance from v is t.

We call a vertex, w, an **eccentric vertex** of v if $d(v, w) = e(v)$. We may think of w as being situated as far as possible from v. In general, this is not a reciprocal relationship; that is, v need not be an eccentric vertex of w. A pair of vertices that happen to be eccentric vertices of one another are called **mutually eccentric**. Sets of such vertices were examined by Buckley and Lau [2]. You may wish to verify that vertex j in Figure 4.1 is not an eccentric vertex of vertex i, even though i is an eccentric vertex of j. However, p and q are mutually eccentric.

4.1 Distance in Graphs

In Figure 4.1, note that $e(c) = e(r) = 4$, while the other vertices all have higher eccentricities. The minimum eccentricity among the vertices of a graph G is called the **radius** of G, usually abbreviated **rad(G)**. So $\text{rad}(G) = 4$. The set of vertices with minimum eccentricity is called the **center**, denoted by **C(G)**. So $C(H) = \{c, d\}$. Let's be clear on this. The radius is a *number* and the center is a *set*.

The center of a graph is important in applications. For example, in facility location problems the graph may model a community where the edges represent roads between locations (the vertices). The center contains the ideal locations for an emergency facility such as a police station, fire station, or hospital. The eccentricity of a given vertex v represents response time from v to a vertex farthest from v. For an emergency facility, we want to minimize the response time.

The **periphery** of a graph G, denoted **P(G)**, is the set of vertices with maximum eccentricity. The maximum eccentricity is called the **diameter** of G, denoted **diam(G)**. For the graph H of Figure 4.1, $\text{diam}(H) = 7$, and $P(H) = \{i, p, q\}$. The periphery of a nontrivial graph G must contain at least one pair of vertices u and v, satisfying $d(u, v) = \text{diam}(G)$. Such a pair of vertices is called **antipodal** or **diametral**. Each vertex is termed an **antipode** of the other. The graph H in Figure 4.1 has two antipodal pairs: i, p and q, p. Since $d(i, q) = 2$, these vertices do not constitute an antipodal pair. Note, furthermore, that antipodal vertices are always mutually eccentric, but the converse is not true (see Exercise 8). A geodesic [necessarily of length rad(G)] joining a central vertex to one of its eccentric vertices is called a **radial path**. A geodesic [necessarily of length diam(G)] joining a diametral pair of vertices is called a **diametral path**.

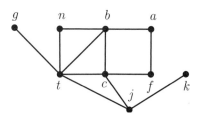

Figure 4.2 A graph G.

Example 4.1 Given the graph G in Figure 4.2,
a. Determine the eccentricity of each vertex.
b. Determine rad(G) and diam(G).
c. Determine $C(G)$ and $P(G)$.
d. Find all radial paths and all diametral paths in G.
e. Find all eccentric vertices of vertex f.
f. Find all pairs of mutually eccentric vertices.

g. Determine which of the pairs of part (f) are antipodal.
h. Which vertices would be the best candidates for an emergency facility if the graph modeled a community in a facility location problem?

Solution
a. $e(a) = 4$, $e(b) = 3$, $e(c) = 2$, $e(f) = 3$, $e(g) = 3$, $e(j) = 3$, $e(k) = 4$, $e(n) = 3$, and $e(t) = 2$.
b. $\text{rad}(G) = 2$ and $\text{diam}(G) = 4$.
c. $C(G)$ and $P(G)$ are the sets of vertices of minimum and maximum eccentricity, respectively. Thus $C(G) = \{c, t\}$ and $P(G) = \{a, k\}$.
d. Since $\text{rad}(G) = 2$ and $C(G) = \{c, t\}$, a radial path in G is a geodesic of length two joining c or t to one of their eccentric vertices. Thus the radial paths in G are c, b, a; c, b, n; c, f, a; c, j, k; c, t, g; c, t, n; t, b, a; t, c, f; and t, j, k. Since $\text{diam}(G) = 4$ and $P(G) = \{a, k\}$, a diametral path in G is a geodesic of length four joining a to k. Thus the diametral paths in G are a, b, c, j, k; a, b, t, j, k; and a, f, c, j, k.
e. $e(f) = 3$, so the eccentric vertices of f are those vertices at distance three from f, namely, g, k, and n.
f. f and g, f and n, and a and k are pairs of mutually eccentric vertices.
g. a and k are antipodal vertices. Not only are they mutually eccentric, but $d(a, k) = 4 = \text{diam}(G)$.
h. The central vertices c and t would serve best as locations for emergency facilities. ◇

If a graph G is connected, its diameter is a nonnegative integer. If G is disconnected, its diameter is defined to be ∞; that is, $\text{diam}(G) = \infty$. When dealing with distance problems, the graphs being considered are usually connected.

Here's a nice theorem about eccentricities that says the eccentricities of adjacent vertices differ by at most 1.

Theorem 4.1 If u and v are adjacent vertices in a connected graph, then $|e(u) - e(v)| \leq 1$.

Proof. Let $e(u) = m$. Then u "reaches" any vertex of G in at most m edges. Now let's look at this from the point of view of vertex v. First, by definition, we have $d(u, x) \leq e(u)$ for any vertex x. Let w be an eccentric vertex of v, so $e(v) = d(v, w)$. Then by the triangle inequality, we have $e(v) = d(v, w) \leq d(v, u) + d(u, w) = 1 + d(u, w) \leq 1 + e(u)$. Hence $e(v) \leq 1 + e(u)$, or, equivalently, $e(v) - e(u) \leq 1$. If we repeat the entire argument with the roles of u and v interchanged, we obtain $e(u) - e(v) \leq 1$. These two inequalities together imply the inequality in the theorem. (Since the left sides of the inequalities are additive inverses of one another, we have an situation similar to $z \leq 1$ and $-z \leq 1$, which implies that $|z| \leq 1$.) ∎

Trees and Distance

In Section 3.1, we examined numerous properties of trees. We now mention some distance properties for trees.

P4.1. Given three vertices u, v, and w of a tree, such that u and v are adjacent, we have $|d(u,w) - d(w,v)| = 1$. In other words, one of u and v is closer to w by precisely one edge. This is not true for arbitrary graphs, as a glance at K_3 shows.

P4.2. All eccentric vertices of a tree are end vertices.

P4.3. Pairs of antipodal vertices of a tree are end vertices.

P4.4. The periphery of a tree consists of end vertices.

P4.5. In any tree T, every diametral path includes all central vertices of T.

The Center of a Tree

Suppose that T is a tree with at least three vertices. What is the effect on T if we delete all of its end vertices? Let's ignore the obvious answer: Make it smaller. The less obvious answer is to decrease the eccentricity of each surviving vertex by exactly 1. This is an immediate consequence of property P4.2. Observe, then, that the new (smaller) tree has the same center as the original. This is because all eccentricities decrease by the same amount.

This process (called **pruning**) can be repeated, but not indefinitely. After several stages we reach one of two graphs—a single vertex, or two adjacent vertices. We have just proved the following theorem originally obtained by Camille Jordan [6] in 1869.

Theorem 4.2 The center of a tree consists of either a single vertex or two adjacent vertices. ∎

It is not hard to show that for any graph G, we have

$$\text{rad}(G) \leq \text{diam}(G) \leq 2\,\text{rad}(G) \qquad (4.2)$$

You will be asked to prove inequality (4.2) in Exercise 9.

Theorem 4.3 For any tree T, if $|C(T)| = 1$, then $\text{diam}(T) = 2\,\text{rad}(T)$, and if $|C(T)| = 2$, then $\text{diam}(T) = 2\,\text{rad}(T) - 1$.

Proof. Suppose that $|C(T)| = 1$ and let $C(T) = \{v\}$. If $V(T) = \{v\}$, then $T = K_1$, so $\text{rad}(T) = \text{diam}(T) = 0$. Thus $\text{diam}(T) = 2\,\text{rad}(T)$. Thus, suppose that T is nontrivial. Then since $|C(T)| = 1$, we get $|V(T)| \geq 3$. There are at least two branches at v. (A **branch at v** is a maximal subtree that has v as an end vertex.) Furthermore, two such branches must contain radial paths. Let x and y be end vertices of radial paths that contain distinct neighbors of v. Then $d(v,x) = d(v,y) = \text{rad}(T)$, and since there is a unique path joining any pair of vertices of a tree, the geodesic joining x and y in T is composed of the radial v–x path together with the radial v–y path. Thus, we find $d(x,y) = d(x,v) + d(y,v) = \text{rad}(T) + \text{rad}(T) = 2\,\text{rad}(T)$. Hence by

(4.2), vertices x and y are at maximum possible distance from one another in T. That is, x and y are antipodes. Hence, diam$(T) = d(x,y) = 2$ rad(T).

Now suppose that $|C(T)| = 2$, and let $C(T) = \{u,v\}$. Then by Theorem 4.2, u and v are adjacent. Let T_u be the set of vertices of T that are closer to u than to v, and let T_v be the set of vertices that are closer to v than to u. Then every radial path emanating from u ends in T_v, and every radial path emanating from v ends in T_u. So all radial paths have edge uv in common. Furthermore, by property (P4.5) every diametral x–y path contains uv. Such a diametral path is composed of a geodesic from x to $C(T)$, followed by edge uv, followed by a geodesic from $C(T)$ to y. Thus a diametral path is composed of two radial paths overlapping at uv. Therefore, diam$(T) = 2$ rad$(T) - 1$. ∎

Thus for all trees, the diameter is either twice the radius or one less than twice the radius. This restores our faith in the terminology, which was, after all, borrowed from circles, for which the diameter is twice the radius.

Self-Complementary Graphs and Distance

Recall that a graph G such that $\bar{G} \cong G$ is called self-complementary. We shall now show that the diameter of a self-complementary graph is at most three. We will need the following interesting theorem.

Theorem 4.4 If diam$(G) \geq 3$, then diam$(\bar{G}) \leq 3$.

Proof. Let G be a graph of diameter at least 3. If $w \in P(G)$, then $e(w) \geq 3$. Partition $V(G)$ into four subsets according to their distance from w as follows:
$$\begin{aligned} W_0 &= \{w\} \\ W_1 &= \{x \in V(G) : d(w,x) = 1\} \\ W_2 &= \{x \in V(G) : d(w,x) = 2\} \\ W_{\geq 3} &= \{x \in V(G) : d(w,x) \geq 3\} \end{aligned}$$

Notice that $V(G) = W_0 \cup W_1 \cup W_2 \cup W_{\geq 3}$, and these four subsets of $V(G)$ are pairwise disjoint; that is, the intersection of any two of them is empty. (This is exactly what is meant by a *partition* of a set.) Figure 4.3 depicts this situation.

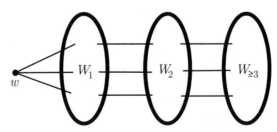

Figure 4.3 A graph G for which diam$(G) \geq 3$. Vertex $w \in P(G)$.

4.1 Distance in Graphs

Note that each edge of G not contained in one of the induced subgraphs on W_1, W_2, or $W_{\geq 3}$, must be incident with vertices in exactly one of the following pairs of subsets:
1. $\{w\}$ and W_1,
2. W_1 and W_2, or
3. W_2 and $W_{\geq 3}$.

To prove that $\text{diam}(\bar{G}) \leq 3$, we must show that given any pair of vertices x and y in $\bar{G}$, $d(x, y) \leq 3$. If x and y were not adjacent in G, then they are adjacent in $\bar{G}$; that is, $d_{\bar{G}}(x, y) = 1 \leq 3$. So suppose instead that x and y were adjacent in G. There are several cases to consider.

Case 1: x and y belong to $w \cup W_1$. Then in $\bar{G}$, x and y are both adjacent to all vertices of $W_{\geq 3}$. So $d_{\bar{G}}(x, y) = 2$.

Case 2: One of x and y, say x, is in W_1 and the other is in W_2. Then let z be any vertex in $W_{\geq 3}$. Then in $\bar{G}$, x and v are adjacent to z and y is adjacent to v. Hence, x, z, v, y is an x–y path in $\bar{G}$. Therefore, $d_{\bar{G}}(x, y) \leq 3$.

Case 3: x and y belong to $W_2 \cup W_{\geq 3}$. Then in $\bar{G}$, x and y are both adjacent to v. So $d_{\bar{G}}(x, y) = 2$. ∎

As a result of Theorem 4.4, a self-complementary graph G couldn't possibly satisfy $\text{diam}(G) \geq 4$, as the theorem would yield $\text{diam}(\bar{G}) \leq 3$, contradicting the assumption that $G \cong \bar{G}$ since isomorphic graphs have the same diameter. Therefore, we have established the following corollary.

Corollary 4.4a If G is self-complementary, then $\text{diam}(\bar{G}) \leq 3$. ∎

Since a nontrivial self-complementary graph can't have diameter 1, Corollary 4.4a can be restated as follows.

Corollary 4.4b The diameter of a nontrivial, self-complementary graph is either 2 or 3. ∎

Note that the self-complementary graphs C_5 and P_4 have diameters 2 and 3, respectively. The self-complementary graph of Figure 4.4 has diameter 3. Take a moment or two to verify that it is self-complementary.

Figure 4.4 A self-complementary graph G of diameter 3.

The Centroid of a Tree

As mentioned earlier, given a vertex v of a tree T, the maximal subtrees that have v as an end vertex are called **branches at v**. The degree of v equals the number of branches at v. If v is an end vertex, there is only one branch with end vertex v, namely, the entire tree. Figures 4.5 and 4.6 show a tree T and the branches of a randomly selected vertex v. The **weight of a vertex** v (in a tree) is the largest number of edges among all of its branches. Vertex v in the tree of Figure 4.5 has weight 5.

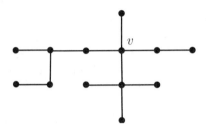

Figure 4.5 A tree T and a chosen vertex v of weight 5.

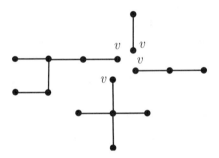

Figure 4.6 The four branches of v.

For the star $K_{1,n}$, the weight of the center vertex is 1 since each of its branches has one edge. On the other hand, the weight of an end vertex is n, which is the total number of edges in the tree. The moral of this story is that adjacent vertices can have weights that differ by any given number. Contrast this with eccentricities. If u and v are adjacent vertices in any graph, then $|e(u) - e(v)| \leq 1$.

The **centroid** of a tree T is the set of vertices with minimum weight. Like the center of a tree, Jordan [6] showed that the centroid consists of either a single vertex or two adjacent vertices. The center and centroid of a tree may be disjoint, as is shown in the tree of Figure 4.7 with center $\{v\}$ and centroid $\{w\}$. In fact, the center and centroid of a tree may be arbitrarily far apart (see Slater [8]).

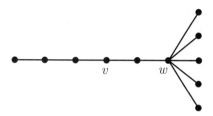

Figure 4.7 A tree with center $\{v\}$ and centroid $\{w\}$.

Exercises 4.1

1. Find the eccentricities of each vertex of the tree T of Figure 4.5. Do the same for the tree of Figure 4.7.

2. Given a vertex v of C_n, find $e(v)$. Show that v has either one or two eccentric vertices, depending on the parity of n.

3. Find the radius and diameter of C_n. Do the same for P_n. Show that the center of P_n consists of one or two adjacent vertices, depending on the parity of n.

4. Let H be a spanning subgraph of a graph G. Given vertices u and v in G, show that their distance from one another in H is at least as big as in G, that is, $d_H(u,v) \geq d_G(u,v)$.

5. Let H be a spanning subgraph of G.

    (a) Use Exercise 4 to show that for any vertex v in G, $e_H(v) \geq e_G(v)$.
    (b) Show that $\text{rad}(H) \geq \text{rad}(G)$ and $\text{diam}(H) \geq \text{diam}(G)$.
    (c) Give an example of G and H for which the two inequalities of part (b) are strict ($>$).
    (d) Give an example of G and H for which we obtain equality for part (b).
    (e) Give an example of G and H for which $\text{rad}(H) = \text{rad}(G)$ and $\text{diam}(H) > \text{diam}(G)$.

6. Show that $C(C_n) = V(C_n)$; that is, show that the center of an n-cycle consists of all of its vertices. A graph with this property is called **self-centered**. (Really!) Find another class of self-centered graphs.

7. Draw graphs that are not self-centered, and with centers consisting of one, two, three, or four vertices. Extend this to n vertices.

8. For the sequential join (see Section 2.4, if necessary) $G = K_1 + K_1 + \bar{K}_2 + K_1 + K_1$, determine $\operatorname{rad}(G)$, $\operatorname{diam}(G)$, $C(G)$, and $P(G)$. Then show that G contains a pair of vertices that are mutually eccentric but not antipodal of one another.

9. Use the definitions of "radius" and "diameter" along with the triangle inequality to prove that for any graph G, $\operatorname{rad}(G) \leq \operatorname{diam}(G) \leq 2\operatorname{rad}(G)$.

10. Prove that all trees have properties P4.1–P4.4 earlier in this section.

11. Prove property P4.5: In any tree T, every diametral path includes all central vertices.

12. Find a graph G that has at least one diametral path that does not contain all central vertices, thereby showing that property P4.5 cannot be extended to graphs in general. (Note: Buckley and Lewinter [3] examined graphs for which *every* diametral path completely avoids the center.)

13. Prove that the wheel $W_{1,n}$ has a spanning tree with one center vertex also has and a spanning tree with two center vertices.

14. Let T be a spanning tree of graph G. Show that if $\operatorname{rad}(T) > \operatorname{rad}(G)$, then $\operatorname{diam}(T) > \operatorname{diam}(G)$. Show that this implication is one way; that is, $\operatorname{diam}(T) > \operatorname{diam}(G)$ does not guarantee that $\operatorname{rad}(T) > \operatorname{rad}(G)$.

15. Find the radius, diameter, center, and periphery for the tree of Figure 4.5. Do the same for the graph of Figure 4.7.

16. Find a spanning tree T of the graph G of Figure 4.2 such that $\operatorname{rad}(T) = \operatorname{rad}(G)$.

17. Suppose that for vertices x, y and z of a graph G, vertices y and z are adjacent, and $d(x,y)$ and $d(x,z)$ have the same parity. Show that $d(x,y) = d(x,z)$.

18. Suppose that for vertices x, y and z of a graph G, the x–y geodesic L and the x–z geodesic M have the same length, and vertex w is the first shared vertex of paths L and M, traveling from y and z toward x. (There may be common vertices other than x, of course). Show that $d(y,w) = d(z,w)$.

19. Consider the following construction. Start with a self-complementary graph A and any graph B. In graph G of Figure 4.4, replace the vertex of degree 2 by a copy of A, replace each vertex of degree 3 by a copy of B, and replace each of the end vertices by a copy of $\bar{B}$. Finally, replace

the edges of G by joins. Show that diam$(G) = 3$, no matter what the diameters of A and B are.

20. Let G and H be graphs, neither of which are complete. Show that diam$(G + H) = 2$. Why must we stipulate that neither of G and H are complete graphs?

21. Show that for the sequential join of any graphs G, H and J, we have diam$(G + H + J) = 2$.

22. Find the weight of each vertex in the graph of Figure 4.5. Find the centroid.

23. Find the weight of each vertex of Figure 4.7.

24. Label the vertices of P_n consecutively from an end vertex using the numbers $1, 2, 3, \ldots, n$. Now determine the weight of vertex i, for $1 \leq i \leq n$.

25. Given integers j and k, with $0 \leq j < k - 1$, construct a graph such that it contains two adjacent vertices whose weights are j and k, respectively.

26. Given a vertex v of Q_3, obtain five spanning trees T_i, $1 \leq i \leq 5$ such that the weight of v in these trees is 3, 4, 5, 6, and 7, respectively; that is, the weight of v in T_i is $i + 2$ for $i = 1, 2, \ldots, 5$. Explain why the weight of v in a spanning tree of Q_3 cannot be 1 or 2.

27. Let v be a vertex of a connected graph G of order n. Show that the weight of v in any spanning tree of G cannot be less than $(n-1)/\deg(v)$. This number is, therefore, a lower bound for the weight of v. Now show that $n - 1$ is an upper bound for the weight of any vertex in G.

4.2 Connectivity Concepts

In some graphs, the removal of one or more vertices will break the graph into several components. In a nontrivial graph, it is always possible to separate a graph into disjoint components by removing edges, provided enough edges are removed. These ideas have received a great deal of study over the years because of their important implications to the reliability of computer networks. Numerous measures of "how connected" a graph is have been developed. We now discuss various connectivity concepts.

Cut Vertices, Bridges, and Connectivity

A vertex in a connected graph G is called a **cut vertex** if its deletion disconnects the graph. If G is a disconnected graph, we define a cut vertex as a vertex whose deletion increases the number of components of G. As we saw in Section 2.4, $G - v$ is the graph obtained by deleting from G vertex v and all edges incident with v. Graph $G - uv$ is the graph obtained by deleting edge uv from G. An edge uv is a **bridge** or **cut edge** if its deletion increases the number of components. Thus, in a connected graph G, edge uv is a bridge if $G - uv$ is disconnected.

While every edge of a tree is a bridge, observe that only those vertices that are not end vertices of the tree are cut vertices. Hence v is a cut vertex of a tree T if and only if $\deg(v) \geq 2$. For $n > 1$, the star $K_{1,n}$ has exactly one cut vertex, for example, while all of its edges are bridges.

If a graph represents a communication or transportation network, such as a network of computers, cut vertices and bridges are items of *vulnerability* that threaten the integrity of the network. Networks welcome redundancy—several ways of traveling between the vertices in the network. A saboteur would have to destroy at least two vertices of a cycle, for example, to disrupt communication between some of the survivors.

Example 4.2 Find all cut vertices and bridges for the graph of Figure 4.8.

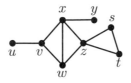

Figure 4.8

Solution
Vertices x, v, and z are cut vertices. Edges uv and xy are bridges. ◇

This brings us to the next definition. The **vertex connectivity** of a graph G, denoted $\kappa(G)$, or just κ (pronounced "*kappa*"), is the minimum number of vertices whose deletion disconnects G or makes G become trivial. The last part of the definition is required for the complete graphs, since K_n will never become disconnected by removing vertices. It follows from the definition that if G is disconnected, then $\kappa(G) = 0$. A set of vertices in a connected graph whose removal disconnects the graph is called a **vertex cutset**. A connected graph G has a cut vertex if and only if $\kappa(G) = 1$. A graph G is called k-**connected** for some positive integer k if $\kappa(G) \geq k$.

This is done for edges, too. The **edge connectivity** of a graph G, denoted $\lambda(G)$, or simply λ (pronounced "*lambda*"), is the minimum number of edges whose deletion disconnects G or makes G trivial. The last part of the definition is needed only for K_1. A set of edges whose deletion causes a connected graph to become disconnected is called an **edge cutset**. A connected graph has a bridge if and only if $\lambda = 1$.

4.2 Connectivity Concepts

The graph G of Figure 4.9 satisfies $\kappa = 1$, $\lambda = 2$, and $\delta = 4$ (recall that δ denotes the minimum degree among the vertices). These three important numbers are usually given in this order because of the content of our next theorem, due to Whitney [9].

Theorem 4.5 Given a connected graph G, we have
$$\kappa(G) \leq \lambda(G) \leq \delta(G) \tag{4.3}$$

Proof. Observe that if v is a vertex of minimum degree, then the removal of all of the edges incident with v disconnects the graph (v is then one of the components). Of course, there may be an edge cutset, elsewhere in the graph, consisting of fewer edges. Thus $\lambda(G) \leq \delta(G)$. Furthermore, if S is an edge cutset consisting of k edges, then the removal of k suitably chosen vertices results in the removal of the edges of S and, therefore, disconnects the graph. (There may be a smaller vertex cutset elsewhere in G.) It follows that $\kappa(G) \leq \lambda(G)$, and combining this with the earlier inequality completes the proof. ∎

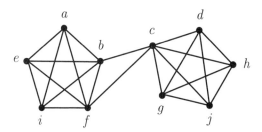

Figure 4.9 A graph G, for which $\kappa = 1$, $\lambda = 2$, and $\delta = 4$.

It is interesting to observe that $\kappa(C_n) = \lambda(C_n) = \delta(C_n) = 2$. We also obtain equality for paths, but the common value is 1 for nontrivial paths and 0 for P_1. In fact, equality of the three values holds for all nontrivial trees, since every nontrivial tree T has end vertices [thus $\delta(T) = 1$].

Example 4.3 Find a minimal vertex cutset of order 1 and of order 2 for graph G in Figure 4.9. Then find minimal edge cutsets of size 2 and 4.

Solution
Vertex c is a cut vertex, so $\{c\}$ is a minimal vertex cutset of order 1. Since the removal of vertices b and f would disconnect G, but neither one alone would (neither is a cut vertex), the set $\{b, f\}$ is a minimal vertex cutset of order 2. There are no others. There is just one edge cutset of size 2—namely, $\{bc, cf\}$. Since neither bc nor cf is a bridge, $\{bc, cf\}$ is a minimal edge cutset of size 2. There are eight minimal edge cutsets of size 4. The set of edges incident with any of the vertices of degree 4 constitutes a minimal edge cutset of size 4. Additionally, $\{cd, cg, ch, cj\}$ is a minimal edge cutset of size 4. ◊

Blocks

A **block of a graph** G is a maximal connected subgraph of G. A block H of G, then, is a subgraph of G such that when it is drawn by itself, it has no cut vertex, but if we expand H by just adding more edges (but no additional vertices), we no longer have a subgraph. Also, if we expand H to a larger connected *subgraph* by adding vertices *and* edges, the resulting subgraph has a cut vertex. Figure 4.10(a) exhibits a graph and Figure 4.10(b) its blocks.

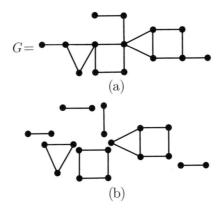

Figure 4.10. A graph G and its seven blocks.

The following theorem of Harary and Norman [5] relates the center of a graph to its block structure.

Theorem 4.6 *The center of a connected graph G belongs to a single block of G.*

Proof. If G is a connected graph without a cut vertex, then there is only one block, namely, G itself, and the theorem is obviously true. Assume then that G has a cut vertex v. Then $G - v$ has at least two components. We now proceed by contradiction. Suppose that $x, y \in C(G)$, such that $x \in V(H)$ and $y \in V(J)$, where H and J are components of $G - v$. Then $e(x) = e(y) = \text{rad}(G)$. Now $e(v) \geq \text{rad}(G)$ implies the existence of a vertex z such that $d(v, z) = e(v) \geq \text{rad}(G)$. Without loss of generality, assume that $z \notin V(J)$. Now this implies that the cut vertex v is on every y–z geodesic. Then $d(y, z) = d(y, v) + d(v, z) \geq 1 + \text{rad}(G) > \text{rad}(G)$. But then we have $e(y) > \text{rad}(G)$, so $y \notin C(G)$, a contradiction. So x and y must be in the same components of $G - v$. This means that no vertex can separate two vertices of $C(G)$, so the center belongs to a single block of G. ∎

We shall refer to the block containing the center as the **central block**.

Example 4.4 Find all blocks of the graph in Figure 4.11. Which block is the central block?

4.2 Connectivity Concepts

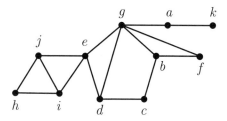

Figure 4.11

Solution
The blocks are displayed in Figure 4.12. The block with six vertices is the central block. The vertices d, e, and g are central vertices of the original graph in Figure 4.11. ◇

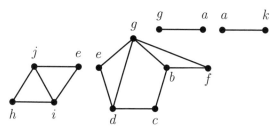

Figure 4.12

Menger's Theorem

In 1927, Menger [7] showed that the connectivity of a graph is related to the number of disjoint paths joining two vertices. His theorem has been extremely useful and has produced many related variations within the theory of graph connectivity. It is also closely related to Hall's theorem (Theorem 3.6) and many other important results throughout discrete mathematics including the max-flow min-cut theorem to be discussed in section 10.2.

Two paths connecting u and v are **internally disjoint u–v paths** if they have no vertices in common other than u and v; they are **edge disjoint** if they have no edges in common. Any internally disjoint paths must also be edge disjoint. A set S of vertices, edges, or both **separates** u and v if u and v are in different components of $G - S$. This means that $u, v \notin S$, and every path connecting u and v passes through S.

Theorem 4.7 (Menger's Theorem) Let u and v be distinct nonadjacent vertices in G. Then the maximum number of internally disjoint paths connecting u and v equals the minimum number of vertices in a set that separates u and v. ∎

For a proof, the interested reader is referred to Buckley and Harary [1, p. 64] or Chartrand and Oellermann [4, p. 155].

Example 4.5 Find a minimum order separating set and a maximum set of internally disjoint paths connecting u and v in the graph of Figure 4.13, thereby illustrating Menger's theorem.

Solution Set $S = \{e, f, d\}$ is a minimum order separating set. Paths u, a, f, v; u, e, i, v; and u, d, v form a maximum collection of disjoint u–v paths. ◇

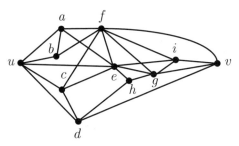

Figure 4.13

Observe that each of the vertices in the minimum order separating set S is contained in one of the disjoint u-v paths. This is always the case, as described in Note 4.2.

Note 4.2
Suppose that S is a minimum order set separating nonadjacent vertices u and v in G, and C is a maximum order collection of disjoint u-v paths. Then each vertex of S is contained in precisely one path of C.

Whitney [9] characterized k-connected graphs as the following corollary to Menger's theorem.

Corollary 4.7a A graph G having at least two vertices is k-connected if and only if for each pair of distinct vertices, there are k internally disjoint paths connecting them. ∎

It was a couple of decades after Menger's theorem that an analogous result was found to hold for edge-separating sets (edge cutsets).

Theorem 4.8 Let u and v be distinct nonadjacent vertices in G. Then the maximum number of edge-disjoint paths connecting u and v equals the minimum number of edges in a set that separates u and v. ∎

Example 4.6 Find a minimum-sized set of edges separating u and v and a maximum set of edge-disjoint paths connecting u and v in the graph of Figure 4.13.

4.2 Connectivity Concepts

Solution
Paths u, a, f, v; u, b, f, i, v; u, e, g, v; u, c, e, h, v; and u, d, v form a maximum collection of edge-disjoint u–v paths. $S = \{ua, ub, uc, ud, ue\}$ is a minimum-sized set of edges separating u and v. ◇

A minimum-sized set of edges separating u and v is referred to as a **minimum cut**. We shall see in Section 10.2 that minimum cuts play a key role in the theory of maximum flows in networks.

Exercises 4.2

1. Find all cut vertices and bridges for the graph in Figure 4.10.

2. Find all cut vertices and bridges for the graph in Figure 4.11.

3. Find a minimal edge cutset of size 5 and two minimal edge cutsets of size 6 for the graph in Figure 4.9.

4. Find three different minimal edge cutsets of size 2 for the graph of Figure 4.8.

5. Determine all blocks for the graph of Figure 4.8.

6. Let x and y belong to different blocks of a graph. What can be said about all x–y geodesics?

7. How many blocks does a tree of order n have? Describe them.

8. Construct a graph G with $\kappa(G) = 2$, $\lambda(G) = 3$, and $\delta(G) = 4$.

9. Construct a graph G with $\kappa(G) = 3$, $\lambda(G) = 3$, and $\delta(G) = 5$.

10. Determine $\kappa(G)$ and $\lambda(G)$ for each of the following graphs:

    (a) The octahedron $\bar{K}_2 + C_4$
    (b) The sequential join $K_2 + K_3 + \bar{K}_2 + \bar{K}_3$
    (c) The cartesian product $P_4 \times C_3$

11. Determine $\kappa(G)$ and $\lambda(G)$ as a function of m and n for the generalized wheel $W_{m,n} = \bar{K}_m + C_n$.

12. Draw the line graph $L(W_{1,4})$. Then find $\kappa(L(W_{1,4}))$ and $\lambda(L(W_{1,4}))$.

13. Prove that every k-connected graph on n vertices has at least $nk/2$ edges.

14. Prove that if G is cubic—that is, 3-regular—then $\kappa(G) = \delta(G)$.

15. Prove that there exists no 3-connected graph having seven edges.

16. Prove that if $\text{diam}(G) \leq 2$, then $\lambda(G) = \delta(G)$.

17. Prove that if G is k-connected, then the join $K_1 + G$ is $(k+1)$-connected.

18. Find a minimum order separating set and a maximum set of internally disjoint paths connecting vertices u and v in the graph of Figure 4.14, thereby illustrating Menger's theorem.

19. Determine $\kappa(G)$ and $\lambda(G)$ for the graph of Figure 4.14.

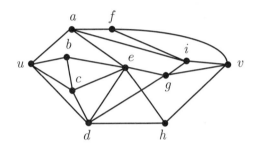

Figure 4.14

4.3 Applications

In this section, we examine two applications, one for distance problems and one for connectivity.

F-Graphs

Recall that the center of a tree consists of either a single vertex or two adjacent vertices. It might be reasonable to expect that if a graph has several center vertices, those vertices would be close to one another. However, this is not the case in general, as the graph G in Figure 4.15 demonstrates. The center $C(G) = \{x, y\}$ and $d(x, y) = 8$. Notice that since $\text{rad}(G) = 8$, the central vertices x and y are situated at maximum possible distance from one another.

This example motivates the following definition. A graph G is an *F*-**graph** ("*F*" for far, since the central vertices are as far apart as possible) if

1. $|C(G)| \geq 2$, and
2. If $x, y \in C(G)$, then $d(x, y) = \text{rad}(G)$.

F-graphs have a very nice property in the context of facility location problems. As we have mentioned, central vertices are the ideal locations

4.3 Applications

for emergency facilities in a community. If the community's road system is modeled by an F-graph, then distinct emergency facilities can be located far apart from one another. For example, when a fire truck and an ambulance are responding to emergencies, an F-graph will decrease the chance of their interfering with one another while on the way to the respective emergency locations.

The F-graph of Figure 4.15 can be modified to yield an infinite class of F-graphs by changing the lengths of the paths x–c, x–d, y–c, y–d, a–c and b–d from 4 to k. Each value of k yields an F-graph with radius $2k$, diameter $4k$, and center $\{x, y\}$.

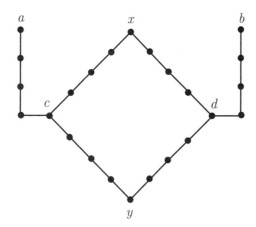

Figure 4.15 Graph G with $C(G) = \{x, y\}$ and $d(x, y) = 8 = \mathrm{rad}(G)$.

For each j, $0 \leq j \leq \mathrm{rad}(G)$, define the **$j$th central distance set**, denoted $\boldsymbol{N_j}$, by
$$N_j(G) = \{x : d(x, C(G)) = j\} \tag{4.4}$$

Thus $N_j(G)$ is the set of vertices at distance j from a nearest central vertex. Note that $N_0(G) = C(G)$, and if $N_k(G) \neq \emptyset$ and $j < k$, then $N_j(G) \neq \emptyset$.

Theorem 4.9 Let G be an F-graph with radius $r \geq 2$ and let $j = \lfloor r/2 \rfloor$. Then $N_j(G) \neq \emptyset$.

Proof. Let $C(G) = \{c_1, c_2, \ldots, c_s\}$. Let z be an internal vertex of a c_1–c_2 geodesic, such that $d(c_1, z) = j$. This is possible because G is an F-graph, implying that $d(c_1, c_2) = r > j$. We claim that $z \in N_j$, implying that $N_j \neq \emptyset$. To prove the claim, suppose to the contrary that $z \notin N_j$. Then since $d(c_1, z) = j$, there must be another central vertex, say c_k, such that $d(c_k, z) < j$. Then $d(c_k, c_1) \leq d(c_k, z) + d(z, c_1) < 2j \leq r$, implying that $d(c_k, c_1) < r$, contradicting the assumption that G is an F-graph. ∎

Before we present the next theorem, let's observe that except for K_n, an F-graph is not self-centered; that is, the diameter of all other F-graphs must exceed the radius. This is so because the distance between any pair of central vertices must equal the radius. If all the vertices are in the center, it follows by selecting a pair of adjacent vertices that their distance from one another is 1, and hence the radius and the diameter are also 1. Therefore, the graph must be K_n.

Theorem 4.10 If G is an F-graph with radius r and diameter d, then any diametral path of G has a subpath of length at least $d - r$, all of whose vertices are in the central block.

Proof. Let G be an F-graph with radius r and diameter d and suppose $x, y \in C(G)$. Let B be the central block; that is, $C(G)$ lies in the block B. Furthermore, let P be a diametral u–v path, where $d(u,v) = \text{diam}(G)$. If $u, v \in B$, then the entire path P is in B and we are done. We shall initially assume then that $u \notin B$ and $v \notin B$ and that $P \cap B = \emptyset$.

Let w be a cut vertex on a geodesic from x to u. [There must be a cut vertex on a u–x geodesic because $x \in C(G) \subseteq B$.] Vertex w must also lie on a geodesic from x to v. Otherwise, there would be another cut vertex, say t, on a v–x geodesic, in violation of the fact that P lies in a different block than x, and these blocks have at most one cut vertex in common—two blocks cannot share two cut vertices!

Since G is an F-graph, $d(x,y) = r$. Using the triangle inequality, we obtain $d(x,w) + d(w,y) \geq r$. Then one of these distances, say $d(x,w)$, satisfies $d(x,w) \geq r/2$. But $d(x,u) \leq r$, from which it follows that $d(u,w) \leq r/2$. By a similar argument, $d(v,w) \leq r/2$. Then $d = d(u,v) \leq d(u,w) + d(w,v) \leq r/2 + r/2 = r$. But $d \leq r$ implies that $d = r$, since for all graphs $d \geq r$. Then G is self-centered; which is only possible if G is K_n. But this is impossible since we assumed that $P \cap B = \emptyset$. It then follows that $P \cap B \neq \emptyset$. In fact, the intersection must have at least two vertices (including two cut vertices) or the same contradiction will ensue.

In light of the foregoing, $P = u, a_1, a_2, \ldots, a_k, P^*, c_1, c_2, \ldots, c_t, v$, where $P_* = w, b_1, b_2, \ldots, b_q, z$ is the subpath of P in B. Note that since P is a geodesic, $d = d(u,v) = d(u,w) + (w,z) + d(z,v)$. Recall that $d(u,w) \leq r/2$, and a similar argument to the one in the previous paragraph shows that $d(z,v) \leq r/2$, since z is a cut vertex on a v–x (or v–y) geodesic. Then $d \leq d(w,z) + r$, or after transposition, $d(w,z) \geq d - r$. This means that the length of the subpath P^* of the diametral path P is at least $d - r$, as claimed.

If only one of u or v, say u, is in B, a similar argument shows that the subpath of P in the central block; that is, the subpath from u to z has length $d(u,z)$, which satisfies $d(u,z) \geq d - r/2 \geq d - r$. ∎

Here is another construction result. This is due to Buckley and Lau [2].

4.3 Applications

Theorem 4.11 For all positive integers r and k with $k \geq 2$, there exists an F-graph with radius r having k central vertices.

Proof. If $r = 1$, then use K_k. For $r > 1$, if r is even, join vertices u and v by k paths of length r. Then attach two pendant paths of length $r/2$, one at u and one at v. For $r > 1$, if r is odd, start with two copies of K_{k+1} with their vertices labeled v_i and w_i, respectively. For $1 \leq i \leq k$, join vertices v_i and w_i by a path of length $r - 1$. Then attach two pendant paths of length $(r-1)/2$, one at v_{k+1} and one at w_{k+1}. ∎

Network Reliability

When we use the word "network" here, you may think of a weighted graph. The weights here are usually probabilities. Those probabilities are assigned to the edges, the vertices, or both. **Network reliability** is concerned with how well a given network, such as a computer or telecommunications network, can withstand failure of individual components of the system. The vertices might represent individual computers, modems, or other pieces of hardware. The edges generally correspond to cables or other such linkages between those pieces of hardware. There are numerous models of network reliability. In one commonly studied model (the **edge-failure model**), the vertices are assumed to be totally immune to failure and all edges are assumed to fail independently with equal probability q. When numerous edges fail, the failures are assumed to happen simultaneously. In such instances, any edge that fails is unavailable for the transmission of information. A common problem in this model is **K-terminal reliability**, which asks us to determine the probability that a subset K of terminal vertices remain connected to one another (that is, they remain within one component).

In a second reliability model (the **vertex-failure model**), the edges are assumed to be perfectly reliable and the vertices fail independently with probability p. When a given vertex fails, the edges incident with that vertex become inoperable. Note that this is similar to vertex connectivity in that when a vertex is removed, the edges incident with it are also removed. In this model, we usually want to know the probability that the network will remain connected.

In a third reliability model (the **edge-neighbor connectivity model**), the edges fail independently with equal probability, but when a given edge fails, not only is the edge inoperable, but its two incident vertices become inoperable as well.

A Little Probability

There are a few basic items about probability that we need to know before proceeding.

1. An **event** (such as a particular edge failing) either occurs or it doesn't. If the probability that an event occurs is q, then the probability that it does not occur is $1-q$. Similarly, if the probability that some compound event A (such as several edges failing) occurs is $\Pr(A)$, then the probability that A does not occur is $1 - \Pr(A)$.
2. If events $A_1, A_2, \ldots, A_k$ occur independently, each with probability q, then they all occur together with probability q^k. That is,
$$\Pr(A_1 \cap A_2 \cap \cdots \cap A_k) = \Pr(A_1) \cdot Pr(A_2) \cdots Pr(A_k).$$
3. $\Pr(A \cup B) = \Pr(A) + \Pr(B) - \Pr(A \cap B)$; that is, the probability that A or B occurs is the probability that A occurs plus the probability that B occurs minus the probability that both A and B occur simultaneously.

Example 4.7 Consider the graph in Figure 4.16. Suppose that each edge fails with probability 0.2. Find the probability that vertices u and v remain in the same component.

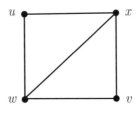

Figure 4.16

Solution
Let A be the event that edges ux and uw are missing; let B be the event that vx and vw are missing; let C be the event that ux, wx, and vw are missing; and let D be the event that uw, wx, and vx are missing. We have $\Pr(A) = \Pr(B) = 0.2^2$ and $\Pr(C) = \Pr(D) = 0.2^3$. Then the probability that vertices u and v remain in the same component equals $1 - \Pr(A \cup B \cup C \cup D)$. *Note*: This uses a combinatorial process called **inclusion-exclusion** to avoid "overcounting." We get
$1 - \Pr(A \cup B \cup C \cup D)$
$$\begin{aligned} &= 1 \; - \Pr(A) - \Pr(B) - \Pr(C) - \Pr(D) + \Pr(A \cap B) + \Pr(A \cap C) \\ &\quad + \Pr(A \cap D) + \Pr(B \cap C) + \Pr(B \cap D) + \Pr(C \cap D) \\ &\quad - \Pr(A \cap B \cap C) - \Pr(A \cap B \cap D) - \Pr(A \cap C \cap D) \\ &\quad - \Pr(B \cap C \cap D) + \Pr(A \cap B \cap C \cap D) \\ &= 1 \; -2(0.2)^2 - 2(0.2)^3 + 5(0.2)^4 + (0.2)^5 - 4(0.2)^5 + (0.2)^5 \\ &= 1 \; -0.08 - 0.016 + 0.008 + 0.0016 - 0.00128 + 0.00032 = 0.91264 \; \diamondsuit \end{aligned}$$

From Example 4.7, we see that the calculations are quite tedious. This is often the case with reliability problems. The situation is somewhat simpler

4.3 Applications

when the graph is a tree because of the uniqueness of the path joining a given pair of vertices.

Exercises 4.3

1. Construct an infinite class of graphs of order $2k$ such that $|C(G)| = k$ and the other k vertices are each adjacent to at least one central vertex. Start with $k = 2$.

2. What are the central distance sets of the 2-mesh $M(3,3)$?

3. Do the four corner vertices of the 2-mesh $M(a,b)$ belong to the same central distance set? If so, which one?

4. Must the induced subgraph of a central distance set be connected? If not, give an example.

5. Given nontrivial graphs G and H, why is $G \times H$ not an F-graph?

6. Construct an F-graph with three central vertices.

7. Using the construction given in the proof of Theorem 4.11, draw an F-graph of radius 3 having four central vertices.

8. Using the construction given in the proof of Theorem 4.11, draw an F-graph of radius 4 having four central vertices.

9. If G and H are F-graphs, determine when $G + H$ is an F-graph.

10. Redo Example 4.7 with probability of edge failure for each edge being 0.1.

11. Suppose that $G = K_{1,4}$ and u and v are two particular vertices at distance 2 from one another. Suppose further that the probability of edge failure is 0.2. Find the probability that u and v end up in the same component.

12. Redo Exercise 11 using the graph $G = K_2 + K_1 + K_1$.

References for Chapter 4

1. Buckley, F., and F. Harary, *Distance in Graphs*, Addison-Wesley, Redwood City, CA (1990).

2. Buckley, F., and W. Y. Lau, Mutually eccentric vertices in a graph. *Ars Combinatoria* (to appear).

3. Buckley, F., and M. Lewinter, Graphs with all diametral paths through distant central nodes. *Mathematical and Computer Modeling* 17, no. 11 (1993) 35–41.

4. Chartrand, G., and O. R. Oellermann, *Applied and Algorithmic Graph Theory*, McGraw-Hill, New York (1993).

5. Harary, F., and R. Z. Norman, The dissimilarity characteristics of Husimi trees. *Annals of Mathematics* 58 (1953) 134–141.

6. Jordan, C., Sur les assemblages de lignes. *J. Reine Agnew. Math.* 70 (1869) 185–190.

7. Menger, K., Zur allgemeinen Kurventheorie. *Fund. Math.* 10 (1927) 96–115.

8. Slater, P. J., Structure of the k-centra of a tree. *Proceedings of the Ninth Southeastern Conference on Combinatorics, Graph Theory, and Computing*, (1978) 663–670.

9. Whitney, H., Congruent graphs and the connectivity of graphs. *American Journal of Mathematics* 54 (1932) 150–168.

Additional Readings

10. Buckley, F., Facility location problems. *College Mathematics Journal* 18 (1987) 24–32.

11. Grimaldi, R. P., and D. R. Shier, Redundancy and reliability of communication networks. *College Mathematics Journal* 27 (1996) 59–67.

12. Gutman, I., More on distance of line graphs. *Graph Theory Notes of New York* 33 (1997) 14–18.

13. Gutman, I., E. Estrada, and O. Ivanciuc, Some properties of the Weiner polynomial of trees. *Graph Theory Notes of New York* 36 (1999) 7–13.

14. Skurnick, R., Extending the concept of branch-weight centroid number to the vertices of all connected graphs via the Slater number. *Graph Theory Notes of New York* 33 (1997) 28–32, (corrigendum, 34 (1998) 54.).

Chapter 5

Eulerian and Hamiltonian Graphs

Numerous theoretic and applied problems in graph theory require us to traverse a graph in a particular way. In some problems, the goal is to find a trail or circuit in order to pass through each edge exactly once. In other problems, we must find a path or cycle that includes each vertex exactly once. We discuss such problems in this chapter.

5.1 Characterization of Eulerian Graphs

For eulerian graphs we want to traverse each edge of a graph and return to our starting point. We first discuss the historical background of this problem. To do so we must consider a generalization of a graph.

Multigraphs

Every now and then we will permit several edges to exist between a given pair of vertices. An airline map might indicate the existence of three flights between Atlanta and Pig's Eye by drawing three edges between them. Several edges incident with the same pair of vertices are called **multiple edges**. If multiple edges are allowed, the resulting structure is called a **multigraph**. Be sure to count each edge incident with a given vertex when computing its degree. Theorem 2.1 remains valid for multigraphs. (The sum of the degrees equals twice the number of edges.) A multigraph is displayed in Figure 5.1. It is important to understand that every graph is a multigraph but not every multigraph is a graph. A graph is a multigraph that has no multiple edges. To distinguish, when no multiple edges are present, some authors use the term **simple graph**.

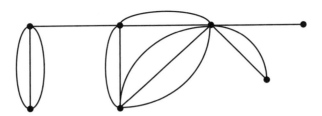

Figure 5.1 An example of a multigraph.

If the number of multiple edges between u and v is large, say k, it is far more convenient to draw only one edge and to weight it by k. Such a structure is called a **weighted graph** or a **network**. The weighted graph of Figure 5.2 does this for the multigraph of Figure 5.1. If the weights are omitted, one obtains the **underlying graph** of the multigraph.

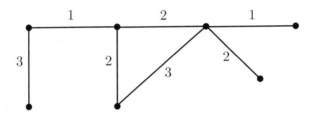

Figure 5.2 The weighted graph corresponding to Figure 5.1.

The Seven Bridges of Königsberg

Let's digress to discuss the origin of graph theory. In the 1730s, an interesting problem intrigued the mathematicians of Europe. The "seven bridges of Königsberg" problem called for a walk that traversed each of the seven bridges of Königsberg in Prussia (now called Kaliningrad in Russia) exactly once and ended where it began. To solve this problem, Euler probably started by drawing a picture of the situation. The drawing of Figure 5.3 will acquaint you with the layout. Here C and D are islands, and A and B are the banks of the river Pregel.

Like a good mathematician, Euler recognized that the sizes of the landmasses (A, B, C, and D) were irrelevant, as were the lengths of the bridges. He redrew the diagram as a multigraph (see Figure 5.4). In graph theoretic terms, the problem requires a closed trail—that is, a circuit that traverses each edge of the multigraph exactly once. The circuit may revisit a vertex as often as needed. Multigraphs with this property are now called **eulerian**, and the associated circuit is called an **eulerian circuit**.

5.1 Characterization of Eulerian Graphs

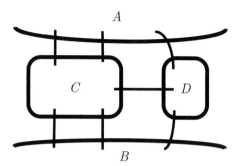

Figure 5.3 The seven bridges of Königsberg and landmasses A, B, C, D.

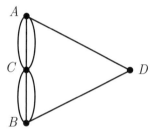

Figure 5.4 The essence of the problem—a multigraph!

Euler observed that each degree in the Königsberg multigraph is odd. Thus no vertex can serve as an initial vertex since the circuit must leave then return, leave then return, and so on, requiring an even degree. Similarly, no vertex can serve as an intermediate vertex since the circuit must enter then leave, enter then leave, and so on, again requiring an even degree.

Euler thereby established a necessary condition for the existence of an eulerian circuit. Every vertex must have even degree. This condition was later shown to be sufficient, producing the following elegant theorem.

Theorem 5.1 A graph (or multigraph) G is eulerian if and only if G is connected and every vertex has even degree.

Proof. The necessity that every vertex has even degree has been described previously. Clearly, it is also necessary that G be connected in order to find an eulerian circuit. Thus, we need to prove sufficiency; that is, we must show that if G is connected and every vertex has even degree, then G is eulerian.

Suppose that G is a connected graph where every vertex has even degree. We construct an eulerian circuit as follows. Begin at any vertex u_1 and start a trail by leaving u_1 along one of its incident edges $u_1 v_1$. Then since the degree of v_1 is even, there is an unused edge at v_1, so leave along one such edge. Each time we arrive at a new vertex, we may leave except eventually when returning to vertex u_1 for the last time (all edges incident with u_1 will then have been used). At this time a circuit C_1 has been found. (For example, in Figure 5.5, we may begin at a and travel subsequently to b, c, d, e, f, g, a along the thin solid edges. Since a's full degree has not yet been achieved, we leave a once again by edge ah and travel along the dashed edges to h, c, f, d, i, j, a. Now a's degree has been exhausted. The resulting circuit is $C_1 = a, b, c, d, e, f, g, a, h, c, f, d, i, j, a$.)

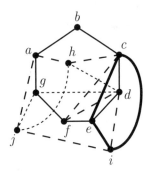

Figure 5.5

If C_1 contains all edges, we are done; C_1 is an eulerian circuit. If not, some vertex u_2 on C_1 is incident with an unused edge (this follows from the fact that G is connected). We begin a new circuit C_2 at u_2 by traveling along unused edges until we can go no farther. Because of the even degree of each vertex, C_2 will end at u_2. (In the graph of Figure 5.5, we may begin a second circuit at d using the dotted edges. This produces $C_2 = d, h, j, g, d$.) If $C_1 \cup C_2$ contains all edges, we are done. If not, we begin anew at a vertex u_3 of $C_1 \cup C_2$ incident with an unused edge and generate another circuit. (In Figure 5.5, we get a third circuit by leaving c along a think solid edge to get $C_3 = c, i, e, c$.) Eventually, we will have generated a sequence of circuits $C_1, C_2, \ldots, C_k$ such that $C_1 \cup C_2 \cup \cdots \cup C_k$ includes all edges of G.

Now, using the sequence of circuits $C_1, C_2, \ldots, C_k$, we generate our eulerian circuit. Begin at u_1, the first vertex of C_1, and travel along C_1 until u_2 is encountered for the first time. Then insert circuit C_2 at that point. In the resulting circuit, find the first time that vertex u_3 appears, and insert C_3 at that point. Proceed with this process until finally finding the first appearance of vertex u_k in the circuit generated from $C_1, C_2, \ldots, C_{k-1}$, and insert C_k at that point. The resulting circuit we have generated is an eulerian circuit for G. ∎

Using the three circuits $C_1 = a, b, c, d, e, f, g, a, h, c, f, d, i, j, a$, $C_2 = d, h, j, g, d$, and $C_3 = c, i, e, c$ related to Figure 5.5 and mentioned parenthetically in the proof of Theorem 5.1, we can construct an eulerian circuit as described there. We insert C_2 where we first see vertex d (the initial vertex of

5.1 Characterization of Eulerian Graphs

C_2) in circuit C_1. Thus, we get $C_1 \cup C_2 = a, b, c, C_2, e, f, g, a, h, c, f, d, i, j, a = a, b, c, d, h, j, g, d, e, f, g, a, h, c, f, d, i, j, a$.

We then insert C_3 into the resulting circuit at the first location where c (the initial vertex of C_3) appears. This gives

$$(C_1 \cup C_2) \cup C_3 = a, b, C_3, d, h, j, g, d, e, f, g, a, h, c, f, d, i, j, a$$
$$= a, b, c, i, e, c, d, h, j, g, d, e, f, g, a, h, c, f, d, i, j, a$$

That is, the resulting eulerian circuit for the graph in Figure 5.5.

The proof of Theorem 5.1 is an example of what is called a **constructive proof**. Such a proof actually gives a method to find (or *construct*) the object that the theorem claims exists. In Theorem 5.1, that object is an eulerian circuit.

Example 5.1 Using the technique in the proof of Theorem 5.1, find an eulerian circuit for the graph given in Figure 5.6.

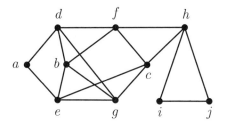

Figure 5.6

Solution
There are many ways to generate an appropriate circuit. Suppose that we begin at vertex a and generate circuit $C_1 = a, d, b, f, d, g, e, a$. There are still unused edges, but none incident with a. If we then begin at, say f, we can generate a circuit $C_2 = f, h, c, g, b, e, c, f$. Then we can generate a third circuit $C_3 = h, i, j, h$. Now all edges have been used. Next we insert C_2 into C_1 at the first place we encounter f, giving the circuit a, d, b, C_2, d, g, e, a or $a, d, b, f, h, c, g, b, e, c, f, d, g, e, a$. Finally, we insert $C_3 = h, i, j, h$ at the first place we encounter h to get $a, d, b, f, C_3, c, g, b, e, c, f, d, g, e, a = a, d, b, f, h, i, j, h, c, g, b, e, c, f, d, g, e, a$, which is an eulerian circuit. ◇

If there are exactly two vertices of odd degree, there is no eulerian circuit but there is a consolation prize: an **eulerian trail**—that is, a trail that traverses each edge of the multigraph exactly once. Such graphs are called **semi-eulerian**. A graph is **traversable** if it is either eulerian or semi-eulerian. Of course, the eulerian trail in a semi-eulerian graph is not closed. It begins and ends at different vertices. This is easy to prove. Let the odd vertices be x and y. Add an edge, e, between them (there may already be several other edges between them). Then in this new graph (or multigraph),

x and y and all other vertices have even degree, implying that the multigraph is eulerian. Produce an eulerian circuit and then delete edge e, yielding the promised eulerian trail. If G has more than two odd degree vertices, then no eulerian trail can exist, because only the first and last vertices of the trail can have odd degree. We have just proved the following result.

Corollary 5.1a A connected graph (or multigraph) G is semi-eulerian if and only if G has precisely two vertices of odd degree. Furthermore, an eulerian trail in G must begin at one of the odd vertices and end at the other. ▮

Eulerian graphs have numerous real-life applications, such as designing appropriate routes for mail delivery, garbage pick-up, snow removal, and so on. A well-known problem related to eulerian graphs, the **Chinese postman problem**, will be discussed in Section 5.3. For now, we consider one such application—namely, snow removal.

Example 5.2 Suppose that the streets of a small town are arranged as in Figure 5.7. Each edge represents a block on a particular street and each vertex is the intersection of where one or more blocks meet. Assume that the town's snowplow garage is located at intersection d. Prove that it is possible for the snowplow to plow each street exactly once without revisiting any block such that the snowplow returns to its garage immediately after plowing the last block. Find a route that the snowplow could take to accomplish the task.

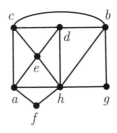

Figure 5.7

Solution
First, note that the graph in Figure 5.6 is connected and each vertex of this graph representing the street layout of the town has even degree. Thus by Theorem 5.1, we are assured that there is an eulerian circuit. This, in turn, implies that it is possible for the snowplow to plow each street exactly once without revisiting any block such that the snowplow returns to its garage immediately after plowing the last block. To find an appropriate route for the snowplow, we piece together circuits as we did in Example 5.1. We start at d, where the snowplow is located, and generate a first circuit, such as $C_1 = d, b, c, d, h, e, d$. Next we may use $C_2 = c, e, a, h, f, a, c$. Then we use $C_3 = b, g, h, b$. We put these together as follows: First, we replace vertex

5.1 Characterization of Eulerian Graphs

c in C_1 by circuit C_2 to get $d, b, C_2, d, h, e, d = d, b, c, e, a, h, f, a, c, d, h, e, d$. Next, we replace vertex b by circuit C_3 to get $d, C_3, c, e, a, h, f, a, c, d, h, e, d = d, b, g, h, b, c, e, a, h, f, a, c, d, h, e, d$. This describes one possible route that the snowplow driver could take to accomplish the task. ◇

Exercises 5.1

1. Verify that the sum of the degrees equals twice the number of edges for the multigraph of Figure 5.1.

2. Find an eulerian circuit for the graph of Figure 5.8. How can you tell in advance that this can be done?

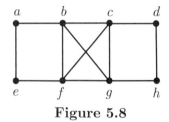

Figure 5.8

3. Show that if G and H are eulerian, so is $G \times H$.

4. Show that if all of the vertices of G and H have odd degree, then $G \times H$ is eulerian.

5. Which hypercubes are eulerian?

6. Prove that if G is a connected graph where every vertex has odd degree, then its line graph $L(G)$ is eulerian.

7. Consider the graphs of Figure 5.9. Which are eulerian? For each such graph, find an eulerian circuit. Which are semi-eulerian? For each such graph, find an eulerian trail.

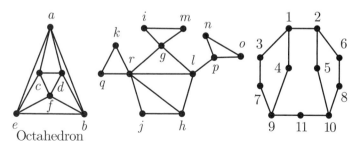

Figure 5.9

8. For which values of m and n is $K_{m,n}$ eulerian?

9. The layout of streets in a certain city is as displayed in Figure 5.10. The traffic department plans to travel along each street and paint new lines down the center of each street. Find a route that can be used to begin at vertex a (the location of their depot), paint the lines on each street, and travel along as few extra streets as possible and return to the depot. Explain why we must travel along at least two streets twice, no matter how we design the route.

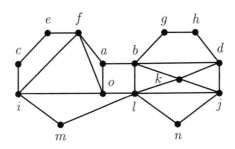

Figure 5.10

10. For a tree T, a **closed eulerian walk** traverses each edge exactly *twice* and ends where it starts. Every tree has such a walk. Prove that an eulerian walk for a tree having q edges has length $2q$.

11. Prove that a connected graph G is eulerian if and only if the edges of G can be partitioned into edge-disjoint cycles. Hint: Use induction on the number of cycles in such a partition and the technique of reconstructing an eulerian cycle used in this section.

12. Prove that a connected graph G is eulerian if and only if every block of G is eulerian.

13. A knight (the one that looks like a horse) can move on a chessboard in an "L shape," either two spaces up or down and one to the side or two spaces to the side and one space up or down. Determine whether it is possible for a knight to tour an 8×8 chessboard making legal moves, landing in each square exactly once and ending in the square it started. *Hint*: You must first determine what the graph of the knight's possible moves looks like.

14. A museum has a floor plan as displayed in Figure 5.11, where each edge represents a hallway along which paintings are displayed. Vertex e is the location of the entrance, and vertex g is the location of the gift shop

through which we exit the museum. Find a route through the museum that begins at the entrance e travels along each hallway exactly once and ends at the gift shop g.

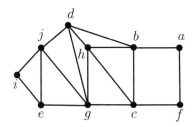

Figure 5.11

5.2 Hamiltonicity

In Section 5.1, we were concerned with traversing a graph so that we include each edge exactly once. In other applications, it is instead important to include each *vertex* just once. That will be the focus of this section.

Hamiltonian Graphs

A graph with a spanning path is called **traceable**. A graph with a spanning cycle is called **hamiltonian** and the associated cycle is a **hamiltonian cycle**. Hamiltonian graphs are traceable, but there is no reverse implication, as P_n demonstrates. A traceable non-hamiltonian graph is called **semi-hamiltonian**. Thus, a semi-hamiltonian graph has a hamiltonian path but no hamiltonian cycle. Unlike the situation for eulerian graphs, it is often extremely difficult to decide whether a given graph is hamiltonian. In fact, there is no known efficient algorithm to do this.

Sometimes we can eliminate the possibility that a graph is hamiltonian by making shrewd observations. If a graph G is bipartite, for example, and the two color sets have different cardinalities, then G cannot be hamiltonian. This follows because a subgraph of a bipartite graph must be bipartite, and a bipartite cycle is equitable—that is, the color sets have the same cardinality. We then have a necessary condition for a bipartite graph to be hamiltonian: It must be equitable. Another simple necessary condition for a graph to be hamiltonian is that it must be 2-connected; that is, the graph must not contain a cut vertex. You will be asked to prove that in the Exercises.

When examining graphs constructed from other graphs, we must always be on the lookout for clever shortcuts when investigating hamiltonicity. The graph $G \times K_2$ consists of two copies of G such that each vertex of one copy is rendered adjacent to its corresponding vertex in the other copy. Call these copies G and G'. If G is traceable, we can obtain a hamilton cycle for $G \times K_2$ by taking a spanning path P for G, with endpoints x and y, and its mirror image path P' in G', with corresponding endpoints x' and y', and then using edges xx' and yy'. We have just proven that if G is a traceable graph, then $G \times K_2$ is hamiltonian. Note that $Q_2 = C_4$, which is hamiltonian, and $Q_n = Q_{n-1} \times K_2$. So it then follows that all hypercubes Q_n are hamiltonian for $n \geq 2$.

Hamilton's Game

Now we give some historical background. Hamiltonian graphs are named for the famous Irish mathematician Sir William Rowan Hamilton (1805–1865), who was the first to give an algebraic (rather that geometric) description of complex numbers. Hamilton also invented the Icosian game, sometimes called the Around the World game, in the mid-1800s. His game involved moving a wooden peg around on the graph of a solid dodecahedron (formerly used as calendar paperweights because of their 12 faces). The dodecahedron has 20 vertices (the corners of the solid), 30 edges (the boundaries of the faces), and 12 faces. Each face is shaped like a pentagon. In the game, each vertex represents a well-known city. The object of the game was to find a round the world tour that visited each city exactly once and ended where it started. In constructing the tour, we are only permitted to "move the peg" along an edge that joins one city to another one.

Example 5.3 The graph of the dodecahedron is displayed in Figure 5.12. Find a hamiltonian cycle in the graph and thereby obtain a round the world tour for Hamilton's Icosian game.

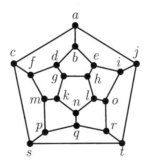

Figure 5.12 The graph of the dodecahedron.

Solution

We begin at a and go around the outer rim to c (we don't want to close off the cycle yet). Next, we go inward to vertex f and then counterclockwise around to e (we must save b as our escape route back to vertex a).

5.2 Hamiltonicity

Finally, we go in toward h, go clockwise around to g, go out to d, go over to b, and then go up to a. Therefore, one possible hamiltonian cycle is $a, j, t, s, c, f, m, p, q, r, o, i, e, h, l, n, k, g, d, b, a$. ◇

That wasn't too difficult. Hamilton sold to his Icosian game to J. Jacques and Sons, makers of high-quality chess sets, for £25 (equivalent to about $125 then; in the year 2003, the corresponding purchasing power would be about £1300, or $2000). The game was patented in London in 1859 but was a failure, perhaps because it was rather easy to solve the problem.

Example 5.4 Draw a connected graph on four vertices (if possible) that is
1. Hamiltonian and eulerian
2. Hamiltonian and semi-eulerian
3. Hamiltonian but not even semi-eulerian
4. Eulerian and semi-hamiltonian
5. Eulerian but not even semi-hamiltonian
6. Semi-eulerian and semi-hamiltonian
7. Neither traceable nor traversable **Solution**

The graphs are displayed in Figure 5.13. Note that (d) and (e) are not possible—that is, no such connected graphs on four vertices exist. ◇

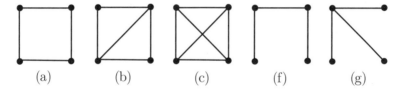

(a)   (b)   (c)   (f)   (g)

Figure 5.13

Sufficient Conditions

Over the years, numerous conditions that guarantee that a given graph is hamiltonian have been proven. We present here one of the earlier ones, due to Dirac [2], both because its proof is rather easy to understand and the proof technique has been used rather widely in studying hamiltonicity.

Theorem 5.2 If G is a graph of order $n \geq 3$ such that $\deg(v) \geq n/2$ for all $v \in V(G)$, then G is hamiltonian.

Proof. If $n = 3$, then $\deg(v) \geq n/2$ implies that $G = K_3$, which is hamiltonian. Thus, suppose that $n \geq 4$, and let $P = v_1, v_2, v_3, \ldots, v_k$ be a path of maximum length in G. Then the only possible additional adjacencies of v_1 and v_k lie within the set $\{v_2, v_3, \ldots, v_{k-1}\}$, or we would be able to find a longer path than P in G. Since $\deg(v_1) \geq n/2$, P must contain at least $1 + n/2$ vertices; that is, $k \geq 1 + n/2$. There must be some vertex v_i, $2 \leq i \leq k$, such that v_1 and v_i are adjacent while v_k and v_{i-1} are adjacent (see Figure 5.13). If this were not the case, then whenever v_1 is adjacent to a vertex v_i, v_k could not be adjacent to v_{i-1}. But since $\deg(v_1) \geq n/2$, this would eliminate at least $n/2$ possible adjacencies for v_k. But since v_k is not adjacent to itself, the elimination of $n/2$ other possible adjacencies would imply $\deg(v_k) < n/2$, a contradiction.

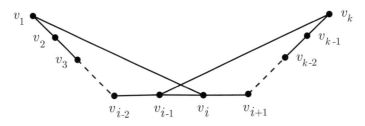

Figure 5.14

Now we know that G contains the cycle $C = v_1, v_2, \ldots, v_{i-1}, v_k, v_{k-1}, v_{k-2}, \ldots, v_i, v_1$ (see Figure 5.14). If C contains all vertices of G, then C is a hamiltonian cycle and we are done. Thus, suppose instead that there is some vertex u not on C. But then since $k \geq 1 + n/2$ and $\deg(u) \geq n/2$, u must be adjacent to at least one vertex on C. But this implies that there is a longer path than P in G, a contradiction. Hence C contains all vertices of G, and G is hamiltonian. ∎

It is important to note that the converse to Theorem 5.2 is obviously false. That is, if a graph is hamiltonian, it need *not* have the property that $\deg(v) \geq n/2$ for all $v \in V(G)$. For example, the cycles C_n are hamiltonian, but when $n \geq 5$, no vertex satisfies the degree condition. The vertices only have degree 2. Similarly, it is important to observe that Theorem 5.2 says that when $n \geq 3$, if *all* vertices satisfy $\deg(u) \geq n/2$, then the graph is hamiltonian. If only *some* of the vertices satisfy the degree condition, there is no guarantee of hamiltonicity. For example, $K_{3,10}$ has three vertices that satisfy the degree condition, but $K_{3,10}$ is definitely not hamiltonian.

Homogeneous and Hamiltonian Connected Graphs

All mathematicians like to generalize. For example, the rational numbers, Q, generalize the integers, Z. An integer n is a special case of a rational number, as can be seen by writing n in the form $n/1$, where $n \in Z$. Multigraphs generalize graphs. A graph is a special type of multigraph—namely, one with no multiple edges.

While we are steeped in the spirit of generalizing, we call a graph G **homogeneous connected** if given any vertex $v \in V(G)$, there exists a hamiltonian path starting at v. This is much stronger than asserting that G is traceable. In fact, we are given the privilege of selecting the initial

5.2 Hamiltonicity

vertex of the spanning path. Note that a hamiltonian graph is homogeneous connected, since we can delete an edge of a spanning cycle incident with any given vertex v, producing a spanning path starting at v. Surprisingly, the reverse implication does not hold. The famous Peterson graph, shown in Figure 5.15, is homogeneous connected but is not hamiltonian.

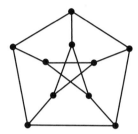

Figure 5.15 The Peterson graph.

The Peterson graph consists of two C_5's and five edges connecting each vertex of the outer C_5 to a vertex of the inner C_5. Notice, however, that the vertices do not correspond with one another, in the sense that if we number the outer cycle's vertices consecutively, the five vertices adjacent to them in the inner cycle are not consecutive. This graph has many interesting properties and is often cited as an example or counterexample for a particular property.

A graph G is called **hamiltonian connected** if, given any pair of vertices x and y, there exists an x–y hamiltonian path for G. This goes one step farther than homogeneous connected, in which we merely have the option to choose one end vertex of the desired spanning path. In a hamiltonian-connected graph, we can find a hamiltonian path with any given pair of specified end vertices. The wheel $W_{1,n}$ is an example of such a graph. Obviously, if G is hamiltonian connected it is also homogeneous connected. Less obviously, when the order is at least three, a hamiltonian-connected graph G is also hamiltonian! To prove this, let x and y be a pair of adjacent vertices in a hamiltonian-connected graph G. Then there exists an x–y hamiltonian path that certainly doesn't include edge xy. Replacing xy, this path becomes a hamiltonian cycle for G. So G is hamiltonian.

Which Meshes Are Hamiltonian?

Since odd meshes are bipartite but not equitable, they are not hamiltonian. What about even meshes?

Theorem 5.3 Every even mesh is hamiltonian.

Proof. The proof follows the maxim "a figure is worth a thousand words of text." See Figure 5.16 for the construction. We assume, without loss of generality, that the number of columns is even. The figure presents a spanning cycle for $M(8,n)$. Larger even meshes are handled similarly. It is clear that this construction will work for any even mesh, no matter how large n is. ∎

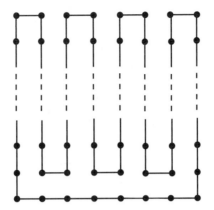

Figure 5.16 A hamiltonian cycle in an even mesh.

All meshes are traceable. If either m or n is even, then Theorem 5.3 implies that $M(m,n)$ is hamiltonian and therefore traceable. If both m and n are odd, then we ca obtain a hamiltonian path for $M(m,n)$ as follows: Delete all horizontal edges except $((1,n),(2,n)), ((3,n),(4,n)), \ldots, ((m-2,n),(m-1,n)); ((2,2),(3,2)), ((4,2),(5,2)), \ldots, ((m-1,2),(m,2))$ and all edges of the form $((k,1),(k+1,1))$, $1 \leq k \leq m-1$. Also, delete the vertical edges $((i,1),(i,2))$, $2 \leq i \leq m$. To picture what the resulting hamiltonian path looks like, delete all vertices and edges to the right of the column seven in Figure 5.16.

Hypercubes

We already observed that all hypercubes Q_n are hamiltonian for $n \geq 2$. Hypercubes are not hamiltonian connected. This can be seen by selecting two vertices, x and y, of the same color in the bipartite hypercube. Hypercubes have an even number of vertices, implying that a hamiltonian path in Q_n has even order. Thus, Q_n has no x–y hamiltonian path because a path of even length in a bipartite graph must have oppositely colored end vertices.

The good news is that hypercubes have the next best thing. They are hamilton laceable. An equitable bipartite graph G is **hamilton laceable** if, given two oppositely colored vertices, x and y, there exists an x–y spanning path for G.

Theorem 5.4 Q_n is hamilton laceable.
Proof. Let's use induction. The theorem is obvious for $n = 1$ and 2. To verify the general case, assume that the statement is true for $n = k$ ($k \geq 2$), and show that the theorem is true for $n = k+1$. To this end, let r and b be

5.2 Hamiltonicity

vertices of Q_{k+1} such that r is red and b is blue. Split Q_{k+1} into two copies of Q_k, which we shall call R and B, such that $r \in R$ and $b \in B$.

Since R is a copy of Q_k, R is hamiltonian, so let C be a spanning cycle of B. Let x be a neighbor of r on C and let x' be the unique vertex of B that is adjacent to x'. Since x' is at distance 2 from r in Q_k, x' has the same color as r; that is, red. So b and x' are in B and are oppositely colored. Since B is a copy of Q_k, we may apply the inductive hypothesis to B. Thus we get a b–x' spanning path L for B. Let M be the spanning path of R from r to x on C. The required r–b hamiltonian path for Q_{k+1} is M, xx', L. Edge xx' is the glue that holds paths L and M together. ∎

Exercises 5.2

1. Draw a connected graph having five vertices that is hamiltonian and eulerian.

2. Draw a connected graph having five vertices that is hamiltonian and semi-eulerian.

3. Draw a connected graph having five vertices that is eulerian and semi-hamiltonian.

4. Draw a connected graph having five vertices that is eulerian but not even semi-hamiltonian.

5. Draw a connected graph having five vertices that is hamiltonian but not even semi-eulerian.

6. Draw a connected graph having five vertices that is neither traceable nor traversable.

7. Draw a connected graph having five vertices that is semi-hamiltonian and semi-eulerian.

8. If G is hamiltonian connected and has order at least 4, show that $G \times K_2$ also has this property. *Hint*: There are two cases to consider.

9. Show that if G and H are traceable, so is $G \times H$.

10. Show that the graph of Figure 5.8 is hamiltonian but is not hamiltonian connected.

11. In Figure 5.17, Theorem 5.2 guarantees that graph G is hamiltonian. Find a hamiltonian cycle in graph G.

12. In Figure 5.17, graphs H and K are semi-hamiltonian but not hamiltonian. Find a hamiltonian path in H and in K. Explain using some structural property of H how we could tell in advance that H could not possibly be hamiltonian.

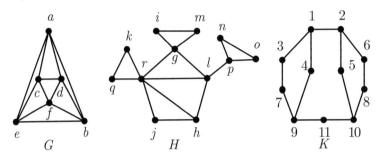

Figure 5.17

13. Graphs M and N of Figure 5.18 are hamiltonian. Find a hamiltonian cycle in M and in N.

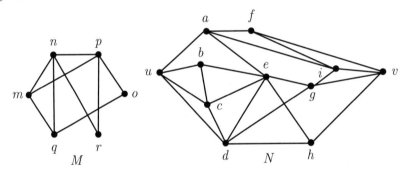

Figure 5.18

14. Prove that $K_{13,20}$ is not hamiltonian.

15. Prove that if G has a cut vertex, then G is not hamiltonian.

16. Show that every even 3-mesh is hamiltonian.

17. Show that the ladder $M(n,2)$, with $n \geq 3$, is not hamilton laceable.

18. Show that $M(4,3)$ is not hamilton laceable.

19. Show that $M(2k,3)$ is not hamilton laceable.

20. Use Theorem 5.2 to show that for $n \geq 6$, the complement of C_n is hamiltonian. Note that C_5 is self-complementary, in which case $\bar{C}_5 = C_5$ is hamiltonian, allowing us to upgrade our assertion to $n \geq 5$.

21. To see that $n/2$ is the "best" result for using $\delta(G)$ to prove hamiltonicity, find a graph G of order 7 for which $\delta(G) = (n-1)/2$, such that G is not hamiltonian.

22. Prove that a bipartite graph G in which each part has order n, and G has at least $n^2 - n + 2$ edges, must be hamiltonian. *Hint*: Examine the degrees and use Theorem 5.2.

23. The following theorem of Chvátal [1] is one of the strongest sufficient conditions for hamiltonicity.

    Theorem 5.5 Let G be a graph with n vertices ($n \geq 3$) with degrees $d_1 \leq d_2 \leq \cdots \leq d_n$. If $d_i \leq i < n/2$ implies that $d_{n-i} \geq n - i$, then G is hamiltonian. ∎

    Informally, the theorem says that if the smallest degrees being small (where small is determined by the given inequality $d_i \leq i < n/2$) always implies that the largest degrees are large (determined by the inequality $d_{n-i} \geq n - i$), then the graph must be hamiltonian. Use Theorem 5.5 to determine whether the graphs having the following degrees are hamiltonian.
    a. $2, 2, 2, 3, 3, 4, 6, 7, 7$ \hspace{1cm} b. $3, 3, 3, 4, 6, 6, 6, 8, 8, 9$

24. Label the vertices of P_4 consecutively using $\{1, 2, 3, 4\}$. Consider the permutation function f for which $f(1) = 2$, $f(2) = 3$, $f(3) = 4$, and $f(4) = 1$. (This is a *cyclic* permutation.) Now draw the related **permutation graph** of P_4, which consists of two copies of P_4 each labeled $1, 2, 3, 4$, with an edge from a vertex i in the first copy to vertex $f(i)$ in the second copy. Is the resulting permutation graph hamiltonian?

25. Repeat the previous exercise for P_3. The function f satisfies $f(1) = 2$, $f(2) = 3$, and $f(3) = 1$.

5.3 Applications

In this chapter, we have considered traversing a graph, first so that each edge is contained exactly once and then so that we include each vertex just once. In the process, we have seen several applications where those ideas were used. In this section, we extend those concepts to provide a simple introduction to two more applications—namely, the Chinese postman problem and the traveling salesman problem.

The Chinese Postman Problem

Suppose that a letter carrier must begin the day's route at the post office and deliver mail along each street in a given neighborhood and return to the post office after the route is complete. It is clear that each block (here we mean a street section between consecutive intersections, not a "graph theoretic" block) must be traversed at least once. Furthermore, it would be possible to devise such a route traversing each block exactly once if and only if the graph modeling the neighborhood is eulerian. By Theorem 5.1, this occurs if and only if the graph is connected and each vertex has even degree. Since not all neighborhoods will satisfy that condition, the problem is relaxed to require that each street block be traversed at least once rather than exactly once, but the minimum number of street blocks should be traversed. That is, the route should be as short as possible.

The **Chinese postman problem** requires us to find a minimum-length closed spanning walk that includes each edge in a network at least once. The network is actually a weighted graph where the weight $w(e)$ of a given edge e is the distance between its endpoints (the length of the block corresponding to edge e). If the total weight of the network is k and each edge has weight $w(e) \geq 0$, then the total distance that the postman must travel is between k and $2k$. The worst case of the problem occurs when the underlying graph of the network is a tree. The name of the problem comes from the fact that it was first described in that way by the Chinese mathematician, Meigu Guan [3] in the early 1960s.

The basic idea needed to find a best possible route in the Chinese postman problem is that when there are odd degree vertices, some edges will have to be traversed more than once. The way that an appropriate route can be found is by first identifying the set S of vertices of odd degree. Then for each pair u, v of vertices in S, find the length of a shortest path joining u and v. (This can be accomplished using Dijkstra's algorithm, which will be discussed in Section 10.2.) We know that there is an even number of vertices of odd degree (this follows directly from Corollary 2.1a), so the next stage is to pair up the vertices in S so that the sum of the distances between those pairs is minimized. The pairs minimizing the sum determine which edges must be traversed more than once—namely, all edges on the shortest paths joining those pairs. We now illustrate the process with a small example.

Example 5.5 Solve the Chinese postman problem for the weighted graph in Figure 5.19.

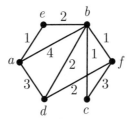

Figure 5.19

Solution

We see that there are four vertices of odd degree—namely, a, b, d, f, so $S = \{a, b, d, f\}$. Now we find the distances between all pairs of vertices in S [there are, of course $\binom{4}{2} = 6$ pairs to consider]. $d(a, b) = 3$, $d(a, d) = 3$, $d(a, f) = 4$, $d(b, d) = 2$, $d(b, f) = 1$, $d(d, f) = 2$. So the pairing that produces the minimum sum of distances is a, d and b, f since $d(a, d) + d(b, f) = 3 + 1 = 4$ minimizes the sum for pairs covering all of set S. Now to solve the problem, we generate a weighted multigraph by inserting additional edges corresponding to those on a shortest path joining the pairs a, d and b, f. This gives the multigraph in Figure 5.20.

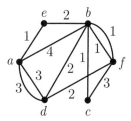

Figure 5.20

Now we find an eulerian tour in the resulting multigraph. For example, $a, e, b, f, c, b, f, d, a, b, d, a$ is one such tour. ◇

We note that in Example 5.5, the shortest paths joining a, d and b, f happened to consist of single edges. That is not always the case. It is important to understand that if the shortest path joining a pair of vertices that is involved in minimizing the sum of additional distances that must be traveled consists of several edges, *all* of those edges must be added to obtain the multigraph that will ultimately be traversed.

The Traveling Salesman Problem

Hamiltonian graphs are related to the traveling salesman problem. In this problem, a salesman plans to visit various cities to show his merchandise. He would like to stop in each city once and return to the home office while minimizing his travel time. To model this problem, we use a weighted graph in which each vertex represents a city, with two vertices joined by an edge if there is a road leading directly from one city to the other. We associate with each edge (road) a positive number, the weight, which represents the travel time between the cities joined by the edge. To solve the traveling salesman problem, therefore, we must find a hamiltonian cycle that has smallest possible total weight in the graph. Such a cycle is called a **minimum-weight hamiltonian cycle**. The traveling salesman problem is related (but not equivalent) to the simple assignment problem of linear programming. Thus, integer programming techniques are used to solve the problems when they are quite large. We shall only discuss the graphical aspects of the problem here.

Example 5.6 Find a minimum-weight hamiltonian cycle for the weighted graph in Figure 5.21.

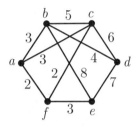

Figure 5.21

Solution
We check all hamiltonian cycles: a, b, c, d, e, f, a has weight 26; a, b, d, e, f, c, a has weight 22; a, b, e, d, c, f, a has weight 28; a, c, b, d, e, f, a has weight 24; and a, c, d, b, e, f, a has weight 26. Thus, we choose a, b, d, e, f, c, a as our minimum-weight hamiltonian cycle. ◇

Example 5.6 illustrates the difficulty of the traveling salesman problem. For the small graph in Figure 5.21, we found five hamiltonian cycles and determined their weights. In applications, the problems are often much larger and more overwhelming. Unfortunately, there is no efficient method for solving the traveling salesman problem. There are, however, approximation techniques and the branch-and-bound method. These are discussed in Sysło, Deo, and Kowalik [4]. We should mention that the traveling salesman problem falls into a class of problems that are intrinsically difficult to solve. This idea is discussed further in Chapter 8 when we discuss algorithms.

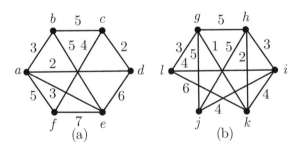

Figure 5.22

5.3 Applications

Exercises 5.3

1. Solve the Chinese postman problem for the weighted graph in Figure 5.21.

2. Solve the traveling salesman problem for the weighted graph in Figure 5.19.

3. Solve the traveling salesman problem for each graph in Figure 5.22.

4. Solve the Chinese postman problem for the graphs in Figure 5.22.

5. Find a hamiltonian path in the Petersen graph (see Figure 5.23).

6. Prove that the Petersen graph is not hamiltonian. *Hint*: You will need to take advantage of the symmetry of the graph as you go along; break the proof into several cases.

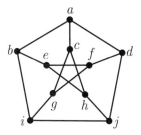

Figure 5.23 The Petersen Graph.

7. Prove that it is not possible to remove edges (but no vertices) from the Petersen graph so that the resulting graph becomes eulerian.

8. Consider the floor plan given in Figure 5.24. Explain why it is impossible to enter through the front door, travel through each internal doorway exactly once, and exit through the rear door.

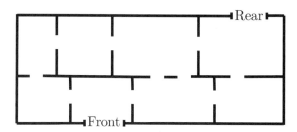

Figure 5.24

9. If a museum is well designed, then there is a route we can take so as to walk through each of its rooms exactly once. Thus its related graph (or multigraph) contains a hamiltonian cycle. Prove that the museum whose floor plan is depicted in Figure 5.25 is *not* well designed.

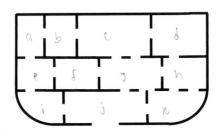

Figure 5.25

References for Chapter 5

1. Chvátal, V., On hamilton's ideals. *Journal of Combinatorial Theory* 12B (1972) 163–168.

2. Dirac, G. A., Some theorems on abstract graphs. *Proceedings of the London Mathematical Society* (3) 2 (1952) 69–81.

3. Guan, M. G., Graphic programming using odd or even points. *Chinese Mathematics* 1 (1962) 273–277.

4. Sysło, M. M., N. Deo, and J. S. Kawalik, *Discrete Optimization Algorithms*, Prentice-Hall, Englewood Cliffs, NJ (1983).

Additional Readings

5. Jelliss, J. P., Generalized knights and hamiltonian tours. *Journal of Recreational Mathematics* 27 (1995) 191–200.

6. McCourt, R. F., A hamiltonian circuit on a box (solution to problem #655, proposed by C. Vanden Eynden). *College Mathematics Journal* 31 (2000) 225–226.

7. McGuirè, G., and F. Ó. Cairbre, A bridge over a hamiltonian path. *Mathematical Intelligencer* 23:3 (Summer 2001) 41–43.

8. Stewart, I., Mathematical recreations: Knight's Tours. *Scientific American* 276:4 (April 1997) 102–104.

Chapter 6

Graph Coloring

Various real-world problems that can be modeled by graphs require that the vertex set or edge set be partitioned into disjoint sets such that items within a given set are mutually nonadjacent. Common problems are scheduling meetings or exams to avoid conflicts and storing of chemicals to prevent adverse interactions. These problems are related to graph coloring, the subject of this chapter.

6.1 Vertex Coloring and Independent Sets

A (proper) **vertex coloring** of a graph G is an assignment of colors to the vertices so that adjacent vertices have distinct colors. A vertex coloring using k colors is a **k-coloring** of the graph. A graph that permits a k-coloring is called **k-colorable**. In this section, we focus our attention on vertex colorings and properties of k-colorable graphs.

The Chromatic Number

The **chromatic number** of a graph G, denoted $\chi(G)$, is the minimum number of colors needed for any proper k-coloring of G. Thus, in particular, a graph G is bipartite exactly when $\chi(G) \leq 2$. Odd cycles, on the other hand, are examples of graphs for which $\chi = 3$. (The symbol χ is a Greek letter, written *chi* and pronounced like the "ky" in "sky.")

Are there graphs on n vertices such that $\chi = n$? In other words, are there graphs for which each vertex needs its own color? Yes. In fact, there are infinitely many such graphs, one for each n. Observe that for the complete graph K_n, we have $\chi(K_n) = n$. As soon as we color a vertex v of K_n, we can never reuse that color since v is adjacent to each of the remaining vertices.

We must use care when trying to achieve a minimum coloring. Consider the graph G in Figure 6.1. If we begin by coloring the vertices v_1, v_2, v_3 by

a, b, and c, respectively, then neither v_4 nor v_5 can be colored c and must be colored differently from one another as well. Suppose that we color v_4 by a and v_5 by b. Next, we color v_6. Since it can't be colored a, we decide to color v_6 by b. But then v_7 is adjacent to v_3, colored c, v_4, colored a, and v_6, colored b. Thus a new color, d, is needed for v_7. Then we can color v_8 with either c or d. A 4-coloring of graph G is displayed in Figure 6.2(a).

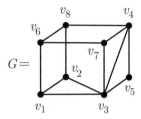

Figure 6.1

With some care, we could have done better and saved one crayon. A 3-coloring of graph G is shown in Figure 6.2(b). Since G contains a triangle, G is not bipartite; that is, $\chi(G) > 2$. But since we have found a 3-coloring, $\chi(G)$ must, therefore, be 3.

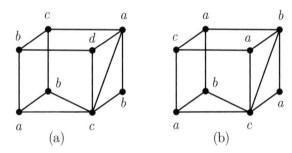

Figure 6.2 A 4-coloring and a 3-coloring of graph G.

Example 6.1 Show that $\chi(W_{1,5}) = 4$.

Solution
The wheel $W_{1,5}$ is equivalent to the join $K_1 + C_5$. When studying bipartite graphs in Chapter 3, we saw that C_5 is not bipartite. Indeed, it is easy to check that three colors are required in a proper coloring of C_5. Since $W_{1,5} = K_1 + C_5$, the additional vertex v is adjacent to every vertex of C_5, so v must use a new color. Thus $\chi(W_{1,5}) = 4$. ◇

A graph is k-colorable if it is possible to assign each vertex a color from a set of k colors such that adjacent vertices receive different colors. The smallest k for which a graph G is k-colorable is its chromatic number $\chi(G)$. So G is k-colorable means that $\chi(G) \leq k$.

6.1 Vertex Coloring and Independent Sets

> **Note 6.1**
> To prove that $\chi(G) = k$, we must show that $\chi(G) > k-1$; that is, it is impossible to $(k-1)$-color G. Then we must show that it is possible to k-color G, which is usually done by exhibiting an actual k-coloring.

A graph G is **k-critical** if (1) $\chi(G) = k$, and (2) $\chi(G - v) = k - 1$ for every vertex $v \in V(G)$; that is, the removal of any vertex of G reduces the chromatic number. (The removal of one vertex can only lower the chromatic number by 1.)

If $\chi(G) = k$, then graph G must contain a k-critical subgraph. We may delete vertices that do not lower χ, one at a time, eventually reaching a graph H for which this is no longer possible. The resulting graph H is k-critical. The next theorem says a mouthful about k-critical graphs.

Theorem 6.1 If G is k-critical, then $\delta(G) \geq k - 1$.

Proof. We proceed by contradiction. Let G be k-critical where $\delta(G) \leq k-2$. Assume that v is a vertex of G satisfying $\deg(v) = \delta(G)$. Since G is k-critical, it follows that $\chi(G - v) = k - 1$.

Now since $\deg_G(v) \leq k - 2$, there is a color, say red, among the $k - 1$ colors used to color $G - v$ that is not used by any neighbor of v in G. Then coloring v in G with red shows that G is $(k-1)$-colorable, which contradicts the assertion that $\chi(G) = k$. ∎

As a consequence of Theorem 6.1, we can make the following chain of observations. First, if $\chi(G) = k$, we know that G has a k-critical subgraph H. Since H also has chromatic number k, it follows that H has at least k vertices. We can then conclude by using Theorem 6.1, that $\delta(H) \geq k - 1$, from which it follows that H has at least k vertices of degree at least $k - 1$. Then these k vertices have degree at least $k - 1$ in G. We have just proved the following beautiful theorem.

Theorem 6.2 If $\chi(G) = k$, then G must have at least k vertices of degree at least $k - 1$. ∎

Corollary 6.2a $\chi(G) \leq \Delta(G) + 1$.

Proof. Suppose, to the contrary, that $\chi(G) = k > \Delta(G)+1$. It would follow that $k - 1 > \Delta(G)$. But this is impossible since Theorem 6.2 guarantees vertices of degree $k - 1$. But G has no vertex of degree greater than $\Delta(G)$. ∎

It is a fact (Brooks [1], 1941), which we will not prove here, that except for two classes of graphs—namely, odd cycles and complete graphs—every connected graph G satisfies $\chi(G) \leq \Delta(G)$.

Chromatic Number and Independence

A subset S of vertices of a graph G is called **independent** if no pair of vertices of S are adjacent. The five cut vertices of the graph of Figure 6.3 form an independent set.

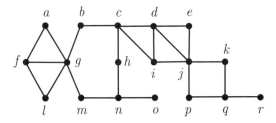

Figure 6.3 The five cutvertices, g, c, j, n,q form an independent set.

Notice, however, that the eight vertices of degree 2 form a larger independent set, setting us to wonder how large we can make an independent set in this graph. This prompts the next exciting definition.

The **independence number** of a graph G, denoted $\beta(G)$, is the maximum size of an independent set. As an example, we have $\beta(P_5) = 3$. To obtain an independent set of vertices of P_5, take the two end vertices and the center vertex. Among all connected graphs on n vertices, the winner is $K_{1,n-1}$, for which $\beta = n - 1$. The loser in this category is K_n, for which $\beta = 1$. Incidentally, for graph G of Figure 6.3, $\beta(G) = 11$, can you find an independent set of size 11 in the graph?

Given an independent set of a graph G, let $X = V(G) - S$; that is, X consists of the vertices of G that are not in S. The vertex set X has an interesting property. Every edge in G is incident with at least one vertex of X. If this were not so—that is, if an edge xy boasted that neither x nor y was in X—we would justifiably deduce that x and y were both in S. But this is impossible since S is an independent set.

Given a graph G of order n such that $\chi(G) = k$, $V(G)$ can be partitioned into k color sets $V_1, V_2, \ldots, V_k$, such that for each $i = 1, 2, \ldots, k$, the vertices of V_i are colored with color i. Since the vertices of V_i all have the same color, they cannot be adjacent with one another. Then each V_i is an independent set, from which we deduce that $|V_i| \leq \beta(G)$, for $i = 1, 2, \ldots, k$. Now since $|V| = |V_1| + |V_2| + \cdots + |V_k|$, we have the inequality $n \leq k \times \beta(G) = \chi(G) \times \beta(G)$, yielding the following rather nice lower bound for $\chi(G)$:

$$\chi(G) \geq \frac{n}{\beta(G)} \qquad (6.1)$$

Encouraged by this inequality, let's obtain an upper bound for $\chi(G)$. Let S be a largest independent set of vertices—that is, $|S| = \beta(G)$. Now consider

6.1 Vertex Coloring and Independent Sets

the graph H with vertex set $V(G) - S$. It follows that $\chi(H) \geq \chi(G) - 1$. In other words, upon deleting S from G, we cannot lower χ by more than one. This is because the addition of S to H requires only one more color (since S is independent).

On the other hand, since $|H| = n - \beta(G)$, we have $\chi(H) \leq n - \beta(G)$. Combining this with the inequality of the previous paragraph yields $\chi(G) - 1 \leq \chi(H) \leq n - \beta(G)$. After eliminating the middle term and solving for $\chi(G)$, we have an upper bound for $\chi(G)$ given by

$$\chi(G) \leq n - \beta(G) + 1 \tag{6.2}$$

The **girth** of a graph G is the length of the smallest cycle (if any) in G. A subgraph of G that is a complete graph is called a **clique**. If a graph G has a large chromatic number, you might naturally think that G has a large clique. It was shown, however, by one of the great twentieth-century mathematicians, Paul Erdős [2], that given any two positive integers a and b, there exists a graph G with $\chi(G) = a$ and whose girth exceeds b. When $b = 3$, this guarantees that there are graphs with very large chromatic number whose largest clique is K_2.

Uniquely k-Colorable Graphs

When dealing with colorings, it is important to remember that a **partition** of a set X is a collection of subsets $X_1, X_2, \ldots, X_k$ such that

$$\bigcup_{i=1}^{k} X_i = X \text{ and } X_i \cap X_j = \emptyset \text{ for } i \neq j$$

Thus a partition of X is a set of disjoint subsets whose union is X.

Let G be a graph for which $\chi(G) = 2$. So G is bipartite and $V(G)$ can be partitioned into subsets V_1 and V_2 such that each edge in $E(G)$ is incident with one vertex of V_1 and with one vertex of V_2. This is the same as stipulating that the vertices incident with any edge of G may not be in the same subset.

Note that this description of a bipartite graph makes no mention of color. It refers to a partition of its vertex set into two subsets and a condition that prohibits incident vertices from being in the same subset. In coloring the vertices of G, we may utilize any two colors we favor. We must, however, assign one color to all of the vertices of V_1, and one color (a different one, of course) to all of the vertices of V_2.

Continuing with this logic, a graph G is k-colorable—that is, $\chi(G) \leq k$—if

1. $V(G)$ can be partitioned into k subsets, and

2. No pair of adjacent vertices of G are in the same subset.

The cycle C_5, shown in Figure 6.4, has chromatic number 3, and $V(C_5)$ has the partition $\{a,c\}, \{bd\}, \{e\}$. A graph G for which $\chi(G) = 3$ is called **tripartite**.

Of course, the colors in Figure 6.4 can be permuted so that vertices a and c are white and b and d are blue and e is red. In fact, the partition $\{a,c\}, \{bd\}, \{e\}$ has 3! or 6 different coloring schemes using the colors red, white, and blue. We are reluctant, however, to call them "different." The partition is the same, after all.

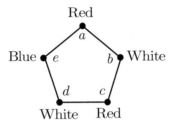

Figure 6.4 A coloring scheme for C_5 using three colors.

On the other hand, the vertex set of the same 5-cycle can be partitioned as in Figure 6.5, using a different partition, $\{a\}, \{bd\}, \{c,e\}$.

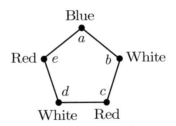

Figure 6.5 Same tripartite graph—different partition.

The existence of more than one partition is not always possible. The tripartite graph G of Figure 6.6, has only one partition—namely, $\{a\}, \{b,d\}, \{c,e\}$—which motivates the following definition. A graph G, with $\chi(G) = k$, is called **uniquely k-colorable** if there is only one partition of $V(G)$ into k subsets such that vertices in the same subset are nonadjacent.

We close this topic with two theorems.

Theorem 6.3 If G is a uniquely k-colorable graph, then $\delta(G) \geq k - 1$.

6.1 Vertex Coloring and Independent Sets

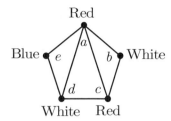

Figure 6.6 A uniquely 3-colorable graph.

Proof. Let the vertices of G be colored using the colors $c_1, c_2, \ldots, c_k$. Then if vertex v is assigned color c_i, it must be adjacent to a vertex of every other color—that is, it must have neighbors assigned each of the $k-1$ colors $c_1, c_2, \ldots, c_{i-1}, c_{i+1}, \ldots, c_k$. If this were not the case—that is, if a color, say c_j, was not used by v or by any neighbor of v—we would be able to change the color of v to c_j, thereby changing the partition of $V(G)$, contradicting the assumption that G is uniquely k-colorable. It follows that each vertex has at least $k-1$ neighbors, implying that $\delta(G) \geq k-1$. ∎

In the next theorem, given a graph G with $\chi(G) = k$, the subset of $V(G)$ with color c_i is denoted V_i, for $i = 1, 2, \ldots, k$.

Theorem 6.4 Let the vertices of a uniquely k-colorable graph G be colored using the colors $c_1, c_2, \ldots, c_k$. Then given two numbers i and j, satisfying $1 \leq i < j \leq k$, the induced graph on $V_i \cup V_j$ is connected.

Proof. We use a proof by contradiction. Assume that $H = \langle V_i \cup V_j \rangle$ is disconnected. Let J and K be components of H. Observe that, in light of Theorem 6.3, $V_i \cap J$ and $V_j \cap J$ are nonempty, as are $V_i \cap K$ and $V_j \cap K$. Then by interchanging the colors of the vertices of $V_i \cap J$ and $V_j \cap J$, we obtain a new partition of $V(G)$, contradicting the fact that G is uniquely k-colorable. ∎

Exercises 6.1

1. Consider graph G in Figure 6.7. Explain why $\chi(G) \geq 3$. Then find a 3-coloring of G, thereby proving that $\chi(G) = 3$.

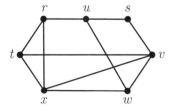

Figure 6.7

2. Prove that $\chi(G + sw) = 4$ for graph G in Figure 6.7.

3. Suppose that no two cycles of G have an edge in common. S[...]
   $\chi(G) \leq 3$.

4. Let H be a subgraph of G. Show that $\chi(G) \geq \chi(H)$.

5. Find $\beta(Q_n)$ for $n = 3$ and 4.

6. Find the independence number for the graphs of Figures 6.3 and 6.7 and for each, state an appropriate independent set of order $\beta(G)$.

7. Find β for the 2-mesh $M(4,3)$.

8. Find the maximum diameter among all connected graphs with $\beta = 3$.

9. Repeat the previous exercise for several other values of β. Then find the maximum diameter among all connected graphs with $\beta = k$.

10. Find β for C_n.

11. Show that $\chi(G \times K_2) = \chi(G)$.

12. Let H be a subgraph of G such that $|V(H)| < |V(G)|$. (H is not a spanning subgraph.) If $\chi(G) = \chi(H) = k$, explain why G is not k-critical.

13. Show using the previous two exercises that if G is nontrivial, $G \times K_2$ is not critical.

14. Show that K_n is n-critical.

15. Which cycles are critical?

16. Show that K_2 is the only connected bipartite graph that is 2-critical.

17. Which graphs of order 4 are critical?

18. If G is disconnected, show that $\chi(G) = \chi(H)$, where H is the component of G with the maximum chromatic number.

19. Use the previous exercise to show that a critical graph must be connected.

20. Prove that if G is a 3-critical graph, then G is an odd cycle.

21. Let G be a bipartite graph of order n. Show in two different ways that $\beta(G) \geq \frac{n}{2}$.

6.2 Edge Coloring

22. In how many ways can the vertices of a labeled C_5 be partitioned into three color sets? We are ignoring the choice of colors, of course.

23. Repeat the instructions of the previous exercise for C_7.

24. Show that $W_{1,4}$ is uniquely colorable while $W_{1,5}$ is not.

25. Show that $\chi(G+H) = \chi(G)+\chi(H)$. If G and H are uniquely colorable, is $G+H$ uniquely colorable?

26. Let G be uniquely 3-colorable. Show that $G \times K_2$ is not uniquely colorable. What is $\chi(G \times K_2)$?

27. Repeat the previous exercise if G is uniquely k-colorable, where $k \geq 3$.

6.2 Edge Coloring

We now consider coloring the edges of a graph. Unlike the situation for vertex coloring, where adjacent vertices had to receive distinct colors, edge colorings generally do not have such a restriction. We shall, however, begin this section by looking at restricted edge colorings and see rather quickly why more general colorings are usually studied.

The Edge-Chromatic Number

An **edge coloring** of a graph G consists of an assignment of colors to the edges of G. Note that there is no restriction on how the edges are colored. If we add the additional restriction that incident edges received distinct colors, then the coloring is called a **proper edge coloring**. The **edge-chromatic number**, $\chi_1(G)$, is the minimum number of colors required in a proper edge coloring of G.

Example 6.2 Find the edge-chromatic number for the graph in Figure 6.8.

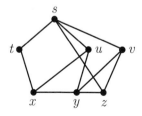

Figure 6.8

Solution

Since the edges must receive distinct colors in a proper coloring, the four edges at s must be colored differently. Without loss of generality, color edges st, sz, su, and sv by red, white, blue, and green, respectively. Then edge vz can only receive color blue or red. So suppose that it is colored blue. Then we can complete our coloring with vy red, uy white, ux red, xy blue, and tx green, among other possible colorings. The coloring described is shown in Figure 6.9. The fact that we needed at least four colors and achieved a coloring using four colors shows that $\chi_1(G) = 4$.   ◇

From the discussion in the solution of Example 6.2, we see that the following simple result holds.

Theorem 6.5 For any graph G, the edge-chromatic number is at least as large as the maximum degree among the vertices of G. That is, $\chi_1(G) \geq \Delta(G)$.

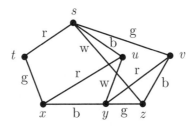

Figure 6.9 A proper edge coloring of the graph in Figure 6.8.

Note that Theorem 6.5 gives a lower bound. There are plenty of graphs that achieve the lower bound. For example, for the complete bipartite graphs, we have $\chi_1(K_{m,n}) = \max\{m,n\} = \Delta(K_{m,n})$. To achieve a proper edge coloring of $K_{m,n}$ using the fewest number of colors, we begin with a coloring of the edges at a vertex v of maximum degree. We then "rotate" the coloring as we move to each successive vertex in the same part as v. The rotation works as follows. Suppose that $m \leq n$ for $K_{m,n}$, and label the vertices of the first part $1, 2, 3, \ldots, m$ and those of the second part $m+1, m+2, \ldots, m+n$. Suppose we color the edge that joins vertex 1 with vertex $m+j$ by color c_j, for $1 \leq j \leq n$. Then for all edges xy in $K_{m,n}$, color xy with c_j if $y - x \equiv (j-1) \bmod m$, for $1 \leq j \leq n$. We illustrate this process on $K_{3,4}$ in Figure 6.10.

The rotation technique is very useful in edge coloring problems. We use it again in the proof of the following theorem.

6.2 Edge Coloring

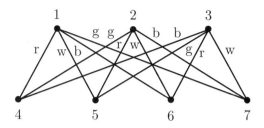

Figure 6.10

Theorem 6.6 For the complete graph K_{2n-1}, we have $\chi_1(K_{2n-1}) = 2n - 1$.

Proof. Graph K_{2n-1} has $\binom{2n-1}{2} = (n-1)(2n-1)$ edges. Since a maximum matching contains $n - 1$ edges, at least $2n - 1$ color sets are needed in a proper edge coloring. We can achieve a proper $(2n - 1)$-edge coloring in the following way. Label the vertices $1, 2, 3, \ldots, 2n-1$ clockwise around a circle. Begin with the matching $(2, 2n-1), (3, 2n-2), (4, 2n-3), \ldots, (n, n+1)$ so that vertex 1 is unmatched. Color the matched edges c_1 [see Figure 6.11(a)]. Now rotate the matching clockwise by one unit so that vertex 2 in unmatched and color the matched edges c_2 [see Figure 6.11(b)]. Continue rotating, and when j is the unmatched vertex, color the matched edges c_j. This defines a proper $(2n - 1)$-edge coloring of K_{2n-1}. ∎

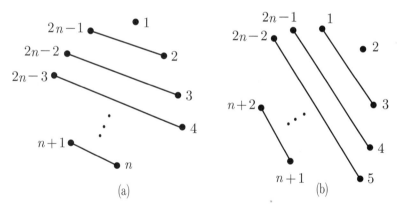

Figure 6.11 Edges colored c_1 in (a) and c_2 in (b).

Note that each rotation in the proof of Theorem 6.6 produces a new maximum matching that determines the edges in one of the color classes. Because of parity, K_{2n-1} has no perfect matching—there is always one unmatched vertex. When we consider K_{2n}, we can use the same matchings as in Theorem 6.6, but each time include a new vertex x. Color the edge joining x to vertex i (the unmatched vertex in the ith matching) with color c_i. That produces a perfect matching of K_{2n} for each i. Thus we have proved the following.

Theorem 6.7 For the complete graph K_{2n}, we have $\chi_1(K_{2n}) = 2n - 1$. ∎

Note that in Theorem 6.7, $\chi_1(K_{2n})$ equals the maximum degree and in Theorem 6.6, $\chi_1(K_{2n-1})$ is one more than the maximum degree. That is no coincidence, as evidenced by the following theorem of Vizing [4].

Theorem 6.8 For any graph G, $\Delta(G) \leq \chi_1(G) \leq \Delta(G) + 1$. ∎

Here is another interesting result about edge colorings, this one [3] due to Denes König (1884–1944), a Hungarian mathematician who, in 1936, wrote the first textbook in graph theory.

Theorem 6.9 If G is a bipartite graph, then $\chi_1(G) = \Delta(G)$.

Proof. By Theorem 3.7, we know that every regular bipartite graph has a perfect matching. We can then show by induction on $\Delta(G)$ that a $\Delta(G)$-regular bipartite graph has $\chi_1(G) = \Delta(G)$. This is clearly true when $\Delta(G) = 1$, since in that case our graph would be n copies of K_2, which has $\chi_1 = 1$. Assume that the statement is true when $\Delta(G) = k$. Then, let G be a $(k+1)$-regular bipartite graph. Then since G is regular bipartite, G has a perfect matching M. Color those edges with a color c_{k+1} and remove them from G. Graph $G - M$ is a k-regular bypartite graph, so by the inductive hypothesis has $\chi_1(G - M) = k$. The coloring of the edges of $G - M$ with k colors, together with the coloring of the edges of M with color c_{k+1}, shows that $\chi_1(G) \leq k + 1$. This together with Theorem 6.5 shows that $\chi_1(G) = k + 1$.

It remains only to prove that if G is a bipartite graph with maximum degree $\Delta(G)$, then G can always be embedded in a $\Delta(G)$-regular bipartite graph. Since a $\Delta(G)$-regular bipartite graph must be equitable, by Lemma 3.7, first add additional vertices, if necessary, to one of the partite sets of G to make the parts have the same cardinality. Then, if the resulting graph is not $\Delta(G)$-regular, there must be a pair of vertices, one from each part, whose degrees are less than $\Delta(G)$. Find such a pair and join them. Continue to find such pairs and join them until the ultimate graph H becomes $\Delta(G)$-regular. H is still bipartite, so we know from the previous paragraph that $\chi_1(H) = \Delta(H)$. So since G is a subgraph of H, we have $\chi_1(G) \leq \chi_1(H) = \Delta(H) = \Delta(G)$. Finally, using the lower bound given in Theorem 6.5, we can conclude that $\chi_1(G) = \Delta(G)$ ∎

Theorem 6.8 tells us that $\chi_1(G)$ will always be one of two possible values, namely, $\Delta(G)$ or $\Delta(G) + 1$. In Theorem 6.9, we see that the value can be determine for regular bipartite graphs. However, except for a few other classes of graphs, it is generally a hard problem to determine which of the two values $\Delta(G)$ or $\Delta(G) + 1$ a graph has for $\chi_1(G)$.

Vizing's theorem (Theorem 6.8) allows for only one of two possibilities for $\chi_1(G)$ once one knows the degrees of the vertices of G. Perhaps this is

6.2 Edge Coloring

part of the reason why graph theorists have studied edge colorings in a more relaxed setting by not insisting the coloring be proper. We now look at the more general situation.

Monochromatic Triangles in K_n

An **edge coloring** of a graph G, is an assignment of a color to each edge of $E(G)$. There is no implication here that edges that share a vertex have distinct color.

Let the complete graph K_n be given an edge coloring using the colors red and blue. Each triangle in this graph will look like one of the triangles in Figure 6.12.

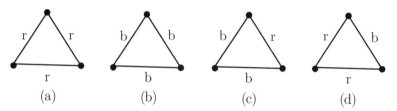

Figure 6.12 The four kinds of triangles in an edge coloring of K_n using the colors red and blue.

Triangles (a) and (b) in Figure 6.12 are called **monochromatic** while triangles (c) and (d) are **bichromatic**. Note that a bichromatic triangle has two vertices that are incident with a red edge and a blue edge. Let's call these vertices **ambivalent**. The third vertex is incident with edges of the same color.

Labeling the vertices of K_n using the numbers $1, 2, \ldots, n$, we define the function $r(i)$ to be the number of red edges incident with vertex i. For each $i = 1, 2, \ldots, n$, it follows that $0 \leq r(i) \leq n - 1$. Moreover, the number of blue edges incident with vertex i is $n - 1 - r(i) = n - r(i) - 1$, since K_n is $(n-1)$-regular.

To summarize, the vertex i is incident with $r(i)$ red edges and $n - r(i) - 1$ blue edges. This makes it easy to determine how many bichromatic triangles there are in an edge coloring of K_n using the two colors red and blue, provided, of course, that we know the values of $r(i)$.

Theorem 6.10 Given an edge coloring of K_n using two colors, say red and blue, and given the function $r(i)$, which yields the number of red edges incident with vertex i, for $i = 1, 2, \ldots, n$, then the number of monochromatic triangles is given by

$$\frac{n(n-1)(n-2)}{6} - \frac{1}{2} \sum_{i=1}^{n} r(i) \times (n - r(i) - 1) \tag{6.3}$$

Proof. Each bichromatic triangle can be counted once at each of its ambivalent vertices as follows. If vertex i is an ambivalent vertex of a bichromatic triangle, then that triangle is formed by selecting a red and a blue edge incident with vertex i. Since there are $r(i)$ ways to choose the red edge and $n - r(i) - 1$ ways to choose the blue edge, the multiplication principle tells us that there are $r(i) \times [n - r(i) - 1]$ bichromatic triangles in which vertex i is ambivalent.

Of course, a bichromatic triangle has a second ambivalent vertex, say vertex k. This triangle will also be counted in the expression $r(k) \times [n - r(k) - 1]$, which yields the number of bichromatic triangles in which vertex k is an ambivalent vertex. Thus each bichromatic triangle is counted twice, because it has two ambivalent vertices.

The number of bichromatic triangles in K_n can now be determined by adding up the expression $r(i) \times [n - r(i) - 1]$ as i varies from 1 to n and then dividing by two to get rid of the double counting.

In K_n, any set of three vertices determines a triangle. It follows that the total number of triangles in K_n is $\binom{n}{3} = \frac{n(n-1)(n-2)}{6}$. We may obtain the number of monochromatic triangles by subtracting the number of bichromatic triangles from this last expression. ∎

The edge coloring of K_4 shown in Figure 6.13 has the value of $r(i)$ written near each vertex. When inserted into expression (6.3), we obtain 1, which indicates that there is exactly one monochromatic triangle in this edge coloring. It is the triangle containing the three vertices satisfying $r(i) \geq 2$.

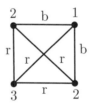

Figure 6.13 An edge coloring with one monochromatic triangle. Numbers indicate red edges at v.

In contrast to the edge coloring of K_4 in Figure 6.13, note that the edge coloring of K_4 in Figure 6.14 contains no monochromatic triangle. It is quite interesting that this can also be done for K_5, as the edge coloring of Figure 6.15 shows. It is easy to verify that there is no monochromatic triangle in this coloring of K_5. The outer spanning cycle has red edges while the inner one has blue edges. Since each triangle has edges in both cycles, no triangle can be monochromatic.

6.2 Edge Coloring

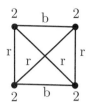

Figure 6.14 An edge coloring of K_4 with no monochromatic triangle.

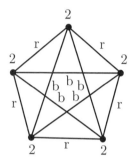

Figure 6.15 An edge coloring of K_5 with no monochromatic triangle.

In light of these edge colorings of K_4 and K_5, free of monochromatic triangles, we might wonder whether there are similar edge colorings using two colors for larger complete graphs that contain no monochromatic triangles. The answer is no! In fact, for $n \geq 6$, an edge coloring of K_n using two colors must have several monochromatic triangles. As this answer (no!) is not very informative, we shall now obtain a lower bound for the number of monochromatic triangles in K_n when n is an even number, that is, when $n = 2k$ and $k \geq 3$. The case for odd n is similar and will be covered in the Exercises.

We will require a little lemma before we continue.

Lemma 6.11 Let m be a given positive number. Then among all pairs of positive numbers x and y such that $x + y = m$, the product xy is maximum when $x = y = m/2$.

Proof. This is equivalent to the inequality $\left(\frac{m}{2}\right)^2 = \frac{m^2}{4} \geq xy$, which we shall now prove. The following says that a square can't be negative: $x^2 - mx + \frac{m^2}{4} = \left(x - \frac{m}{2}\right)^2 \geq 0$. We then eliminate the middle expression, and then transpose and factor, yielding $\frac{m^2}{4} \geq mx - x^2 = x(m - x)$. Since $m - x = y$, we have $\frac{m^2}{4} \geq xy$, thereby completing the proof. ∎

Corollary 6.11a When m, x, and y are positive *integers* where $x + y = m$, we maximize the product xy when $x = \lfloor m/2 \rfloor$ and $y = \lceil m/2 \rceil$.

Proof. When m is even, this is the situation of the lemma, and the rounding functions are not required. When m is odd, it is not hard to see that we come as close as possible to the highest point, $(\frac{m}{2}, \frac{m}{2})$ on the parabola $f(x) = mx - mx^2$ by selecting the point with integer coordinates $(\lfloor \frac{m}{2} \rfloor, \lceil \frac{m}{2} \rceil)$, thereby maximizing the product xy. ∎

To continue our quest for a lower bound for the number of monochromatic triangles in an edge coloring of K_n using red and blue when $n = 2k$, we must minimize the expression $\frac{n(n-1)(n-2)}{6} - \frac{1}{2} \sum_{i=1}^{n} r(i) \times (n - r(i) - 1)$, obtained earlier. This is done by maximizing the sum, since it is being subtracted from the fixed number $n(n-1)(n-2)/6$. The variable summand $r(i) \times (n-r(i)-1)$ is the product of two integers whose sum is the constant, $n-1$. By Corollary 6.11a, the maximum product can be achieved in the odd case ($n = 2k$ is even, so $n-1$ is odd), when $r(i) = k$ and $n-r(i)-1 = k-1$ for each $i = 1, 2, \ldots, n$. Since the summand is then constant, we obtain, after substituting $2k$ for n,

$$\frac{2k(2k-1)(2k-2)}{6} - \frac{1}{2} \sum_{i=1}^{2k} k(k-1) = \frac{k(2k-1)(2k-2)}{3} - \frac{1}{2} \cdot 2k \cdot k(k-1) = \frac{k(k-1)(k-2)}{3},$$

proving the following beautiful theorem.

Theorem 6.12 Let $n = 2k$, with $k \geq 3$. Then in any edge coloring of K_n using two colors, there are at least $\frac{1}{2}\binom{k}{3} = \frac{k(k-1)(k-2)}{3}$ monochromatic triangles. ∎

In particular, when $k = 3$, we find that an edge coloring of K_6 with two colors has at least two monochromatic triangles. An asymptotic estimate of the expression in this theorem yields the interesting fact that for large k, the number of monochromatic triangles in K_n is of the order $k^3/3$. Since $k = n/2$, this becomes $n^3/24$. As an example, K_{10}, which contains $\binom{10}{3} = 120$ triangles, must have at least $\frac{5 \cdot 4 \cdot 3}{3} = 20$ monochromatic triangles when its edges are colored using two colors.

Exercises 6.2

1. Produce an edge coloring, with two colors, for K_6 containing exactly two monochromatic triangles.

2. Find a proper edge coloring of $K_{4,5}$ using five colors.

3. Prove that for all odd cycles C_{2k+1}, we have $\chi_1(C_{2k+1}) = 3$.

4. Show that $\chi_1(G) = 4$ for the graph in Figure 6.7.

5. Prove that the edge chromatic number of the Peterson graph is four. *Hint*: Up to isomorphism, there is just one way to 3-color the edges of

6.2 Edge Coloring

the outer cycle. Try to proceed from there and show that eventually a fourth color is needed. Then complete the 4-edge coloring.

6. Prove that $\chi_1(W_{1,n}) = n$.

7. Prove that there are 48 possible proper 4-edge colorings of $W_{1,4}$ whose vertices are labeled a, b, c, d, e.

8. Explain why it is true that for all graphs G, $\chi_1(G) = \chi(L(G))$.

9. Prove that if G is k-regular of odd order, then $\chi_1(G) = k + 1$.

10. Prove that if G is k-regular and contains a cut vertex, then $\chi_1(G) = k + 1$.

11. If $\langle r(i) \rangle = (3,1,1,2,3,2)$ is the number of red edges incident with vertex i, $(r(1) = 3$ is for the center) for wheel $W_{1,5}$. Find a 2-edge coloring with one monochromatic triangle and a different one with two monochromatic triangles.

12. Suppose that the vertices of C_5 are labeled u, v, w, y, z. How many distinct edge colorings of C_5 are possible? How many of these are proper edge colorings? How many are proper 3-edge colorings?

13. Prove by mathematical induction that $\chi_1(Q_n) = n$.

14. Determine the edge chromatic number for the graphs in Figure 6.16.

15. A graph is **uniquely k-edge colorable** if the partition of its edges into k sets using a proper k-edge coloring is unique. Show that graph H of Figure 6.16 is uniquely $\chi_1(H)$–edge colorable, but G is not uniquely $\chi_1(G)$–edge colorable.

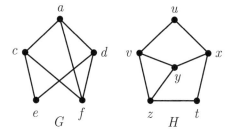

Figure 6.16

16. Derive a lower bound for the number of monochromatic triangles for an edge coloring, using two colors, for K_n when $n = 2k + 1$ and $k \geq 3$.

17. Let the edges of K_{2k} be colored either red or blue. Compute a lower bound for the probability that a randomly selected triangle is monochromatic. *Hint*: There are $\binom{n}{3} = \frac{n(n-1)(n-2)}{6}$ triangles in K_n. Replace n by $2k$. Now use Theorem 6.12 to determine the fraction of triangles that are monochromatic.

18. Use calculus to prove that the product of two positive numbers, x and y, with fixed sum m is maximum when $x = y$. *Hint*: Observe that $y = m - x$. Then the product $xy = x(m - x) = mx - x^2$.

19. (For readers who have studied multivariable calculus.) Show that the product of three positive numbers, x, y, and z, whose sum is m is maximum when $x = y = z$. *Hint*: Observe that $z = m - x - y$, in which case the product $f(x, y) = xy(m - x - y) = mxy - x^2y - xy^2$. Now find the partial derivatives f_x and f_y and set them equal to zero, and solve.

6.3 Applications of Graph Coloring

Graph coloring can be used as an aid in solving several real-world problems. This is particularly the case in problems where certain items must be separated from one another because of basic properties of those items such as possible chemical interactions or predator-prey relationship among animals. We consider several examples here to give the flavor of such applied problems.

Example 6.3 (Designing a Zoo) Modern zoos permit the animals to run free as much as possible. However, enclosures are still needed for two main reasons. First, a basic enclosure is required to keep all animals within the zoo. Second, since some animals find others quite tasty, enclosures are needed to separate a predator from its prey. Since building the enclosures is a major expense, how do we determine the fewest number of enclosures that are needed?

Solution
We can set up a graph to model this problem. Specify a vertex to represent each type of animal. Put an edge between two vertices if one of the corresponding animals is a predator of the other. The minimum number of enclosures needed turns out to be the chromatic number of the resulting graph. Additionally, a possible safe arrangement of the animals in the enclosures is given by a minimum coloring. Animals whose vertices have the same color may be placed within the same enclosure. ◇

Example 6.4 Zambula Safari-Land Inc. is planning a new zoo. They plan initially to have the following animals: baboon (b), fox (f), goat (g), hyenas

6.3 Applications of Graph Coloring

(h), kudu (k), lion (l), porcupine (p), rabbit (r), shrew (s), wildebeest (w), and zebra (z). They have found from past experience the following eating habits. Baboons like to eat goats, kudu (before they get too big), rabbits, and shrews; foxes like to eat goats, porcupines, rabbits, and shrews; hyenas like to eat goats, kudu, wildebeest, and zebra. Lions prefer goat, kudu, wildebeest, and zebra; porcupine eat shrews and rabbits; and the rest prefer bugs or lizards, or grass and other vegetation. Zambula will feed the animals and would like them to run free but does not want the animals eating one another because of the great expense in acquiring them. Find a grouping of animals that would require the minimum number of enclosures to meet Zambula's needs.

Solution

We represent each animal by a vertex labeled with its corresponding letter and place an edge between vertices u and v if one is a predator of the other. By doing so, we arrive at the graph G in Figure 6.17. We must now determine a minimum coloring of the vertices of G; that is, a coloring using $\chi(G)$ colors.

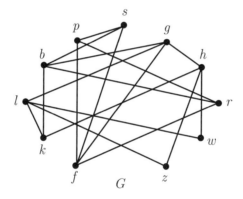

Figure 6.17

Since G contains K_3 (see vertices f, p, s), we know that $\chi(G)$ is at least three, and vertices f, p, and s must receive distinct colors. It is easy to verify that the vertex coloring in Figure 6.18 is proper. Thus three enclosures are required, and Zambula will enclose the baboons, hyenas, porcupines, and lions within the first enclosure; the foxes, kudu, wildebeest, and zebra in the second; and the goats, rabbits, and shrews in the third. ◇

Another useful application of graph coloring is in scheduling of meetings or exams. When a corporation, government agency, or university schedules meetings, certain individuals need to be at more than one of the meetings. Of course, for that to happen, the meetings that a given individual must attend have to occur at different times. If you are expecting the different times to have something to do with the different colors of the vertices in the partition of the vertex set of a graph, you are thinking in the right direction. We illustrate the idea in Example 6.5.

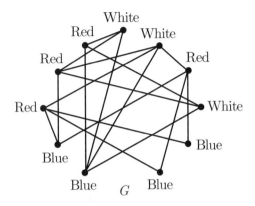

Figure 6.18 A 3-coloring of graph G.

Example 6.5 (Scheduling Final Exams) Department X at Complex U. has eight faculty members, each of whom is teaching four courses this semester. Their schedules are shown in Table 6.1.

Professor	Courses Taught
Agnesi	132, 136, 211
Bernoulli	127, 131, 153
Cauchy	131, 132, 211
Descartes	127, 131, 205
Euler	131, 138, 154
Frobenius	132, 136, 201
Gauss	127, 131, 138
Hamilton	153, 154, 205

Table 6.1

In arranging the final exams, it has been decided that each course will have a department final exam such that all sections of a given course will have their finals at the same time. Each professor must proctor his or her own exam. There is plenty of classroom space available, but everyone would like to get the exams over as soon as possible so that they can rush off for summer vacation and prove lots of new theorems. What is the fewest number of time slots that can be used to give the exams?

6.3 Applications of Graph Coloring

Solution

Construct the graph where each vertex represents a given course and label the vertex by the course number. Next, place an edge between two vertices if some professor is teaching both of those courses. See Figure 6.19(a).

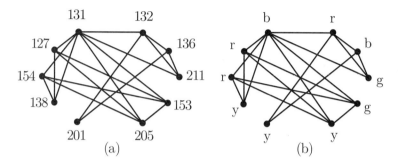

Figure 6.19

A schedule of final exams that uses the fewest number of time slots is equivalent to a coloring of graph G in Figure 6.19(a) using $\chi(G)$ colors. Since vertices 127, 131, 153, and 205 are mutually adjacent, it follows that $\chi(G) \geq 4$. A 4-coloring is displayed in Figure 6.19(b), thereby establishing that $\chi(G) = 4$. Thus four time slots are required for the scheduling of exams. Just to confirm that each professor can be present to proctor his or her own exam, we display the complete schedule in Table 6.2. ◇

Time Slot	Exams Scheduled	Professors Proctoring Exams
1 (red)	127, 132, 154	Agnesi (132), Bernoulli (127), Cauchy (132), Descartes (127), Frobenius (132), Euler (154), Gauss (127), Hamilton (154)
2 (blue)	131, 136	Agnesi (136), Bernoulli (131), Cauchy (131), Descartes (131), Frobenius (136), Euler (131), Gauss (131),
3 (yellow)	138, 201, 205	Descartes(205), Euler (138), Frobenius (201), Gauss (138), Hamilton (205)
4 (green)	153, 211	Agnesi (211), Bernoulli (153), Cauchy (211), Hamilton (153)

Table 6.2

Here's another example. This one involves storage of chemicals.

Example 6.6 (Storage of Chemicals) An architect has been asked to design storage cabinets for the chemical supplies of a certain laboratory. Because of budgetary constraints, the lab would like to construct as few storage cabinets as possible. It has been determined that certain chemicals may interact with other ones and should not be stored together. For simplicity, we will refer to the 14 chemicals by letter as a through n. The possible interaction between chemicals u and v is indicated by an edge in the graph H in Figure 6.20. Determine the minimum number of storage cabinets needed, and state which chemicals will be stored in those cabinets.

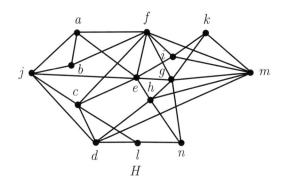

Figure 6.20 Possible chemical interactions.

Solution
Although the graph looks very complicated and it appears that there might be an explosion any minute, it is rather easy to confirm that $\chi(H) = 3$. This is because of the pleasant surprise that graph H is uniquely 3-colorable. Once you color the vertices of a given K_3 with three distinct colors, you can successively move to a vertex v adjacent to one of the colored vertices such that the color of v is forced. The unique 3-coloring of H is displayed in Figure 6.21, where the colors pink (p), tan (t), and white (w) are used. Thus we can see that three storage cabinets are required and the chemicals that will be stored together are as follows. In the pink cabinet, we store chemicals f, h, j, k, and l; in the tan cabinet, we store chemicals b, d, e, m, and n; and in the white cabinet, we store chemicals a, c, g, and i. ◇

As a final application, we mention the **timetabling problem**. It is a scheduling problem but of a somewhat different type than that given in Example 6.5. A typical situation in the timetabling problem is that there are teachers, each of whom will be teaching certain classes. A given class may meet several times in a week, as any student is well aware. The problem is to schedule all classes for all the teachers by using the minimum total number of time periods. At first this problem may seem very much like the

6.3 Applications of Graph Coloring

scheduling problem, but it is quite different. First of all, in this problem we are not dealing with a graph but a multigraph. Furthermore, the multigraph is bipartite, and we solve the timetable problem by using edge colorings instead of vertex colorings.

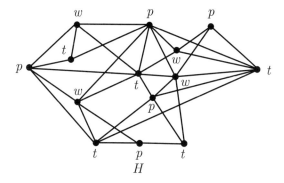

Figure 6.21 The coloring of the uniquely 3-colorable graph H

Example 6.7 Suppose that teachers t_1, t_2, t_3, t_4, t_5 must meet with classes $c_1, c_2, c_3, \ldots, c_7$, and that t_i must meet with class c_j a total of w_{ij} times per week. Note that class c_j is a fixed set of students who are in the same classes together all day long. We represent the situation with a weighted bipartite graph. The weight w_{ij} on a given edge corresponds to the number of edges there would be joining t_i and c_j in the related bipartite multigraph. Suppose that the required class meetings are as depicted in Figure 6.22. Determine the minimum number of periods per week that must appear in the timetabling grid. Then determine an appropriate schedule of classes.

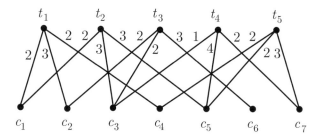

Figure 6.22 The weighted bipartite graph for the timetabling problem

Solution
Since each teacher can meet with only one class during a given time period, and each period a given class will have just one teacher, we must find a proper edge coloring. Each color class will match a teacher with a given

class during one of the time periods. Since the vertices do not all have the same degree, there may be some time periods during which a given class will not meet with a teacher, as well as some time periods during which some teachers may not be meeting with a class. Since the maximum degree (the maximum of the sum of the weights at a given vertex) is nine, at least nine time periods are necessary. It follows from Theorem 6.9 that $\chi_1(G) = \Delta(G)$, so nine time periods are also sufficient. A timetable is displayed in Table 6.3. The entries within the table tell which class meets with which teacher during the given time period. Note that each time period corresponds to a different color class in the edge coloring. A class c_j appears at most one time in a given row because it can meet with just one teacher during that period. The number of times that c_j appears in the column t_i corresponds to the weight w_{ij} on edge $t_i c_j$. ◇

Period ↓ \ Teacher →	t_1	t_2	t_3	t_4	t_5
1	c_1	c_3	c_2	c_5	c_7
2	c_2	c_5	c_3	c_7	c_4
3	c_4	c_5	c_6	c_3	c_7
4	c_1	c_3	c_2	c_5	c_4
5	c_2	c_1	c_3	c_7	c_5
6	c_4	c_3	c_6	c_5	c_7
7	c_2	c_1	c_6	-	c_5
8	-	c_5	-	-	-
9	-	-	-	c_5	-

Table 6.3

Exercises 6.3

1. In Example 6.4, if Zambula introduced wombats that are plant eaters, which enclosure would be safest for them?

2. In Figure 6.17, find an edge e not in G whose addition to G would increase the chromatic number. Then find a missing edge whose addition to G would not alter the chromatic number.

3. Repeat Exercise 2 but using graph H in Figure 6.20 this time.

4. Find an edge xy of graph H in Figure 6.20 such that $H - xy$ is not uniquely 3-colorable.

5. Suppose that the chemical lab in Example 6.6 decides that they will no longer need or store chemicals e, h, j, and m.

6.3 Applications of Graph Coloring

(a) Draw the new graph of chemical interactions.

(b) Is the resulting graph uniquely k-colorable?

(c) Determine an arrangement of the remaining chemicals in as few storage cabinets as possible.

6. Suppose that it is decided for the timetabling problem of Example 6.7 that each class that meets with a given teacher three times a week will now meet with that teacher four times per week. Also, teacher t_4 will begin meeting with c_3 three times a week instead of once and with c_7 three times a week instead of twice.

    (a) Draw the new weighted bipartite graph for this timetabling problem.

    (b) Determine the minimum number of periods per week that must appear in the timetabling grid.

    (c) Determine an appropriate schedule of classes.

7. Suppose that it is decided for the timetabling problem of Example 6.7 that each class that meets with a given teacher twice a week will now meet with that teacher three times per week. Also, each class that is meeting with a given teacher three times a week will now meet with that teacher just twice per week.

    (a) Draw the new weighted bipartite graph for this timetabling problem.

    (b) Determine the minimum number of periods per week that must appear in the timetabling grid.

    (c) Determine an appropriate schedule of classes.

References for Chapter 6

1. Brooks, R. L., On colouring the nodes of a network. *Proc. Cambridge Phil. Soc.* 37 (1941) 194–197.

2. Erdős, P., Graph theory and probability II. *Canadian Journal of Mathematics* 13 (1961) 346–352.

3. König, D., Über Graphen und ihre Anwendung auf Determinantentheorie und Menenlehre. *Math. Annalen* 77 (1916) 453–465.

4. Vizing, V. G., On an estimate of the chromatic class of a p-graph. *Diskrete Analiz.* 3 (1964) 25–30.

Additional Readings

5. Chung, F., and R. Graham, *Erdős on Graphs: His Legacy of Unsolved Problems.* A. K. Peters, Wellesley, MA (1998).

6. Coulson, D., An 18-coloring of 3-space omitting distance one. *Discrete Mathematics* 170 (1997) 241–247.

7. Donini, D., A 4-regular, 3-chromatic, self-complementary graph. (Problem #632, proposed by F. Schmidt) *College Mathematics Journal* 30 (1990) 320–321.

8. Hoffman, P., *The Man Who Loved Only Numbers: The Story of Paul Erdős and the Search for Mathematical Truth.* Hyperion Press, New York (1998).

9. Schechter, B., *My Brain Is Open: The Mathematical Journeys of Paul Erdős.* Simon & Schuster, New York (1998).

10. Temple, D. W., Colored polygon triangulations. *College Mathematics Journal* 29 (1998) 43–47.

Chapter 7

Matrices

A **matrix** is a rectangular table of numbers, which is usually denoted using a capital letter. The numbers within the matrix are called **entries** of the matrix. The horizontal lists of numbers are **rows** of the matrix and the vertical lists are **columns**. The **dimension** of a matrix with m rows and n columns is denoted $m \times n$ (read "m by n"). If $m = n$, the matrix is **square**. The entry that is in row i and column j is called the (i,j)**-entry**, and its **address** is (i,j). If the matrix is represented by A, we denote its (i,j)-entry by a_{ij}.

The following is an abstract representation of an $m \times n$ matrix, A.

$$A = \begin{bmatrix} a_{11} & a_{12} & \cdots & a_{1n} \\ a_{21} & a_{22} & \cdots & a_{2n} \\ \vdots & \vdots & \ddots & \vdots \\ a_{m1} & a_{m2} & \cdots & a_{mn} \end{bmatrix}$$

Graph theory makes effective use of matrices. We will first review some basic properties of matrices and then consider some uses of matrices within graph theory.

7.1 Review of Matrix Concepts

In this section, we will consider basic matrix operations. We begin with a simple real-world example.

Example 7.1 Alice and Bob work at a car dealership. Their car sales figures for the first through fourth quarters are given in matrix C, where the first row represents the sales for Alice and the second row the sales for Bob.

$$C = \begin{bmatrix} 3 & 2 & 7 & 5 \\ 1 & 4 & 8 & 2 \end{bmatrix}$$

a. What is the meaning of the $(2,3)$-entry of C?

b. How many cars did Alice sell during the fourth quarter?

c. How many cars did Bob sell altogether during the year?

d. What is the total number of cars that Alice and Bob sold during the second quarter?

Solution

a. The $(2,3)$-entry of C tells us that Bob sold eight cars during the third quarter.

b. Five

c. Add the entries in row two to get fifteen.

d. Add the entries in column two to get six.

Matrix Arithmetic

We will first illustrate the simplest matrix operations with an example that uses matrix C from Example 7.1 together with another matrix T.

Example 7.2 The dealership mentioned in Example 7.1 also sells pick-up trucks. Matrix T is the sales matrix, by quarter, of trucks sold by Alice and Bob, respectively.

$$T = \begin{bmatrix} 1 & 0 & 3 & 1 \\ 2 & 1 & 4 & 0 \end{bmatrix}$$

a. Construct a matrix that lists sales of vehicles (both cars and trucks) by quarter for Alice and Bob.

b. The sales manager is very optimistic about truck sales next year based on the new models that are coming out. He would like Alice and Bob to each double their sales output during next year. Write the resulting sales matrix for truck sales next year if Alice and Bob were to achieve that goal.

c. The main office has issued a new directive. They want sale matrices listed so that the sales representatives correspond to the columns and the rows correspond to the quarters in order, with row one corresponding to the first quarter. Rewrite matrix T as a new matrix called D according to the directive of the main office.

Solution

a. Since we want total sales of cars and trucks by quarter for Alice and Bob, we must add corresponding entries—that is, entries that have the same address, in matrices C and T. We get what is known as the sum of the two matrices:

7.1 Review of Matrix Concepts

$$C+T = \begin{bmatrix} 3 & 2 & 7 & 5 \\ 1 & 4 & 8 & 2 \end{bmatrix} + \begin{bmatrix} 1 & 0 & 3 & 1 \\ 2 & 1 & 4 & 0 \end{bmatrix}$$

$$= \begin{bmatrix} 3+1 & 2+0 & 7+3 & 5+1 \\ 1+2 & 4+1 & 8+4 & 2+0 \end{bmatrix} = \begin{bmatrix} 4 & 2 & 10 & 6 \\ 3 & 5 & 12 & 2 \end{bmatrix}$$

b. To get the new sales matrix, we must multiply each entry in matrix T by two. We obtain the matrix

$$2T = 2\begin{bmatrix} 1 & 0 & 3 & 1 \\ 2 & 1 & 4 & 0 \end{bmatrix} = \begin{bmatrix} 2 & 0 & 6 & 2 \\ 4 & 2 & 8 & 0 \end{bmatrix}$$

c. The new matrix, called the transpose of T, and is denoted T^t. Thus the matrix we get is

$$D = T^t = \begin{bmatrix} 1 & 0 & 3 & 1 \\ 2 & 1 & 4 & 0 \end{bmatrix}^t = \begin{bmatrix} 1 & 2 \\ 0 & 1 \\ 3 & 4 \\ 1 & 0 \end{bmatrix} \qquad \diamondsuit$$

Now we give some formal definitions for the operations encountered so far. For starters, matrices **A and B are equal**—that is, $\boldsymbol{A} = \boldsymbol{B}$ provided they have same dimension, say $m \times n$, and $a_{ij} = b_{ij}$, for $1 \leq i \leq m$ and $1 \leq j \leq n$. In other words, their corresponding entries are equal.

To add two matrices, their dimensions must be the same; that is, they must both be both $m \times n$ matrices. In that case, **the sum of matrices $\boldsymbol{A}$ and $\boldsymbol{B}$** is a matrix C whose entries satisfy the equation $c_{ij} = a_{ij} + b_{ij}$—that is, the sum of A and B is obtained by adding corresponding entries. Their **difference** is obtained similarly. Thus, if we subtract matrix B from matrix A, we get a new matrix D whose entries satisfy $d_{ij} = a_{ij} - b_{ij}$. Since addition of real numbers is commutative, it should be obvious from the definition that matrix addition is commutative; that is, $A + B = B + A$. Equally obvious is the fact that, in general, $A - B \neq B - A$.

Since multiplication is repeated addition ($3 \cdot 17 = 17 + 17 + 17$), it makes sense to define the product of a number r and a matrix M. **Scalar multiplication** consists of a scalar (number) r times a matrix M, which produces the new matrix $\boldsymbol{rM}$, whose (i,j)-entry is $r \cdot m_{ij}$. In other words, we distribute r to each entry of M. Scalar multiplication was used in Example 7.2(b).

The **transpose** of an $m \times n$ matrix A is the $n \times m$ matrix, denoted $\boldsymbol{A^t}$, obtained from A by interchanging its rows and columns. That is, the first row of A becomes the first column of A^t, the second row of A becomes the second column of A^t, and so on. Similarly, the first column of A becomes the first row of A^t, and so on. This is equivalent to defining A^t as the matrix whose (i,j)-entry is the (j,i)-entry of A. The transpose was illustrated in Example 7.2(c).

The **diagonal** of an $n \times n$ matrix A consists of the entries a_{ii}, $1 \leq i \leq n$; that is, the entries whose row and column numbers are identical. The $n \times n$ matrix whose diagonal entries each equal 1 and whose remaining entries each equal 0 is called the **identity matrix** and is denoted I. We will see why it is named that way shortly when we consider matrix multiplication.

As an example, consider the matrix

$$A = \begin{bmatrix} 5 & 8 & 6 \\ 7 & 2 & 4 \\ 9 & 0 & 3 \end{bmatrix}$$

Then the diagonal consists of the entries 5, 2, and 3. Note that the diagonal entries remain in place when we find the transpose. For matrix A, we find its transpose to be

$$A^t = \begin{bmatrix} 5 & 7 & 9 \\ 8 & 2 & 0 \\ 6 & 4 & 3 \end{bmatrix}$$

A $1 \times n$ matrix is called a **row matrix** and an $n \times 1$ matrix is called a **column matrix**. In either of those cases, the word "vector" is sometimes used in place of "matrix." That is, a **vector** is a matrix that has a single row or a single column. A matrix is a rectangular array of numbers, while a vector is a linear array of numbers. A vector is sometimes thought of as an ordered n-tuple of numbers. The order matters. Of course, the entries of a vector need only one coordinate as an address. Also, a vector is usually named using a lowercase letter. So the entries of an $n \times 1$ vector v would be $v_1, v_2, v_3, \ldots, v_n$. We call v_i the ith coordinate or the ith entry of v.

There are many uses for matrices. Imagine three countries #1, #2, and #3 that trade solely amongst each other. The (i,j)-entry of a 3×3 matrix might yield the value of the exports from country i to country j. The diagonal entries of this matrix are obviously 0. The sum of the entries of row i represents the value of all exports from country i, while the sum of the entries of column j represents the value of all imports to country j.

Here's another nice example: Suppose that a particle has four possible energy levels. Every second it changes levels (or remains at the same level). Let p_{ij}, where $1 \leq i \leq 4$ and $1 \leq j \leq 4$, be the probability that a particle at energy level i will, in the next second, be at energy level j. Then the matrix P, whose (i,j)-entry is p_{ij}, yields insight into the behavior of the particle over time. Notice that the sum of the entries in any row of such a matrix must be 1, since the particle must be in some energy level at any point in time. Think about why the column sums do not have to be 1.

A distance chart between various cities provides one more example. The (i,j)-entry of this matrix (even if its just called a chart) is the distance between cities i and j. Notice that the (i,j) and (j,i)-entries are the same,

7.1 Review of Matrix Concepts

from which it follows that the matrix and its transpose are the same. A square matrix A for which $A = A^t$ is called **symmetric**.

Here's an example of a symmetric 3×3 matrix:

$$A = \begin{bmatrix} 2 & 3 & 6 \\ 3 & 0 & 5 \\ 6 & 5 & 13 \end{bmatrix}$$

Notice how the entries reflect through the diagonal, which serves as a line of symmetry for this matrix. Note also that although A is symmetric, it cannot be a distance chart. A distance chart is symmetric but also has the property that the diagonal entries are zero. Why must it have this last property?

Matrix Multiplication

One of the authors has been known to say, "Multiplication of two matrices is, at first, a real pain. After a little practice, however, it is merely a nuisance." Matrix multiplication is certainly far more complicated than the operations that we considered so far. It may help to understand how it works by considering an example.

Example 7.3 Harvey's Health Foods packages two different raisin-nut mixtures, each in 10-oz bags. Alluring Almonds contains 5 oz of almonds, 4 oz of raisins, and 1 oz of pecans. Pecan Pleasures contains 2 oz of almonds, 3 oz of raisins, and 5 oz of pecans. Harvey obtains the ingredients for the mixtures from two different suppliers. The Nut House charges him 12¢ per oz for almonds, 10¢ per oz for raisins, and 15¢ per oz for pecans. We Are Nuts charges him 13¢ per oz for almonds, 9¢ per oz for raisins, and 14¢ per oz for pecans. We display this information in matrix form in matrices N and P as follows:

$$N = \begin{bmatrix} 5 & 4 & 1 \\ 2 & 3 & 5 \end{bmatrix} \quad P = \begin{bmatrix} 12 & 13 \\ 10 & 9 \\ 15 & 14 \end{bmatrix}$$

Note that it is usual to list the entries so that each row of the first matrix corresponds to a different mixture. On the other hand, each ingredient for the mixtures corresponds to a column in the first matrix. In the second matrix, however, the rows represent the different ingredients, and column one represents the respective prices charged by The Nut House, while column two the prices charges by We Are Nuts for those ingredients. We now construct a matrix that lists the total cost for the ingredients of each mixture if purchased from each of the suppliers. For example, we determine the cost of the first mixture from the first supplier by multiplying the respective entries in row 1 of N by those in column 1 of P and adding. (By "respective," we mean, first×first, second×second, and so on.) We do similarly for the other rows and columns and obtain the following:

$$NP = \begin{bmatrix} 5 & 4 & 1 \\ 2 & 3 & 5 \end{bmatrix} \begin{bmatrix} 12 & 13 \\ 10 & 9 \\ 15 & 14 \end{bmatrix}$$

which yields

$$\begin{bmatrix} 5(12) + 4(10) + 1(15) & 5(13) + 4(9) + 1(14) \\ 2(12) + 3(10) + 5(15) & 2(13) + 3(9) + 5(14) \end{bmatrix}$$

Combining, we get

$$\begin{bmatrix} 60 + 40 + 15 & 65 + 36 + 14 \\ 24 + 30 + 75 & 26 + 27 + 70 \end{bmatrix} = \begin{bmatrix} 115 & 115 \\ 129 & 123 \end{bmatrix}$$

Thus if Harvey were to purchase all his ingredients from one supplier, he would be better off buying from We Are Nuts, represented by column two. From them, the ingredients for his Alluring Almond mixture would cost him $1.15 (remember that the costs within the matrix are in cents) and for his Pecan Pleasures it would cost him $1.23. ◇

Now we formally describe this process. Given an $m \times r$ matrix A and an $r \times n$ matrix B, the **product $C = AB$** is an $m \times n$ matrix. Furthermore, the (i,j)-entry of C, c_{ij}, is given by

$$c_{ij} = a_{i1}b_{1j} + a_{i2}b_{2j} + a_{i3}b_{3j} + \cdots + a_{ir}b_{rj} \tag{7.1}$$

This is more impressively written as

$$c_{ij} = \sum_{k=1}^{r} a_{ik}b_{kj}$$

We may describe this process as multiplying row i of A into column j of B pairwise and adding. The "pairwise" means that we pair the first entry of row i of A with the first entry of column j of B, the second entry of row i of A with the second entry of column j of B, and so on.

Note, carefully, the prerequisite sizes of A $(m \times r)$ and B $(r \times n)$ if we are to multiply them. The number of columns of A (that is, r) must equal the number of rows of B (also r). This ensures that we can form the r products of the entries of row i of A by the corresponding entries of column j of B and add them to obtain the (i,j)-entry c_{ij} of the product matrix C.

Example 7.4 Determine the missing dimensions in Table 7.1.

Solution

In the first row, AB has dimension 3×7. In the second row, B must have dimension 2×3. In the third row, the product AB is undefined; the number of columns of A is not equal to the number of rows of B. In the fourth row, the dimension of A must be 6×1. ◇

7.1 Review of Matrix Concepts

A	B	AB
3×4	4×7	
4×2		4×3
2×4	2×2	
	1×5	6×5

Table 7.1

One of the more surprising things about matrix multiplication is that it is not commutative. First, we can see this from analyzing the dimension of the product. Sometimes, if A and B are compatible for the multiplication AB, the reverse product BA might not even be defined. For example, the product BA would be undefined for the matrices whose dimensions are given in row one of Table 7.1 even though AB is defined. But even if both AB and BA exist, there still is usually a problem. For example, if A is a 2×3 matrix and B is a 3×2, then AB is a 2×2 matrix, but BA is a 3×3 matrix. So, clearly, $AB \neq BA$. If A and B are both square matrices of the same dimension, then both AB and BA are square matrices of that dimension, but *still*, usually $AB \neq BA$.

Example 7.5 Calculate both AB and BA for the matrices
$A = \begin{bmatrix} 1 & 2 \\ 8 & -5 \end{bmatrix}$ and $B = \begin{bmatrix} 4 & 3 \\ -2 & 0 \end{bmatrix}$

Solution

$$AB = \begin{bmatrix} 1 & 2 \\ 8 & -5 \end{bmatrix} \begin{bmatrix} 4 & 3 \\ -2 & 0 \end{bmatrix} = \begin{bmatrix} 1(4)+2(-2) & 1(3)+2(0) \\ 8(4)+(-5)(-2) & 8(3)+(-5)(0) \end{bmatrix}$$

$$= \begin{bmatrix} 4-4 & 3+0 \\ 32+10 & 24+0 \end{bmatrix} = \begin{bmatrix} 0 & 3 \\ 42 & 24 \end{bmatrix}$$

$$BA = \begin{bmatrix} 4 & 3 \\ -2 & 0 \end{bmatrix} \begin{bmatrix} 1 & 2 \\ 8 & -5 \end{bmatrix} = \begin{bmatrix} 4(1)+3(8) & 4(2)+3(-5) \\ -2(1)+0(8) & -2(2)+0(-5) \end{bmatrix}$$

$$= \begin{bmatrix} 4+24 & 8-15 \\ -2+0 & -4+0 \end{bmatrix} = \begin{bmatrix} 28 & -7 \\ -2 & -4 \end{bmatrix}$$

We see from this example that the commutative law ($ab = ba$) of ordinary arithmetic breaks down with matrices. In general, we have $AB \neq BA$. ◇

Since matrix multiplication is rather complicated, it would be nice if there were an easier way. Fortunately, now there is, thanks to graphic calculators. For example, with the Texas Instruments TI-89 calculator, we would calculate the product AB from Example 7.5 as follows: $[1, 2; 8, (-)5] * [4, 3; (-)2, 0]$ *enter*. It is important to note that both minus signs use the additive inverse key (-), not the subtraction sign key −. Some older TI models also can do matrix multiplication, but the rows are separated somewhat differently. On those, each row is enclosed in brackets. Thus you would enter instead $[[1,2],[8,-5]] * [[4,3],[-2,0]]$ *enter*. If you have a graphic calculator, explore how to do matrix multiplication on your calculator. Try calculating the products AB and BA from Example 7.5 on your calculator.

Here are some other properties of matrix multiplication. Matrix multiplication is associative. Assuming that the matrices are compatible for multiplication, it is the case that $(AB)C = A(BC)$.

We define **powers of a square matrix** exactly as you might expect, $A^2 = AA$, $A^3 = AAA$, and so on. It should be easy for you to verify the following laws.

> **Note 7.1** Assume that A is $n \times n$ matrix and I is the $n \times n$ identity matrix. Then
> 1. $A^r A^s = A^{r+s}$
> 2. $(A^r)^s = A^{rs}$
> 3. $IA = AI = A$

It should now be clear why the identity matrix has the name it does. The $n \times n$ identity matrix plays the role of a multiplicative identity for $n \times n$ matrices in the same way that the number 1 does for real numbers. It makes sense, in light of the preceding rules, to define A^0 as I. Notice also that the first law implies that $A^r A^s = A^s A^r$; that is, powers of A commute.

Exercises 7.1

For Exercises 1–10, let

$$A = \begin{bmatrix} 2 & 1 & 5 \\ 0 & 5 & 2 \\ 6 & 1 & 4 \end{bmatrix}, B = \begin{bmatrix} 4 & 1 & 0 \\ 0 & 3 & 3 \\ 1 & 2 & 2 \end{bmatrix}, C = \begin{bmatrix} 2 & 0 & 4 \\ 1 & 1 & 0 \\ 0 & 5 & 2 \end{bmatrix},$$

$$D = \begin{bmatrix} 3 & 6 \\ 1 & 4 \\ 0 & 6 \end{bmatrix}, \text{ and } v = \begin{bmatrix} 3 \\ 2 \\ 5 \end{bmatrix}.$$

1. Find AB, AC, and BC.

2. Show that $AB \neq BA$.

3. Find $(AB)^t$ and show that it equals $B^t A^t$.

4. Verify that $(AB)C = A(BC)$.

5. Find Av. Why can't we find Dv?

6. Show that $(A+B)v = Av + Bv$.

7. Find v^t.

8. Find $A + B$, $3C - 2B$, and $I + A + 10B$.

9. Which of the following are undefined: DA, $C + D$, CD, $A + v$, D^2 and ID?

10. Find $(A + B)^2$. Is this equal to $A^2 + 2AB + B^2$? If not, how do you explain this?

11. Given two arbitrary $n \times n$ matrices M and N, show that $(MN)^t = N^t M^t$.

12. A **diagonal matrix** is a square matrix all of whose nondiagonal entries are zero. Show that if D is a diagonal matrix, then D^k is also diagonal, and its diagonal entries are the kth powers of the corresponding diagonal entries of D.

13. If A is a symmetric matrix—that is, $A = A^t$—show that A^k is also symmetric. Then show that $A^k = (A^k)^t$. *Hint*: Use Exercise 11 and mathematical induction of k.

14. Show that $(A + B)^t = A^t + B^t$.

7.2 The Adjacency Matrix

Many computer algorithms use graphs. To use those algorithms, we need a convenient way to store and manipulate graphs within the computer. There are a variety of ways of storing graphs, some of which are more efficient than others. We now consider one of the simplest ways of storing a graph, that is, as a matrix.

A Simple Example

Consider the airline map in Figure 7.1 of Take-a-Chance Airlines. Two cities in the map are adjacent if there is a (direct) flight connecting them. Otherwise, the cities nonadjacent.

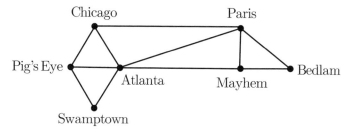

Figure 7.1 An airline map.

Let's number these cities in alphabetical order. So the numbers are Atlanta 1, Bedlam 2, Chicago 3, Mayhem 4, Paris 5, Pig's Eye 6, and Swamptown 7. We shall create a 7×7 matrix, A, by defining its entries as follows.

$$a_{ij} = \begin{cases} 1 \text{ if cities } i \text{ and } j \text{ are adjacent} \\ 0 \text{ if cities } i \text{ and } j \text{ are not adjacent} \end{cases}$$

Using this information, we get

$$A = \begin{bmatrix} 0 & 0 & 1 & 1 & 1 & 1 & 1 \\ 0 & 0 & 0 & 1 & 1 & 0 & 0 \\ 1 & 0 & 0 & 0 & 1 & 1 & 0 \\ 1 & 1 & 0 & 0 & 1 & 0 & 0 \\ 1 & 1 & 1 & 1 & 0 & 0 & 0 \\ 1 & 0 & 1 & 0 & 0 & 0 & 1 \\ 1 & 0 & 0 & 0 & 0 & 1 & 0 \end{bmatrix}$$

Matrix A is an example of a **binary matrix**, since each entry is an element of the set $\{0, 1\}$. This important matrix, which tells us the adjacencies of the airline map, is an **adjacency matrix**. Notice that the diagonal entries are zero, indicating that flights go between *distinct* cities. Observe, also, that an adjacency matrix is symmetric; that is, $a_{ij} = a_{ji}$ or, equivalently, $A = A^t$. This follows from a general assumption that air service between two given cities is always two way. When this is not the case, the flight routes are modeled using directed graphs, the subject of Chapter 10.

If we square the adjacency matrix A, we get the following:

$$A^2 = \begin{bmatrix} 5 & 2 & 2 & 1 & 2 & 2 & 1 \\ 2 & 2 & 1 & 1 & 1 & 0 & 0 \\ 2 & 1 & 3 & 2 & 1 & 1 & 2 \\ 1 & 1 & 2 & 3 & 2 & 1 & 1 \\ 2 & 1 & 1 & 2 & 4 & 2 & 1 \\ 2 & 0 & 1 & 1 & 2 & 3 & 1 \\ 1 & 0 & 2 & 2 & 1 & 1 & 2 \end{bmatrix}$$

which is no longer a binary matrix.

What does the (i, j)-entry of this matrix signify? Let's recall how it is obtained. It is found by multiplying row i of A into column j of A pairwise and adding. Since A is symmetric, the successive entries of column j are the same as those of row j. Thus, the (i, j)-entry of A^2 can be found by multiplying each entry of row i by its corresponding entry of row j and adding. The answer counts the number of *corresponding* 1's in these rows. This means that if there is a 1 in the third place, say, of both rows, it will contribute a 1 to the sum of the products. If either (or both) third entries are zero, there will be no contribution from the third place.

Put another way, the product of rows i and j gives the number of cities that are adjacent both to city i and to city j. Then the (i, j)-entry of

7.2 The Adjacency Matrix

A^2, denoted $(A^2)_{ij}$, counts the number of ways to go from city i to city j in exactly two flights—namely, flights that involve a stop at a common adjacency.

It is interesting to observe that the diagonal entry $(A^2)_{ii}$, by this logic, counts the number of cities adjacent to city i. This is the same as the sum of the entries in row i of A. The number of trips, involving a single stop, from a city to itself involves flying to a neighboring city and then flying right back; that is, a round trip. The $(1, 1)$-entry of A^2 for the Take-a-Chance airline map, for example, is 5, indicating that Atlanta has flights to five cities: Chicago, Mayhem, Paris, Pig's Eye, and Swamptown. This is also the sum of the entries of the first row of A – the row corresponding to Atlanta.

It is harder to show that $(A^3)_{ij}$ [the (i, j)-entry of A^3] counts the number of trips from city i to city j involving two stops; that is, involving three flights. More generally, the (i, j)-entry of A^k counts the number of trips from city i to city j involving k flights.

Be aware that this concept extends to other kinds of situations, such as bus transportation, commerce between industries, friendship, trade relations, and so on. Imagine a graph that shows alliances between nations. An edge connecting two vertices (representing two nations) indicates that they are allies. Two nations might not be allies but might have a common ally. Failing that, they may each have an ally and these allies might be allies.

The Adjacency Matrix of a Graph

Given a graph G of order n, with $V(G) = \{v_1, v_2, \ldots, v_n\}$, we define the $n \times n$ **adjacency matrix** A as follows:

$$a_{ij} = \begin{cases} 1 \text{ if } v_i v_j \in E(G) \\ 0 \text{ if } v_i v_j \notin E(G) \end{cases}$$

Note that $a_{ii} = 0$, for $i = 1, 2, \ldots, n$. Furthermore, since $a_{ij} = a_{ji}$, it follows that $A^t = A$, that is, A is a symmetric binary matrix with zeros on the diagonal.

We denote by J the $n \times n$ matrix with all entries equal to 1. This enables us to write the adjacency matrix of $\bar{G}$ as $J - A - I$. Subtracting A from J changes each 0 of A into a 1 and each 1 into a 0. This is a good thing since adjacencies and nonadjacencies are reversed in going from G to $\bar{G}$. What about the diagonal entries? We want them to be zero in the adjacency matrix of $\bar{G}$, requiring that we subtract I from $J - A$.

It follows from our discussion of the airline map problem that $(A^k)_{ij}$ [the (i, j)-entry of A^k] counts the number of v_i-v_j walks of length k in G. Moreover, $(A^2)_{ii}$, in counting walks of length 2 from v_i to itself, yields the degree of v_i since a walk of length 2 from v_i to itself requires visiting a neighbor and coming right back. In symbols, we have $(A^2)_{ii} = \deg(v_i)$.

What does $(A^3)_{ii}$ count? A walk of length 3 from v_i to itself requires two adjacent neighbors, say v_j and v_k, which means there will be two such walks for each triangle (the common word for K_3) to which v_i belongs. One walk goes clockwise and the other goes counterclockwise around the triangle. The triangle with vertices v_i, v_j and v_k will then be counted twice in each of the diagonal entries of A^3 corresponding to these three vertices, bringing the total to six. This yields a nice theorem, in which we use the term **trace** to denote the sum of the diagonal entries.

Theorem 7.1 The number of triangles in a graph G is the trace of A^3 divided by 6. ∎

Note that the adjacency matrix of a multigraph that is not a graph is not binary; that is, 0 and 1 are not the only entries. This is because for a multigraph, its adjacency matrix has a_{ij} equaling the number of edges joining vertex i to vertex j. If G is a multigraph but not a graph, then some a_{ij} must be at least two.

The Incidence Matrix

Let G have n vertices and m edges, such that $V(G) = \{v_1, v_2, \ldots, v_n\}$ and $E(G) = \{e_1, e_2, \ldots, e_m\}$. The **incidence matrix** of G is an $n \times m$ binary matrix M where $m_{ij} = 1$ if v_i is incident with e_j, and $m_{ij} = 0$ otherwise.

Example 7.6 Find the incidence matrix for the labeled graph displayed in Figure 7.2.

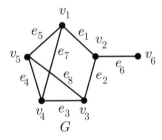

Figure 7.2

Solution
Graph G has six vertices and eight edges. Its 6×9 incidence matrix M is as follows:
$$M = \begin{bmatrix} 1 & 0 & 0 & 0 & 1 & 0 & 1 & 0 \\ 1 & 1 & 0 & 0 & 0 & 1 & 0 & 0 \\ 0 & 1 & 1 & 0 & 0 & 0 & 0 & 1 \\ 0 & 0 & 1 & 1 & 0 & 0 & 1 & 0 \\ 0 & 0 & 0 & 1 & 1 & 0 & 0 & 1 \\ 0 & 0 & 0 & 0 & 0 & 1 & 0 & 0 \end{bmatrix}$$ ◊

7.2 The Adjacency Matrix

Note that the sum of the entries in row i of M is $\deg(v_i)$. This is also true, of course, for the adjacency matrix. Moreover, the sum of the entries in each column of M is 2, because each edge is incident with exactly two vertices. Here is an interesting relationship between M and A. Let D be the $n \times n$ diagonal matrix (all nondiagonal entries are 0) such that $d_{ii} = \deg(v_i)$.

Theorem 7.2 Given a graph G with $V(G) = \{v_1, v_2, \ldots, v_n\}$ and $E(G) = \{e_1, e_2, \ldots, e_m\}$. Then for the adjacency matrix A, the incidence matrix M and the diagonal matrix D, where $d_{ii} = \deg(v_i)$, we have $MM^t = A + D$.

Proof. Let v_i and v_j be adjacent vertices of G, so $i \neq j$, and let edge $e_k = v_i v_j$. Then the (i,j)-entry of MM^t, obtained by multiplying row i by row j of M, will be 1 since entry k in both rows will be 1, and this will happen only in entry k since v_i and v_j share only one edge, e_k. Since v_i and v_j are adjacent (and $i \neq j$), the (i,j)-entry of the right side of the equation, $A + D$, will also be 1.

On the other hand, if v_i and v_j are not adjacent, the (i,j)-entry on both sides will be 0. On the left side, rows i and j of M have no ones in corresponding entries, since v_i and v_j share no edge. Finally, the diagonal entries of both sides agree because multiplying row i of M by itself yields $\deg(v_i)$. ∎

It should be clear that both M and A determine the graph. In other words, if we know either matrix M or matrix A, we are able to draw the graph. Unfortunately, the same is not true in reverse. The graph G can have a multitude of adjacency matrices because we can permute the indices of the vertices. In the case of M, we can also permute the indices of the edges.

The Adjacency Matrices of Classes of Graphs

Certain classes of graphs are easily recognizable from their adjacency matrices if (and that's a big if) the labeling conforms to a specified method. Consider the following adjacency matrix A for the path P_5 with consecutively labeled vertices.

$$A = \begin{bmatrix} 0 & 1 & 0 & 0 & 0 \\ 1 & 0 & 1 & 0 & 0 \\ 0 & 1 & 0 & 1 & 0 \\ 0 & 0 & 1 & 0 & 1 \\ 0 & 0 & 0 & 1 & 0 \end{bmatrix} \quad (7.2)$$

Note that the entries immediately above and below the diagonal are 1's, while all other entries are 0's. Stated elegantly, $a_{ij} = 1$ if and only if $|i - j| = 1$. Thus consecutively labeled paths are easily recognized from the adjacency matrix.

What about cycles? The adjacency matrix B for C_5, again labeled consecutively, should convince you that cycles, too, are easily spotted (if the vertices going around the cycle are labeled consecutively).

$$B = \begin{bmatrix} 0 & 1 & 0 & 0 & 1 \\ 1 & 0 & 1 & 0 & 0 \\ 0 & 1 & 0 & 1 & 0 \\ 0 & 0 & 1 & 0 & 1 \\ 1 & 0 & 0 & 1 & 0 \end{bmatrix} \tag{7.3}$$

You may have noticed a connection between the matrices (7.2) and (7.3). The latter has two additional ones—the $(1,5)$-entry and the $(5,1)$-entry. Of course! A cycle can be obtained from a path by rendering the first and last vertices adjacent.

One graph that can be recognized from its adjacency matrix is K_n. No matter how the vertices of K_n are labeled, $A(K_n) = J - I$. Thus, there are zeros on the diagonal and ones everywhere else.

Submatrices and Matrix Blocks

A **submatrix** B of a given $m \times n$ matrix M is an $r \times s$ matrix, where $r \leq m$ and $s \leq n$, obtained by retaining r (partial) rows and s (partial) columns of M. The retained rows and columns are obviously shortened to s and r entries, respectively. You may think of the submatrix B as consisting of the entries there the specific r rows intersect the selected s columns of M. A submatrix is called a *block* of the original matrix when the r rows and the s columns are consecutive. If all entries of a block are equal to 0, we may write a large 0 (the same goes for 1), as in the 8×6 matrix of Figure 7.3. The 0 and 1 blocks are each 4×3 submatrices.

$$\begin{bmatrix} & & & 1 & 2 & 5 \\ & \mathbf{0} & & 8 & 2 & 1 \\ & & & 5 & 4 & 8 \\ & & & 3 & 7 & 4 \\ 3 & 2 & 9 & & & \\ 5 & 8 & 8 & & \mathbf{1} & \\ 0 & 0 & 1 & & & \\ 3 & 8 & 2 & & & \end{bmatrix}$$

Figure 7.3 An 8×6 matrix with 4×3 blocks.

Given a bipartite graph G with m red and n blue vertices and such that the red vertices are labeled v_1 through v_m and the blue vertices are labeled v_{m+1} through v_{m+n}, the adjacency matrix A will consist of four blocks. The

7.2 The Adjacency Matrix

upper left and lower right blocks will be zero blocks of dimensions $m \times m$ and $n \times n$, respectively. This is because vertices of the same color in a bipartite graph are not adjacent. The upper right block, called B, is an $m \times n$ submatrix that contains enough information to draw the entire graph! The lower left block is B^t, since A is symmetric.

C_8 is bipartite. If we label the red vertices $1, 2, 3, 4$ and the blue vertices $5, 6, 7, 8$, then the adjacency matrix A looks as in (7.4).

$$A = \begin{bmatrix} 0 & 0 & 0 & 0 & 1 & 0 & 0 & 1 \\ 0 & 0 & 0 & 0 & 1 & 1 & 0 & 0 \\ 0 & 0 & 0 & 0 & 0 & 1 & 1 & 0 \\ 0 & 0 & 0 & 0 & 0 & 0 & 1 & 1 \\ 1 & 1 & 0 & 0 & 0 & 0 & 0 & 0 \\ 0 & 1 & 1 & 0 & 0 & 0 & 0 & 0 \\ 0 & 0 & 1 & 1 & 0 & 0 & 0 & 0 \\ 1 & 0 & 0 & 1 & 0 & 0 & 0 & 0 \end{bmatrix} \quad (7.4)$$

As you may have expected, A consists of upper left and lower right 0 blocks, an upper right 4×4 submatrix B, and B^t in the lower left corner. Of course, we no longer have the easily recognizable cycle format typified by the matrix (7.3) for C_5.

For the important complete bipartite graphs $K_{n,n}$ and $K_{1,n}$, the adjacency matrices, with appropriate vertex labeling, have the forms given in (7.5), where each submatrix J is the $n \times n$ matrix of all ones. In $A(K_{1,n})$, the upper right block is a $1 \times n$ matrix of zeros, and the lower left block is an $n \times 1$ block of zeros.

$$A(K_{n,n}) = \begin{bmatrix} J & 0 \\ 0 & J \end{bmatrix} \text{ and } A(K_{1,n}) = \begin{bmatrix} 1 & 0 \\ 0 & J \end{bmatrix} \quad (7.5)$$

Let G be a disconnected graph with two components H and K of orders r and s, respectively. Label the vertices of the first component v_1 through v_r and those of the second component v_{r+1} through v_{r+s}. Let the adjacency matrices of H and K be B and C, respectively. Then the adjacency matrix of graph G is given by

$$\begin{bmatrix} B & 0 \\ 0 & C \end{bmatrix}$$

where B is $r \times r$, C is $s \times s$, the upper right 0 block is $r \times s$, and the lower 0 block is $s \times r$.

Finally, we can often easily determine the adjacency matrix for a graph constructed using certain graph operations if we know the adjacency matrices of the graphs involved in the operation. For example, given a graph G of order n with adjacency matrix A, the adjacency matrix of the cartesian product $G \times K_2$ is very simple. Suppose that we label the vertices of the first copy of G by v_1 through v_n, and the second copy by v_{n+1} through v_{2n}, such that v_i in the first copy of G is adjacent to v_{n+i} in the second copy within $G \times K_2$. Then the adjacency matrix of $G \times K_2$ (each block has dimension $n \times n$) is as follows.

$$\begin{bmatrix} A & I \\ I & A \end{bmatrix}$$

Exercises 7.2

1. What can be said about an airline map with adjacency matrix A if the matrix $A + A^2$ has a zero entry?

2. Prove that the (i,j)-entry of the cube of the adjacency matrix of an airline map counts the number of trips from city i to city j consisting of three flights. *Hint*: Write A^3 as $A^2 A$, and use the fact that the (r,s)-entry of A^2 counts the number of trips from city r to city s consisting of two flights.

3. Why are there 7! different ways to label the cities of the airline map for Take-a-Chance Airlines of Figure 7.1 using the numbers from 1 to 7? Find the adjacency matrix for several labelings other than the one used in this section. Why may they differ from one another? Can two different labelings for the same airline map ever produce the same adjacency matrix?

4. Find the adjacency matrix A for Q_3 if consecutive vertices of the outer cycle are labeled 1, 2, 3, 4, vertex 4 is adjacent to vertex 5 on the inner cycle, and consecutive vertices of the inner cycle are labeled 5, 6, 7, 8. Relabel the vertices of Q_3 to obtain a different adjacency matrix that more clearly shows that Q_3 is bipartite.

5. Draw the adjacency matrix for $K_4 - e$, where the vertices are labeled v_1, v_2, v_3, v_4 and $v_3 v_4$ is the missing edge.

6. What can be said about two vertices v_i and v_j in G if rows i and j of the adjacency matrix are identical?

7. Using the graph G from Figure 7.4,

    (a) Find A, A^2, and A^3.
    (b) Determine the number of walks of length of length 3 in G.
    (c) Determine the trace of A^3 and the number of triangles (K_3's) in G. How do your answers coincide with what should happen according to Theorem 7.1?

7.3 The Distance Matrix

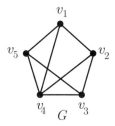

Figure 7.4

8. Let the edges of graph G in Figure 7.4 be labeled so that $e_1 = v_1v_2$, $e_2 = v_1v_4$, $e_3 = v_1v_5$, $e_4 = v_2v_3$, $e_5 = v_2v_4$, $e_6 = v_3v_4$, $e_7 = v_3v_5$, $e_8 = v_4v_5$.

   (a) Write the incidence matrix M for G.

   (b) Verify Theorem 7.2 for graph G.

9. Characterize graphs for which the adjacency matrix A equals the incidence matrix M for some labeling of the graph.

10. Prove that if G is bipartite, then $(A^3)_{ii} = 0$ for all i.

11. Show that if A is a symmetric binary matrix with all zeros on the diagonal, then A is the adjacency matrix of some graph G.

12. Prove that if the incidence matrix M of a graph G is a square matrix, then G must contain a cycle.

7.3 The Distance Matrix

We now consider another matrix, the distance matrix, that stores distance between vertices rather than adjacencies or incidences with edges. We shall see that the distance matrix can be generated with an algorithm that makes use of the adjacency matrix.

A Simple Example

Given a graph G of order n, with $V(G) = \{v_1, v_2, \ldots, v_n\}$, we define the $n \times n$ distance matrix D as the matrix with $d_{ij} = d(v_i, v_j)$. Note that $d_{ii} = 0$, for $i = 1, 2, \ldots, n$. Furthermore, since the distance function is a metric, $d(v_i, v_j) = d(v_j, v_i)$, so $d_{ij} = d_{ji}$. It follows that $D^t = D$; that is, D is a symmetric matrix with zeros on the diagonal. Note, however, that unlike the adjacency matrix, the distance matrix of a graph is generally not binary.

Example 7.7 Consider the graph G in Figure 7.5. Write the distance matrix D for G.

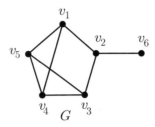

Figure 7.5

Solution
The distance matrix D is given by

$$D = \begin{bmatrix} 0 & 1 & 2 & 1 & 1 & 2 \\ 1 & 0 & 0 & 2 & 2 & 1 \\ 2 & 1 & 0 & 1 & 1 & 2 \\ 1 & 2 & 1 & 0 & 1 & 3 \\ 1 & 2 & 1 & 1 & 0 & 3 \\ 2 & 1 & 2 & 3 & 3 & 0 \end{bmatrix}$$

◇

Obtaining D from A

There are numerous algorithms for generating the distance matrix of a graph with a given labeling. Although there are more efficient techniques, we will show here one that is easy to understand based on our work with the adjacency matrix in Section 7.2.

Let us denote by S_k the sum $I + A + A^2 + \cdots + A^k$. Notice that a nonzero (i,j)-entry in A^r, for some r such that $0 \le r \le k$, guarantees that $(S_k)_{ij} \ne 0$. Recall that $A^0 = I$. In fact, $(S_k)_{ij} \ne 0$ if and only if $(A^r)_{ij} \ne 0$ for some r satisfying $0 \le r \le k$. This observation gives us several exciting algorithms (easily implemented with a computer program) for determining various facts about G. The next theorem will get the ball rolling. Recall that $e(v)$ is the eccentricity of v.

Theorem 7.3 If G is a graph with vertex set $V(G) = \{v_1, v_2, \ldots, v_n\}$ and adjacency matrix A, then $e(v_i)$ is the minimum value of k such that row i of S_k has no zero entries.

Proof. Let $e(v_i) = m$. Then given any vertex $v_j \in V(G)$, it follows that $d(v_i, v_j) \le m$. Then for some $r \le m$, we have $(A^r)_{ij} \ne 0$. This says that there is at least one v_i-v_j walk of length $r \le m$. Then $(S_m)_{ij} \ne 0$ for each j satisfying $1 \le j \le n$. In other words, the row i of S_m has no zero entries. How do we know, however, that m is the minimum index of S for which this

7.3 The Distance Matrix

happens? Given $t < m$, there is an eccentric vertex, say v_p, of v_i such that $d(v_i, v_p) > t$. Then $(S_t)_{ip} = 0$, and row i of S_t has a zero entry, proving that m is indeed the minimum index for which S_m has no zero entries. ∎

It should be noted that it is possible that S_k never has a row with no zero entries. When can this happen? When G is disconnected! In that case, no matter which vertex v_i we consider, there is always a vertex v_j for which there is no $v_i - v_j$ walk of any length. Thus, the (i,j)-entry of A^k will be 0 for all k, and thus so will the (i,j)-entry of S_k. How do we know when to stop; that is, how do we know that G is disconnected if we do not have a drawing of G but only have its matrix representation A? Since a shortest path is a walk, and the diameter of a connected graph with n vertices is at most $n-1$, and A^{n-1} will list the number of walks of length $n-1$ in G, we have the following simple result.

Theorem 7.4 A graph G on n vertices is connected if and only if S_{n-1} has no zero entries. ∎

We shall now restrict our attention to connected graphs, where distances are all finite. Now we are equipped to determine $\text{rad}(G)$.

Theorem 7.5 For any connected graph G, $\text{rad}(G)$ is the minimum k such that at least one row of S_k has no zero entries.

Proof. The radius of G is the minimum eccentricity among its vertices. This requires at least one vertex $v_i \in V(G)$ such that $\text{rad}(G) = e(v_i)$. Then, if $e(v_i) = k$, this number will be the first index for which S_k will have an entire row of nonzero entries (row i, of course). ∎

As a byproduct of the last two theorems, we have a method of determining the center $C(G)$. We evaluate the running matrix sums S_m, for $m = 1, 2, 3, \ldots$, and stop as soon as we reach the smaller of a number k such that S_k has at least one row with no zero entries, or we reach $m = n - 1$. In the first case, the vertices corresponding to the rows without zeros constitute the center of G. In the latter case, if there are rows without zeros, the corresponding vertices constitute the center of G. If, however, there are no such rows, then the graph is disconnected and the center consists of $V(G)$.

Our next theorem yields a way to find $\text{diam}(G)$.

Theorem 7.6 If G is a connected graph, then $\text{diam}(G)$ is the minimum k such that S_k has no zero entries.

Proof. Let $\text{diam}(G) = m$. Then for any two vertices v_i and v_j in G, we have $d(v_i, v_j) \leq m$. Then for some $r \leq m$, $(A^r)_{ij} \neq 0$, in which case $(S_m)_{ij} \neq 0$. Then all the entries of S_m are nonzero. To show that m is the minimum index, let $p < m$. Then there exists a pair of vertices, say v_s and v_t, such that $d(v_s, v_t) > p$. Then clearly $(S_p)_{st} = 0$, and we see that m is indeed the minimum index. ∎

Note that if G is disconnected, there is no k such that S_k has no zero entries, and Theorem 7.4 would alert us to this fact. In that case, of course, we have $\text{diam}(G) = \infty$.

Distance Matrix Realizability

An interesting yet very difficult problem concerning distance matrices is the following. Given a symmetric matrix D of nonnegative integers with all zeros on the diagonal, is D the distance matrix of some graph? First, to ensure that the distance function is a metric, we must verify that the triangle inequality holds; that is, $d(v_i, v_j) \leq d(v_i, v_k) + d(v_k, v_j)$ for all i, j, k. We consider a small example.

Example 7.8 Consider the following matrix. Determine whether it is the distance matrix of some graph G, and if so, find graph G.

$$D = \begin{bmatrix} 0 & 2 & 1 & 2 & 1 & 2 \\ 2 & 0 & 1 & 2 & 3 & 2 \\ 1 & 1 & 0 & 1 & 2 & 1 \\ 2 & 2 & 1 & 0 & 1 & 1 \\ 1 & 3 & 2 & 1 & 0 & 1 \\ 2 & 2 & 1 & 1 & 1 & 0 \end{bmatrix}$$

Solution

Immediately, we see the immense difficulty even for a small graph. If we were to check the triangle inequality for all possible ordered triples, we would have to check $P(6,3) = 6 \cdot 5 \cdot 4 = 120$ inequalities. However, because D is symmetric, this is cut in half, so there are 60 inequalities to check, still a burdensome task. Instead of checking all those, we will plow forward in a different way, constructing the graph in steps by looking for recognizable subgraphs in blocks of the distance matrix. Remember that, by definition, if $d_{ij} = k$, then $d(v_i, v_j) = k$. In particular, if $d_{ij} = 1$, then v_i and v_j are adjacent. From the lower right 3×3 block of D, we see that v_4, v_5, and v_6 are mutually adjacent. Additionally, we see from row 3 that v_3 is adjacent to both v_4 and v_6, but at distance two from v_5. Based on those observations alone, we obtain the induced subgraph in Figure 7.6(a). Since adjacencies are easiest to deal with, we consider v_1 next because there are more ones in row 1 than row 2. We see that v_1 is adjacent to v_3 and v_5, but at distance 2 from both v_4 and v_6. This forces the induced subgraph in Figure 7.6(b). Finally, row 2 of D indicates that v_2 is adjacent to v_3, at distance 2 from v_1, v_4, and v_6, and at distance 3 from v_5. The only possible way that can happen is if $v_3 v_1$ is a pendant edge. This forces G to be the graph in Figure 7.6(c). ◇

7.3 The Distance Matrix

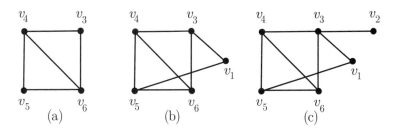

Figure 7.6

When the number of vertices is small and we are looking for a graph G that realizes a given distance matrix D (that is, a graph G for which D is its distance matrix), the problem can usually be solved in a reasonable amount of time. We have found the best approach to be by building successively larger induced subgraphs in steps, as was done in Example 7.8. Distance matrix realizability problems can be much more difficult, however, because in applications, we are often looking for a weighted graph rather than a graph. Also, the weights on the edges are not restricted to integer values. The interested reader is referred to Section 6.3 of Buckley and Harary [1].

Exercises 7.3

1. Find the distance matrix for C_{2n} where the vertices are labeled consecutively $v_1, v_2, v_3, \ldots, v_{2n}$ as you go around the cycle.

2. Find the distance matrix for C_{2n+1} where the vertices are labeled consecutively $v_1, v_2, v_3, \ldots, v_{2n+1}$ as you go around the cycle.

3. Find the distance matrix for the wheel $W_{1,n}$ where the central vertex of degree n is labeled v_1, and the vertices of the outer cycle are labeled consecutively $v_2, v_3, \ldots, v_{n+1}$ as you go around the cycle.

4. Write the distance matrix D for the graph G displayed in Figure 7.7.

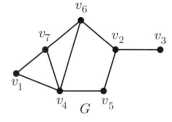

Figure 7.7

For Exercises 5–7, use the given matrices D_1, D_2, and D_3. Without trying to draw the graph, but just by using the theorems from this section, answer each question.

$$D_1 = \begin{bmatrix} 0 & 2 & 1 & 2 & 1 & 1 \\ 2 & 0 & 1 & 3 & 2 & 3 \\ 1 & 1 & 0 & 2 & 1 & 2 \\ 2 & 3 & 2 & 0 & 1 & 2 \\ 1 & 2 & 1 & 1 & 0 & 1 \\ 1 & 3 & 2 & 2 & 1 & 0 \end{bmatrix} \quad D_2 = \begin{bmatrix} 0 & 1 & 2 & 2 & 3 & 3 \\ 1 & 0 & 1 & 1 & 2 & 2 \\ 2 & 1 & 0 & 1 & 1 & 2 \\ 2 & 1 & 1 & 0 & 2 & 1 \\ 3 & 2 & 1 & 2 & 0 & 1 \\ 3 & 2 & 2 & 1 & 1 & 0 \end{bmatrix}$$

$$D_3 = \begin{bmatrix} 0 & 1 & 2 & 3 & 3 & 4 & 4 \\ 1 & 0 & 1 & 2 & 2 & 3 & 3 \\ 2 & 1 & 0 & 1 & 1 & 2 & 2 \\ 3 & 2 & 1 & 0 & 2 & 3 & 3 \\ 3 & 2 & 1 & 4 & 0 & 1 & 1 \\ 4 & 3 & 2 & 3 & 1 & 0 & 2 \\ 4 & 3 & 2 & 3 & 1 & 2 & 0 \end{bmatrix}$$

5. Determine each of the following for the graph corresponding to D_1, without drawing the graph: the order, the eccentricity of v_2 and v_5, the radius, the diameter, and the center.

6. Determine each of the following for the graph corresponding to D_2, without drawing the graph: the order, the eccentricity of v_2 and v_5, the radius, the diameter, and the center.

7. Determine each of the following for the graph corresponding to D_3, without drawing the graph: the order, the eccentricity of v_3 and v_6, the radius, the diameter, and the center. Also, determine the periphery for the graph.

8. Using the technique exhibited in Example 7.8, construct a graph G_1 whose distance matrix is D_1.

9. Using the technique exhibited in Example 7.8, construct a graph G_2 whose distance matrix is D_2.

10. Using the technique exhibited in Example 7.8, construct a graph G_3 whose distance matrix is D_3.

11. Determine a class of graphs for which the distance matrix always equals the adjacency matrix.

12. Find the distance matrix of the paths P_n, where the vertices are labeled consecutively $v_1, v_2, v_3, \ldots, v_n$ along the path.

7.3 The Distance Matrix

13. Prove that the following matrix D cannot be the distance matrix of a graph if the edges are unweighted; that is, if each edge has length 1. What basic property of a distance function fails, and where does it fail?

$$D = \begin{bmatrix} 0 & 1 & 2 & 2 & 3 \\ 1 & 0 & 1 & 4 & 2 \\ 2 & 1 & 0 & 2 & 1 \\ 2 & 4 & 2 & 0 & 3 \\ 3 & 2 & 1 & 3 & 0 \end{bmatrix}$$

References for Chapter 7

1. Buckley, F., and F. Harary, *Distance in Graphs*, Addison-Wesley, Redwood City, CA (1990).

Additional Readings

2. Bivins, I., The linear transformation associated with a graph. *College Mathematics Journal* 24 (1993) 76–78.

3. Kędzierawski, A., and O. Nicodemi, Image reconstruction in linear algebra. *College Mathematics Journal* 32 (2001) 128–134.

Chapter 8

Graph Algorithms

There are many connections between graph theory and computer science. Thus it should not be surprising that algorithms have played a strong role in recent graph theory research, so much so that several books have been devoted to algorithmic graph theory. In this chapter, we give an introduction to graph algorithms and indicate some of their uses. We shall not examine efficiency of algorithms or NP-completeness here, but refer the interested reader to Buckley and Harary [3], Chartrand and Oellermann [5], or Gould [6] for discussions of those topics.

8.1 Graph Searching

Numerous graph theory problems require us to explore a graph in search of a particular structure such as a spanning tree, a hamiltonian cycle, an eulerian circuit, a cut vertex, a matching, and so on. In this section, we discuss two of the most important search techniques: breadth-first search and depth-first search.

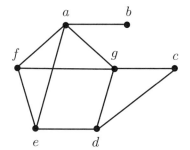

Vertex v	Adjacency list $N(v)$
a	b, e, f, g
b	a
c	d, g
d	c, e, g
e	a, d, f
f	a, e, g
g	a, c, d, f

Figure 8.1 A graph and its adjacency lists

In describing the two search algorithms, we will assume that the graph is stored using adjacency lists, a generally more efficient method than using an adjacency matrix. For each vertex, its neighbors are listed in some given order. We illustrate this concept in Figure 8.1.

Breadth-First Search

Given a graph G and a vertex $v \in V(G)$, we shall now present an algorithm for constructing a spanning tree T such that $d_T(v, x) = d_G(v, x)$ for all $x \in V(G)$. In other words, vertex v, called the **root**, has all of its distances preserved. Such a vertex is called **reach preserving**. The algorithm we shall now present is called a **breadth-first search** and is very important in computer science. We abbreviate it **BFS**.

BFS will construct a spanning tree in stages, keeping two growing lists, a list S of vertices and a list T of edges. The initial vertex list is quite short; it contains only the root. The initial list of edges is even shorter. It is empty. In the algorithm, we assign two labels to each vertex x. The first, $l(x)$, indicates the order in which x was reached. The second, $p(x)$, describes a shortest path from the root to x.

> **Algorithm 8.1 (Breadth-First Search)**
> 1. Let $S = \{v\}$ [v is the root]; $T = \emptyset$ [initially there are no edges in the tree].
> 2. $C = N(v)$ [C is the current set of vertices being processed].
> 3. $l(v) = 0$ [label root as vertex 0]; $p(v) = v$; $b^* = v$ [keeps track of current vertex we are branching from]; $i = 1$ [initializes variable i used to help label the vertices]; remove b^* from all adjacency lists.
> 4. For each $w \in N(b^*)$, place w in S and place edge $b*w \in T$; assign successive labels $l(w) = i$ and $p(w) = p(b^*), w$ to vertices of C. Add one to i after each vertex is labeled; remove w from all adjacency lists [this ensures that a vertex w gets labels $l(w)$ and $p(w)$ just once].
> 5. Define a new b^* to be the vertex x in C such that $l(x)$ is minimum; remove b^* from C, and return to step 4. If, however, C is empty, stop. If every vertex of G has been labeled, a spanning tree has been found. If not, then G is disconnected, but a spanning tree of the component containing the root has been found.

Recall that Prim's algorithm generates a minimum spanning tree in a weighted graph. When a graph is unweighted and ties are suitably broken, implementing Prim's algorithm corresponds to performing a breadth-first search.

8.1 Graph Searching

Example 8.1 Illustrate Algorithm 8.1 for the graph in Figure 8.1, first using vertex a as the root and then using vertex d as the root. Draw the resulting spanning trees.

Solution

Begin with $S = \{a\}$ and $T = \emptyset$. At step 2, C becomes $\{b, e, f, g\}$. At step 3, we get $l(a) = 0$, $p(a) = a$, $b^* = a$, $i = 1$, and a is removed from all adjacency lists. In Table 8.1, we show the results of steps 4 and 5 in each subsequent pass through the algorithm.

After two more passes through steps 4 and 5, C will be empty and the algorithm will terminate. The process when we begin with vertex d as the root is done similarly, but we will get a different spanning tree. The spanning trees with a and d as the root are shown in Figures 8.2(a) and (b), respectively. ◊

Pass	Step	b^*	$x \in S$	$e \in T$	$y \in C$	i	New Labels
1	4	a	$a, b, e,$ f, g	$ab, ae,$ af, ag	b, e, f, g	5	$l(b) = 1,\ p(b) = a, b,$ $l(e) = 2,\ p(e) = a, e,$ $l(f) = 3,\ p(f) = a, f,$ $l(g) = 4,\ p(g) = a, g$
	5	b			e, f, g		
2	4#, 5	e			f, g		
3	4		$a, b, d,$ e, f, g	$ab, ae,$ $af, ag,$ ed	f, g, d	6	$l(d) = 5,\ p(d) = a, e, d$
	5	f			g, d		
4	4#, 5	g			d		
5	4		$a, b, c, d,$ e, f, g	$ab, ae,$ $af, ag,$ ed, gc	d, c	7	$l(c) = 6,\ p(c) = a, g, c$

Table 8.1

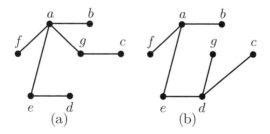

Figure 8.2

It should be noted that there are often many BFS spanning trees for a given unlabeled graph. For a labeled graph, the BFS spanning tree is determined by the ordering given to the labels. We used alphabetical order in Example 8.1(a) but d, a, b, c, e, f, g for 8.1(b). A different ordering affects the root, the ordering of the adjacency list, and ultimately the BFS spanning tree.

With some practice, you will easily be able to generate BFS spanning trees without having to use the whole labeling procedure as illustrated in Table 8.1. As another example, Figure 8.3 illustrates the BFS algorithm in stages for the 2-mesh $M(5,3)$ using the center vertex as the root. This time we are working on an unlabeled graph. BFS is modified to a labeling procedure where the k neighbors of the root are labeled $1, 2, \ldots, k$. Subsequent labels $p(x)$ in Algorithm 8.1 are then based on that initial assignment of labels. Labels on x shown in Figure 8.3 correspond to $p(x)$.

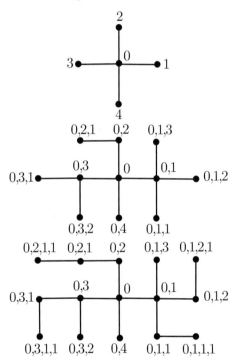

Figure 8.3 A spanning tree of the mesh $M(5,3)$.

A spanning tree T of a graph G satisfies the inequality $\mathrm{rad}(T) \geq \mathrm{rad}(G)$. (See Section 4.1, Exercise 14). Armed, however, with the BFS algorithm, we have the following heartwarming guarantee.

Theorem 8.1 Every connected graph G has a spanning tree T such that $\mathrm{rad}(T) = \mathrm{rad}(G)$.

8.1 Graph Searching

Proof. Select a vertex $v \in C(G)$ and produce a BFS spanning tree with v as root. Since v is reach-preserving, it follows that $\text{rad}(T) = e_T(v) = e_G(v) = \text{rad}(G)$. ∎

A spanning tree T of a graph G satisfying $\text{rad}(T) = \text{rad}(G)$ is called a **radius-preserving spanning tree (RPST)**. Theorem 8.1 guarantees that every connected graph has one. Unfortunately, it is not true that every connected graph has a **diameter-preserving spanning tree (DPST)**. The cycle C_n has only the path P_n as spanning tree and the path approximately doubles the diameter of the cycle.

Our next theorem has something to say about preserving the diameter of a graph.

Theorem 8.2 If T is a diameter-preserving spanning tree of a graph G, then $\text{rad}(T) = \text{rad}(G)$; that is, T is also a radius-preserving spanning tree.

Proof. We shall prove this by contradiction. Assume that T is a DPST of G such that $\text{rad}(T) > \text{rad}(G)$. Let $\text{rad}(G) = r$. Then $\text{rad}(T) \geq r+1$. Now since T is a tree, $\text{diam}(T) \geq 2\,\text{rad}(T) - 1 \geq 2(r+1) - 1 = 2r+1$. Since T is a DPST, it follows that $\text{diam}(G) \geq 2r+1$. This cannot be the case since for any connected graph G, $\text{rad}(G) \leq \text{diam}(T) \leq 2\,\text{rad}(T)$. (See Section 4.1, Exercise 9.) ∎

Depth-First Search

Depth-first search (DFS) is an important searching algorithm. The way that it works on a graph G is to begin at some vertex, the **root**, and travel as far as possible along a path P emanating from v. It then backtracks toward v until it finds a first vertex u from which there is an edge to a vertex w not yet reached. It then proceeds to generate a path P' by moving out along the edge uw toward additional vertices not yet reached. Path P' together with P is a subtree of G. The algorithm again continues visiting new vertices on a path until it can go no farther and then again backtracks, searching for another vertex that it can branch out from. In this way, DFS will generate a spanning tree T of G and will have traversed each edge of T exactly twice, once forward along a path of new vertices and once backtracking in search of a new branch location. We describe this process more formally in Algorithm 8.2 that follows.

Note that in Algorithm 8.2, step 2 is called a **while loop**, a standard item used in computer programming. When reaching step 2, the condition following "while" is tested. If the condition is true, the statements between "begin" and "end" will be performed. Each time "end" is reached, the condition will be retested. If the condition is still true, we stay in the loop and process the statements again. The first time that we find the condition to be false—that is, $N(b^*) \cap U = \emptyset$—we move on to step 3. As with BFS, Algorithm 8.2 detects whether G is connected.

220   Chapter 8 Graph Algorithms

Example 8.2 Illustrate Algorithm 8.2 using the graph in Figure 8.4, first with the vertex ordering a, b, c, d, e, f, g and then with the ordering d, g, c, b, a, f, e. Draw the resulting DFS spanning trees.

Algorithm 8.2 (Depth-First Search)
1. Let $S = \{v\}$ [v is the root], $T = \emptyset$ [initially there are no edges in the tree]; $b^* = v$ [our initial branch vertex]; $p^* = v$ $l(v) = 0$ [label root as vertex 0]; $i = 1$ [initializes variable i used to help label the vertices]; $U = V(G) - \{v\}$ [keeps track of unlabeled vertices].
2. While $N(b^*) \cap U \neq \emptyset$ [b^* has unlabeled neighbors] do
   begin
      label the next unlabeled neighbor w of b^* by i; place b^*w in T; remove w from U; let $p^* = b^*$ [helps in backtracking]; $b^* = w$ [new branch vertex]; $i = i + 1$ [increment i for future labeling].
   end.
3. $b^* = p^*$ [backtrack].
4. If $b^* = v$ and $N(b^*) \cap U = \emptyset$ [v has no unlabeled neighbors], stop. A spanning tree for the component containing the root v has been found. Otherwise, repeat step 2. If $U = \emptyset$, then the tree found is a spanning tree of all of G. If $U \neq \emptyset$, then G is disconnected.

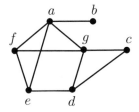

Figure 8.4

Solution
The vertex ordering determines the root and the order in which neighbors appear on the adjacency list of a given vertex. That will affect the order in which the vertex and an incident edge will enter the tree. Using the ordering a, b, c, d, e, f, g, we obtain the following sequence of events. Vertex a gets label 0, then b gets label 1. We then backtrack to vertex a since b has no unlabeled neighbors. Then e gets label 2, d gets label 3, c gets label 4, g gets label 5, and f gets label 6. Then there is successive backtracking to g, c, d, e, a. Upon finally arriving back at a, we find that even a has no unlabeled neighbors, and the algorithm terminates. The resulting spanning tree is shown in Figure 8.5(a).

8.1 Graph Searching

Using the vertex ordering d, g, c, b, a, f, e, the algorithm proceeds as follows. Vertex d gets label 0, g gets label 1, c gets label 2. We then backtrack to g since c has no unlabeled vertices. Vertex a then gets label 3, b gets label 4, and we backtrack to vertex a. Then f gets label 5, e gets label 6, and we backtrack to f, a, g, d. Ending at the root d, we find it has no unlabeled neighbor, and the algorithm terminates. The resulting DFS tree is displayed in Figure 8.5(b). ◇

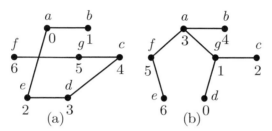

Figure 8.5 Two DFS spanning trees.

BFS and DFS are useful in many graph theory problems. In general, BFS produces spanning trees that are more compact and have smaller diameter than those produced by DFS. Thus DFS is often more useful in problems where one is trying to find long paths emanating from a given vertex. BFS can easily be adapted to aid in finding eccentricities, the radius, the diameter, the center, and the periphery of a graph, among other items. In section 10.1, we show how to use DFS to turn a 2-connected graph into a directed graph (there are directions on the edges) where it is possible to get from any vertex u to any other vertex v while following the directions on the edges. DFS can also be used to help find the blocks in a graph (see Chartrand and Oellermann [5, pp.79–85]).

Exercises 8.1

1. Use the breadth-first search algorithm to obtain three different spanning trees for $M(5,5)$.

2. Show that every odd 2-mesh has a diameter preserving spanning tree. *Hint*: Use the BFS algorithm.

3. Find a radius-preserving spanning tree for the hypercube Q_3.

4. Show that Q_n has no diameter-preserving spanning tree for $n \geq 2$.

5. Find all spanning trees of $W_{1,7}$ that are produced using the BFS algorithm.

6. Find a spanning tree of $M(5,6)$ such that there are six reach-preserving vertices in the tree.

7. Given vertices $v = (i,j)$ and $w = (h,k)$ of a 2-mesh such that $i \neq h$ and $j \neq k$, show that there is no spanning tree in which both v and w are reach preserving.

8. Use the preceding exercise to prove that no spanning tree of the mesh $M(m,n)$ can have more than k reach-preserving vertices, where $k = \max\{m,n\}$. Show how to construct a spanning tree with this maximum number of reach-preserving vertices.

9. Generate a spanning tree using Algorithm 8.1 (BFS) for the graph in Figure 8.6 with vertex ordering a,b,c,d,e,f,g,h. (This ordering determines both the root and the ordering of vertices on the adjacency lists.)

10. Repeat Exercise 9 but with vertex ordering b,e,a,d,c,g,f,h.

11. Apply Algorithm 8.2 (DFS) to the graph in Figure 8.6 with vertex ordering a,b,c,d,e,f,g,h.

12. Repeat Exercise 11 but with vertex ordering b,e,a,d,c,g,f,h.

13. Repeat Exercise 11 but with vertex ordering c,b,h,e,g,a,f,d.

14. Explain precisely how you would modify Algorithm 8.1 to determine the radius upon completion.

15. Explain how BFS could be used to find the eccentricities of all vertices and therefore both the center and periphery.

16. If G is disconnected, explain how you would modify DFS to find a spanning forest.

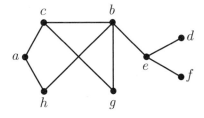

Figure 8.6

8.2 Graph Coloring Algorithms

Coloring the vertices of a graph to obtain a proper coloring with a minimum number of colors is known to be intrinsically difficult. In fact, even the problem of determining whether an arbitrary graph of order n is 3-colorable is NP-complete. Some hope of not going too far off when attempting a minimum coloring is given by the following result due to Brooks [2].

Theorem 8.3 (Brooks' Theorem) If G is a connected graph and is neither complete nor an odd cycle, then $\chi(G) \leq \Delta(G)$. ∎

Unfortunately, even though standard proofs (see, for example, Gould [6, p. 228] or Chartrand and Oellermann [5, p. 289]) indicate an algorithm that will achieve a $\Delta(G)$-coloring, the difference $\Delta(G) - \chi(G)$ may be huge. For example, consider $K_{n,n}$ or $W_{1,n}$ when n is large. Thus, various approximation algorithms have been devised. We now discuss a couple of them.

Sequential Coloring

This coloring algorithm assumes that the vertices are labeled and vertex labels are ordered. The possible colors are also ordered from lowest to highest. For simplicity, we use letters for the vertex labels, ordered alphabetically, and positive integers for the colors, ordered numerically. At each stage, the lowest possible color is assigned to the next vertex while maintaining a proper coloring. We describe the procedure with the following algorithm.

Algorithm 8.3 (Sequential Coloring)
1. Color the first vertex by 1.
2. Color each subsequent vertex by the lowest color not used by any of it neighbors that have already been colored. Do this until all vertices have been colored.

In our next example, we illustrate Algorithm 8.3 on a single graph with two different initial labelings.

Example 8.3 Find sequential colorings using Algorithm 8.3 for the labeled graphs in Figure 8.7.

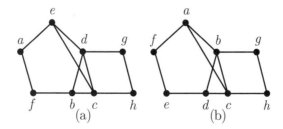

Figure 8.7

Solution
In graph (a) we color a by 1, then b, c, and d by 1, 2, and 3, respectively. Then e must get color 4, f gets color 2, g gets color 1, and h gets 3. The resulting 4-coloring is shown in Figure 8.8(a).

For graph (b) of Figure 8.7, we color a, b, and c by 1, 2, and 3, respectively. Then d, e, and f get colors 1, 2, and 3, respectively. Finally, g gets color 1, and h gets 2. The resulting 3-coloring is shown in Figure 8.8(b). ◊

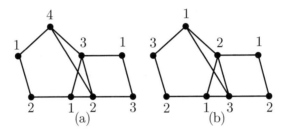

Figure 8.8 Sequential colorings.

Since we achieved a 3-coloring in Figure 8.8(b), and G has triangles, we know that $\chi(G) = 3$. Note, however, that the ordering of the vertices did make a difference. One of the factors that makes achieving a minimum coloring difficult is that there are $N!$ ways that the labels can be assigned, all of which may, in fact, produce distinct labeled graphs. To understand this, consider $K_4 - e$, which has four vertices. Although there are $4! = 24$ ways to assign labels a, b, c, d, only $\binom{4}{2} = 6$ distinct labeled graphs are produced. Once we choose the two labels for the nonadjacent vertices, the labeled graph is uniquely determined. The interested reader is referred to the book by Harary and Palmer [7] on graphical enumeration.

Maximum Color-Degree Coloring

This approach to coloring is due to Brelaz [1]. The idea is to try to color vertices of high degree early because they may create lots of difficulties if colored later in the process. It also simultaneously tries to color a vertex whose neighbors have already received many distinct colors. The reasoning here is the same: Postponing the decision on such a vertex could create more trouble later, so perhaps it would be wise to color it earlier.

In order to present the algorithm, we need a definition. Suppose that a partial coloring of G has been obtained; that is, some of the vertices have been assigned colors. Then the **color degree** of vertex v is the number of distinct colors that have been assigned to neighbors of v. It is important to understand that the color degree is not the number of neighbors that have been colored, but is instead the *number of different colors* that have been used to color those neighbors.

8.2 Graph Coloring Algorithms

Algorithm 8.4 (Maximum Color-Degree Coloring)
1. Sort vertices in order from largest degree to smallest onto a list U [uncolored vertices].
2. Color first vertex v of U by 1 [$\deg(v) = \Delta(G)$]; $i = 1$ [keeps track of how many vertices have been colored]; delete v from U [U keeps track of uncolored vertices].
3. while $i < N$ [not all vertices have been colored] do
   begin
       $j = 1$; found = "no";
       select a vertex w from U with maximum color-degree. When
         there is a tie, choose the vertex that appears earliest on U
         [w has maximum color-degree in U and has maximum
         degree among such vertices].
       while found = "no" do
       begin
         if some $x \in N(w)$ has color j then
           $j = j + 1$
         else
           begin
             found = "yes"
             color w by j
             $i = i + 1$[another vertex has been colored]
             remove w from U.
           end {else}
       end {while found}
   end {while $i < N$}
4. All vertices have been colored; output the coloring.

Example 8.4 Apply Algorithm 8.4 to the graph in Figure 8.7(a).
Solution
First sort by maximum degree to get the ordered list $U = c, d, b, e, a, f, g, h$. Then we color c, d, b, e by $1, 2, 3, 3$, respectively. Next, a gets color 1 and f gets color 2. Finally, g gets color 1 and h gets color 2. The resulting 3-coloring is shown in Figure 8.9. ◇

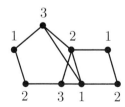

Figure 8.9 A Maximum color-degree coloring.

We now consider a larger graph in order to show better some of the features of Algorithm 8.4.

Example 8.5 Find a coloring of the vertices of the graph in Figure 8.10 using Algorithm 8.4.

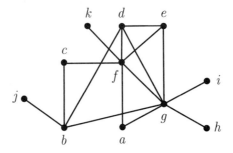

Figure 8.10

Solution
We sort the vertices by degree and get $g, f, b, d, e, a, c, h, i, j, k$. We color g by 1, then color f by 2. Here's where it gets interesting. Although b has the next largest degree, its color degree is only 1, while the color degrees of a, d, and e are each 2. Among those three vertices, we color the one with maximum degree, namely, d by 3. Now b and a have color degree 2, but e has color degree 3, so e is colored next. We color e by 4. The coloring thus far is shown in Figure 8.11.

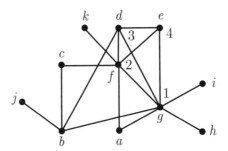

Figure 8.11 A partial coloring.

Now b and a each have color degree 2, so we color b, which has higher degree, with the lowest available color, namely 2. Then a gets color 3, c gets color 1, h and i get color 2, j gets 1, and k gets 1. The final coloring is shown in Figure 8.12. ◇

Chartrand and Oellermann [5] and Gould [6] discuss efficiency measures of graph coloring algorithms. Gould also presents two additional graph coloring algorithms. The smallest last algorithm of Matula, Marble, and Isaacson

8.2 Graph Coloring Algorithms

[8] recursively removes vertices of smallest degree and indicates these as being colored latest. After a vertex v is removed, degrees are recalculated for $H = G - v$ to see which vertex w will be removed next from H. Then degrees are recalculated for $H - w$, and so on. Gould also discusses an interchange coloring algorithm that is slightly more complex.

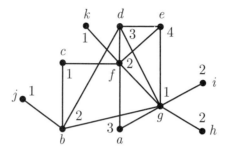

Figure 8.12 Another maximum color-degree coloring.

Exercises 8.2

1. Find a sequential coloring of the graph of Figure 8.4 if the vertices are ordered alphabetically.

2. Redo Exercise 1 but with vertex ordering a, e, f, d, c, g, b this time.

3. Find a maximum color-degree coloring of the graph in Figure 8.4.

4. Show that the vertex ordering given in Exercise 2 affects the initial ordering of vertices by maximum degree for the graph in Figure 8.4 but does not affect the actual coloring using Algorithm 8.4.

5. Find a sequential coloring of the graph in Figure 8.6.

6. Find a maximum color-degree coloring for the graph in Figure 8.6.

7. Prove that a graph with $\chi(G) = 2$ will not always be 2-colored if Algorithm 8.3 is used.

8. Prove that a graph with $\chi(G) = 2$ will always be 2-colored if Algorithm 8.4 is used.

9. Show that there will be seven vertices assigned color 1 when Algorithm 8.3 is used to color the graph of Figure 8.10.

10. Find a labeling of the Peterson graph G that would cause G to be 4-colored using Algorithm 8.3 assuming the vertices are ordered alphabetically. Then show that $\chi(G) = 3$ by finding a 3-coloring.

11. Show that Algorithm 8.4 always 3-colors the Peterson graph.

12. Rewrite Algorithm 8.3 more precisely using while loops and counters similar to the way Algorithm 8.4 was written.

8.3 Tree Codes

In this section, we consider two problems concerning tree codes. The first one assigns a code to a tree on n vertices that uniquely describes the tree. That is, if we know the code, then there is a unique tree that we can construct from it. The second coding scheme is called an addressing problem. In this problem we label each vertex with a binary vector so that the distance between two vertices can be ascertained directly from their labels.

Prüfer Codes

Given a tree T on n vertices, such that $V(T) = \{1, 2, \ldots, n\}$, the **Prüfer code** of T is an ordered $(n-2)$-tuple of integers $(a_1, a_2, \ldots, a_{n-2})$ with $1 \leq a_i < n$, obtained by an algorithm soon to be described, from which T may be reconstructed. In other words, the Prüfer code completely characterizes the tree T, implying a one-to-one correspondence between labeled trees and ordered $(n-2)$-tuples with entries from the set $\{1, 2, \ldots, n\}$.

Since each of the $n-2$ entries in the Prüfer code can be chosen in n ways, it follows by the multiplication principle that there are n^{n-2} Prüfer codes.

The Prüfer code $(a_1, a_2, \ldots, a_{n-2})$ of a labeled tree T is constructed simultaneous with a sister code $(b_1, b_2, \ldots, b_{n-2})$ as follows. The second code, by the way, will not be needed to reconstruct the tree T. In fact, we will show how to derive the b_i's from the Prüfer code $(a_1, a_2, \ldots, a_{n-2})$.

Algorithm 8.5
1. Let b_1 be the end vertex of T with the smallest label. Let a_1 be the unique neighbor of b_1.
2. Delete vertex b_1 from T, yielding a new tree T_1.
3. Let b_2 be the end vertex of T_1 with the smallest label. Let a_2 be the neighbor of b_2.
4. Delete vertex b_2 from T_1, yielding a new tree T_2.
5. Continue in this manner until the Prüfer code is complete.

Note that each time we deleted a vertex b_i in Algorithm 8.5, edge $a_i b_i$ of T is also deleted. This idea will be important in the reconstruction process. Algorithm 8.5 will be easier to follow with an example. Let T be the labeled tree of Figure 8.13.

8.3 Tree Codes

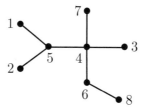

Figure 8.13 A labeled tree T.

The end vertices of T are 1, 2, 3, 7, and 8. Since the smallest of these numbers is 1, we set $b_1 = 1$ and b_1's neighbor, $a_1 = 5$. Now delete b_1, obtaining the tree T_1 with end vertices 2, 3, 7, and 8. Since the smallest end vertex is 2, set $b_2 = 2$ so $a_2 = 5$. Delete b_2, obtaining the tree T_2 shown in Figure 8.14.

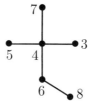

Figure 8.14 The tree T_2, obtained from T after two deletions.

Since the end vertices of T_2 are 3, 5, 7, and 8, set $b_3 = 3$, so $a_3 = 4$. (Note that 5 is an end vertex of T_2 even though it is not an end vertex of T.) Deleting b_3, we get the tree T_3 with end vertices 5, 7, and 8. Set $b_4 = 5$, so $a_4 = 4$. Upon deletion of 5, we get the tree T_4 shown in Figure 8.15.

Figure 8.15 The tree T_4. We are almost finished.

Now set $b_5 = 7$, so $a_5 = 4$, obtaining the tree T_5 (which is the path P_3 with end vertices 4 and 8). Finally, setting $b_6 = 4$, we obtain $a_6 = 6$ and the tree T_6, which is K_2. We have the Prüfer code $(5, 5, 4, 4, 4, 6)$ and the sister code $(1, 2, 3, 5, 7, 4)$.

Note that for any given nontrivial tree, we obtain K_2 when the algorithm terminates. One of the survivors of the "deleting frenzy"—that is, one of the vertices of K_2, "the last of the trees"—must be n. At no stage of the algorithm is n the smallest end vertex. Thus n is never deleted.

Given a Prüfer code, how do we reconstruct the tree? The first thing to do is to let $a_{n-1} = n$. In other words, lengthen the Prüfer code (which has $n-2$ entries) by letting n be the $(n-1)$-st entry. Thus we have the sequence $(a_1, a_2, \ldots, a_{n-1})$. If we were to know the sister sequence $(b_1, b_2, \ldots, b_{n-1})$, we would be able to construct the tree since we would know all of the $n-1$ edges, $a_1 b_1, a_2 b_2, \ldots, a_{n-1} b_{n-1}$. Merely add these edges to the vertex set $\{1, 2, \ldots, n\}$.

The following algorithm yields the sequence $(b_1, b_2, \ldots, b_{n-1})$.

Algorithm 8.6
1. Let b_1 be the smallest vertex not in $(a_1, a_2, \ldots, a_{n-1})$.
2. Let b_2 be the smallest vertex not in $(a_2, a_3, \ldots, a_{n-1})$ and not b_1. (In forming the Prüfer code, b_2 is deleted after b_1.)
3. Let b_3 be the smallest vertex not in $(a_3, a_4, \ldots, a_{n-1})$ and not b_1 or b_2. (In forming the Prüfer code, b_3 is deleted after b_1 and b_2.)
4. Continue in this manner until the sequence $(b_1, b_2, \ldots, b_{n-1})$ is obtained. For each $k = 1, 2, \ldots, n-1$, b_k is the smallest vertex not in $(a_k, a_{k+1}, \ldots, a_{n-1})$, and $b_k \neq b_i$, for $i = 1, 2, \ldots, k-1$.

Example 8.6 Precisely one tree has Prüfer code $(7, 4, 9, 9, 2, 9, 1, 7)$. Find that tree.

Solution

Since the code is an 8-tuple, we know that our tree has 10 vertices. Add the number 10 to the end of the code to get $(7, 4, 9, 9, 2, 9, 1, 7, 10)$. Now construct the sister code by using Algorithm 8.6. b_1 is the smallest number (from $1, 2, 3, \ldots, 10$) not on the list, so $b_1 = 3$. Then we delete the first number, 7. Then b_2 is the smallest number not on this shorter list and not equal to b_1, so $b_2 = 5$. Continuing in this way, using Algorithm 8.6, we get the sequence $(3, 5, 4, 6, 8, 2, 9, 1, 7)$. From the Prüfer code and the sister code, we obtain the edge set $\{(7, 3), (4, 5), (9, 4), (9, 6), (2, 8), (9, 2), (1, 9), (7, 1), (10, 7)\}$. The resulting labeled tree is displayed in Figure 8.16. $\diamond$

Since there is a one-to-one correspondence between labeled trees on n vertices and Prüfer codes of length $n-2$—that is, $(n-2)$-tuples composed from the set $\{1, 2, \ldots, n\}$—we have the following theorem first proved by Cayley [4].

8.3 Tree Codes

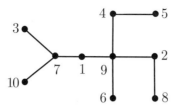

Figure 8.16 The unique tree with Prüfer code $(7, 4, 9, 9, 2, 9, 1, 7)$.

Theorem 8.4 (Cayley's Theorem) *The number of labeled trees on n vertices using n distinct labels is n^{n-2}.*

Corollary 8.4a *The complete graph K_n has n^{n-2} distinct labeled spanning trees.*

Binary Addressing of a Tree

The **Hamming distance** between two binary vectors (each entry is 0 or 1) of length n is the number of bit positions that differ in the two vectors. For example, the binary vectors $(0, 1, 0, 0, 1, 1, 0)$ and $(1, 0, 0, 1, 0, 1, 0)$ have Hamming distance 4 because they differ in positions 1, 2, 4, and 5. Hamming distance is used in the analysis of **error-correcting codes**. Is it possible to label each vertex of a graph G with a binary vector with n entries such that the distance between any pair of vertices equals the Hamming distance of their labels? If this is possible, what is the smallest n for which it can be achieved?

Note that this labeling scheme, where the Hamming distance of the labels equals the *graph* distance between the vertices, called a **binary addressing**, is employed in the hypercube Q_n, where the binary vectors have length n. It is not hard to show that "shorter" binary vectors cannot do the job for Q_n.

An attempt to find a binary addressing for K_3 is futile for the following reason. Call the vertices x, y, and z. Without loss of generality, let two of the vertices, say x and y, have the binary vectors $(b_1, b_2, \ldots, b_k, 0, b_{k+1}, \ldots, b_t)$ and $(b_1, b_2, \ldots, b_k, 1, b_{k+1}, \ldots, b_t)$, respectively, that disagree precisely in the ith place, as indicated. The vectors differ in just one place because x and y are adjacent; that is, $d(x, y) = 1$. Now the third vertex z must have either a 0 or 1 in the ith place.

Once again, we invoke the "no loss of generality" license and assume that, like vertex x, vertex z has a 0 in the ith place, thereby requiring a disagreement in, say, the jth place with both x and y. This implies, however, that the Hamming distance between y and z is 2, but we know that $d(y, z) = 1$ since y and z are adjacent. We conclude that K_3 has no binary addressing. Graphs that have a binary addressing are called **binary Hamming graphs**. Thus Q_n is a binary Hamming graph, while K_3 is not.

The star $K_{1,4}$ has a labeling using binary 4-tuples, as is exhibited in Figure 8.17.

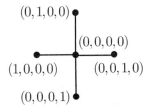

Figure 8.17 A binary addressing for the star $K_{1,4}$.

The next theorem settles the addressing problem for trees.

Theorem 8.5 Let T be a nontrivial tree of order n. Then the vertices of T binary addressing using $(n-1)$-tuples.

Proof. The proof uses induction on n. When $n = 2$, T is K_2 and the vertices may be labeled using the binary 1-tuples (0) and (1). Assume that the theorem is true for $n = k$. Then we must show that it is true when $n = k+1$.

Let T be a tree on $k+1$ vertices. Let v be an end vertex of T. Let w denote its neighbor. Now consider $T - v$. Since $T - v$ has order k, we apply the inductive hypothesis to produce a binary addressing using $(k-1)$-tuples. Now to obtain a labeling for T, assign k-tuples to its vertices as follows. If $x \in V(T-v)$ has the binary $(k-1)$-tuple $(x_1, x_2, \ldots, x_{k-1})$, then in T, label x with the k-tuple $(0, x_1, x_2, \ldots, x_{k-1})$. On the other hand, to label v, use the k-tuple of its neighbor w but change the initial 0 to 1, thereby labeling v by $(1, w_1, w_2, \ldots, w_{k-1})$. Since the k-tuple of w in T is $(0, w_1, w_2, \ldots, w_{k-1})$, the addressing conforms to the requirement that $d(v, w) = 1$. Also, for $x \neq w$ in $T - v$, $d_T(v, x) = d_T(v, w) + d_T(w, x) = 1 + d_T(w, x)$. But the addressing for v differs from the addressing for x in precisely the $d_T(w, x)$ positions that the addressing for w does, plus the first position since x has 0 in position 1, but v has 1 in position 1. So $d(v, x)$ equals the Hamming distance between v and x for all x. ∎

Theorem 8.5 assures us that binary $(n-1)$-tuples suffice in the addressing of any tree!

It is probably clear to most readers that the constructive proof of Theorem 8.5 indicates an algorithm for a binary addressing of any tree. It works as if we are constructing the tree edge by edge and increasing the length of the k-tuples that label all existing vertices at each stage by 1.

8.3 Tree Codes

Algorithm 8.7 (Binary Addressing for a Nontrivial Tree)
1. Begin with any edge uv of tree T. Label u by 0 and v by 1. Call the current tree T_1, and set $i = 1$.
2. If $T_i = T$, stop. We have a binary addressing.
3. If $T_i \neq T$, let xy be any edge that is incident with, but not in, T_i. One of x or y, say x, is in T_i. Let its current label be $(x_1, x_2, \ldots, x_i)$. Then for each w in T_i with label $(w_1, w_2, \ldots, w_i)$, assign the new label $(0, w_1, w_2, \ldots, w_i)$ and label y by $(1, x_1, x_2, \ldots, x_i)$. Let $i = i + 1$ [increment i, we have an additional vertex labeled]. Return to Step 2.

Let's clarify the labeling procedure of Algorithm 8.7 with an example.

Example 8.7 Find a binary addressing for the tree in Figure 8.18.

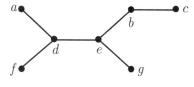

Figure 8.18

Solution
We may begin with any edge, so start with ad and label a by 0 and d by 1. Next, we consider df. We alter a's label to $(0, 0)$ and d's label to $(0, 1)$. Vertex f then gets label $(1, 1)$. Edge de is next. We alter existing labels of a, d, and f to $(0, 0, 0)$, $(0, 0, 1)$, and $(0, 1, 1)$, respectively. Since e is adjacent to d, we label e by $(1, 0, 1)$. This process continues. You should verify that the final binary addressing if we proceed to label b, c, and g (in that order) next is given in Figure 8.19. ◇

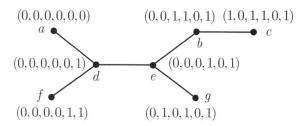

Figure 8.19 A binary addressing of the tree in Figure 8.18.

Exercises 8.3
1. Find the Prüfer code for the paths P_6 and P_7, using a consecutive labeling.

2. Find the Prüfer code for the path P_n, using a consecutive labeling.

3. Find the Prüfer code for the star $K_{1,n}$ if the center is labeled 1.

4. Find the tree whose Prüfer code is $(3, 6, 3, 5, 2, 1)$.

5. Find the tree whose Prüfer code is $(2, 6, 3, 7, 3, 4, 5)$.

6. How many labeled trees are there of order 5?

7. How many nonisomorphic trees are there of order 5?

8. Why does the Prüfer code algorithm terminate with K_2?

9. Why must n be one of the vertices of the K_2 in the previous exercise?

10. Find an addressing using binary $(n-1)$-tuples for the star $K_{1,n}$. Prove that no addressing exists for this graph using binary k-tuples such that $k < n - 1$.

11. Find an addressing using binary $(n-1)$-tuples for the path P_n.

12. Draw six nonisomorphic trees of order 6 and find an addressing for each of them.

13. Find the Prüfer code for the tree of Figure 8.18 if "smallest" vertex means earliest alphabetically.

14. Find a binary addressing for the tree in Figure 8.13.

References for Chapter 8

1. Brelaz, D., New methods to color the vertices of a graph. *Communications of the Association for Computing Machinery* 22 (1979) 251–256.

2. Brooks, R. L., On coloring the nodes of a network. *Cambridge Philosophical Society* 37 (1941) 194–197.

3. Buckley, F., and F. Harary, *Distance in Graphs*, Addison-Wesley, Redwood City, CA (1990).

4. Cayley, A., A theorem on trees. *Quarterly Journal of Mathematics* 23 (1889) 376–378.

5. Chartrand, G., and O. R. Oellermann, *Applied and Algorithmic Graph Theory*, McGraw-Hill, New York (1993).

6. Gould, R., *Graph Theory*, Benjamin Cummings, Menlo Park, CA (1988).

7. Harary, F., and E. M. Palmer, *Graphical Enumeration*. Academic Press, New York (1973).

8. Matula, D. W., G. Marble, and J. D. Isaacson, Graph coloring algorithms. *Graph Theory and Computing*, Academic Press, New York (1972) 109–122.

9. J. W. Moon, Various proofs of Cayley's formula for counting trees. *A Seminar on Graph Theory*, Holt, Rinehardt, and Winston, New York (1967) 70–78.

Additional Readings

10. Brigham, R. C., R. D. Dutton, P. Z. Chinn, and F. Harary, Realization of parity visits in walking a graph, *College Mathematics Journal* 16 (1985) 76–78.

11. Buckley, F., On graphs with unique breadth-first spanning spanning trees, *Graph Theory Notes of New York* 17 (1989) 23–26.

12. Reiter, H., and I. Sonin, The "join the club" interpretation of some graph algorithms, *College Mathematics Journal* 27 (1996) 54–58.

13. Vandeskam, J. M., Path-sequential labelings of cycles. *Discrete Mathematics* 162 (1996) 239–249.

Chapter 9

Planar Graphs

A graph is **planar** if it can be drawn in the plane with no crossing edges. Otherwise it is nonplanar. You may think of the plane on which we are drawing the graph as a sheet of paper having infinite length and width. Planar graphs have received a great deal of attention over the years because of a long-standing problem, the four-color conjecture, that took over one hundred years to prove. Planar graphs remain important today because of their applications. They are crucial in problems such as facility layout in operations research and in the design of printed circuit boards in computer science. We discuss planar graphs and their properties in this chapter.

9.1 Planarity

It is easy to see that all trees are planar. One way to see this is to draw the tree as a rooted tree. Since cycles are obviously planar, so are **unicyclic** graphs, which are connected graphs containing exactly one cycle.

Caution: A graph might be planar even though it *looks* nonplanar. Figure 9.1 depicts two drawings of K_4, the one on the right demonstrating its planarity.

Figure 9.1 Two drawings of the planar graph K_4.

Thus, planarity is a potentiality. If there is *some* way to draw a graph without edges crossing, the graph is planar, no matter how we actually draw it. If a planar graph is small, a bit of creative doodling will soon reveal a planar drawing. Finding such a drawing for a large graph can be quite difficult. Furthermore, if a graph is *nonplanar*, it is not always easy to figure out how to draw it with the fewest number of crossings. This number is called the **crossing number** of the graph.

A plane graph is a planar graph that has achieved its potential. Thus a **plane graph** is a graph that *is* drawn in the plane with no crossing edges. Of course, that is possible only if the graph is planar. In Figure 9.1 both drawings depict the planar graph K_4, but only the second one is a plane graph.

Example 9.1 Redraw the graph G in Figure 9.2 as a plane graph, thereby showing that G is planar.

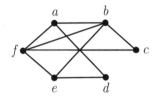

Figure 9.2

Solution
In redrawing the graph, we may move the vertices around and we may bend the edges. We like to think of this process in the following way. Think of the graph as being posted on a bulletin board where the vertices are thumbtacks and edges are rubber bands joining thumbtacks. We may move a thumbtack to a different location on the bulletin board, but any rubber band joined to the thumbtack remains attached. Also, we may stretch a rubber band while leaving the pair of thumbtacks joined by that rubber band in place. Two drawings of G as a plane graph are shown in Figure 9.3. ◇

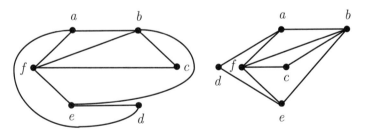

Figure 9.3 Two planar drawings of graph G.

Incidentally, there is a theorem, proven independently by Wagner [10] and Fáry [3], that says if a plane graph can be drawn with curved edges, it can also be drawn with straight edges.

9.1 Planarity

We shall now see that the complete graph K_5 has the distinction of being the nonplanar graph of smallest order. To determine whether or not it is planar, we could start with its spanning subgraph, C_5. So far so good—cycles are planar. The remaining five edges must be drawn either outside the cycle or inside. There is no difference between these two options. The infinite extent of the outer region is of no concern. So, without loss of generality, we should be able to draw three of the five remaining edges inside the cycle. But this cannot be done, since the first edge eliminates one vertex from further consideration and permits only one more edge to be drawn.

By the way, the crossing number of K_5 is 1—that is, K_5 can be drawn with one pair of edges intersecting, but K_5 cannot be drawn as a plane graph.

> **Note 9.1**
> If a graph A is nonplanar and A is a subgraph of B, then B is also nonplanar. That is, the impossibility of drawing A as a plane graph forces the impossibility of drawing B as a plane graph.

It follows from Note 9.1 that K_n is nonplanar when $n \geq 5$, since K_5 is a subgraph of all of those graphs. It is easy to see that K_1, K_2, K_3, and K_4 are all planar. You might be wondering which complete bipartite graphs are planar. For starters, $K_{1,n}$ is a tree so it is planar, no matter how large n is. Interestingly enough, $K_{2,n}$ is also planar for all n, even though $K_{2,n}$ is not a tree when $n > 1$. Figure 9.4 depicts a crafty way to draw $K_{2,n}$, by putting the two white vertices on either side of a column of black vertices. The three dots imply that n can be arbitrarily large.

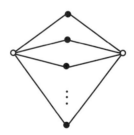

Figure 9.4 $K_{2,n}$ is planar.

On the other hand, we will now state an interesting theorem about $K_{3,n}$.

Theorem 9.1 $K_{3,n}$ is nonplanar for $n \geq 3$.

Proof. By Note 9.1, it suffices to show that $K_{3,3}$ is nonplanar. Start with a plane drawing of the bipartite graph C_6, which is a spanning subgraph of $K_{3,3}$. The three white vertices are labeled a, b, and c, while their black eccentric vertices are labeled A, B, and C, respectively, in Figure 9.5.

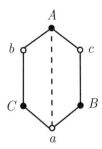

Figure 9.5 C_6 is a spanning cycle of $K_{3,3}$.

Since vertices a and A must be adjacent, we draw aA (see the dashed line) *inside* the cycle, without loss of generality. This forces us to draw the two remaining edges (bB and cC) *outside* the cycle, which cannot be done without their crossing. ∎

This problem is sometimes presented as a tale of the plight of three mutually antagonistic farmers (A, B, and C) who wish to build roads to each of three stores (a, b, and c) such that the nine roads do not intersect. The farmers will then not need to greet each other. Well, too bad for them—it can't be done.

Corollary 9.1a The crossing number of $K_{3,3}$ is 1. ∎

Corollary 9.1b The complete bipartite graph $K_{m,n}$ is nonplanar when m and n are each at least 3.

Proof. The result follows directly from Theorem 9.1 by using Note 9.1. ∎

Euler's Formula

The great Euler noticed that a plane graph divides the plane into regions, or **faces**, as they are usually called. One face, the **outer face**, is of unbounded extent and would be the region of choice for a frisky dog if the edges of the graph were fences. The four faces of the graph of Figure 9.6 are labeled a, b, c, and d. The outer face is d, of course.

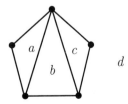

Figure 9.6 A planar graph with four faces.

Euler then discovered a relation between the number of vertices (n), the number of edges (q), and the number of faces (f). Here is his theorem.

9.1 Planarity

Theorem 9.2 (Euler's Formula) Let G be a connected plane graph with n vertices, q edges, and f faces. Then
$$n - q + f = 2. \tag{9.1}$$

Proof. Let us begin the proof with a spanning tree T of G. For a tree, $q = n - 1$ and $f = 1$, so (9.1) holds for T. Now G may be obtained from T by adding edges. Each time we add an edge, we create a new face. So the value of q goes up by 1, as does the value of f. This causes the value of $n - q + f$ to remain unchanged at 2 no matter how many new edges (and resulting faces) are added. ∎

Corollary 9.2a Let G be a plane graph with n vertices, q edges, f faces, and k components. Then $n - q + f = k + 1$. ∎

The proof of Corollary 9.2a is left as an exercise.

The Number of Edges in a Planar Graph

What is the maximum number of edges a planar graph G on n vertices can have? Clearly, when the number of edges is maximum and G is drawn as a plane graph, each face of G must be a triangle (3-cycle). Otherwise, we would be able to add more edges. We can then rephrase the question: How many edges does a plane graph have if every face (even the outer one) is a triangle?

Let's assume that G is such a graph. Let the number of vertices, edges, and faces of G be denoted n, q, and f, respectively. Since each face of G has a boundary of three edges and each edge is on exactly two faces, it follows that $2q = 3f$, or $f = 2q/3$. We now apply Euler's formula to G, substituting $2q/3$ for f, obtaining $n - q + 2q/3 = 2$. By multiplying both sides by 3, this becomes $3n - 3q + 2q = 6$, or $q = 3n - 6$. So for all planar graphs with $n \geq 3$ vertices, we have
$$q \leq 3n - 6. \tag{9.2}$$

This yields a necessary condition for planarity. A graph that violates (9.2) is nonplanar. Observe that for K_5, $n = 5$ and $q = 10$, which proves rather nicely that K_5 is nonplanar. Condition (9.2) is not sufficient, however, to prove that a graph is planar. The nonplanar complete bipartite graph $K_{3,3}$, for example, has $n = 6$ and $q = 9$. Since $9 \leq 12$, condition (9.2) is satisfied but to no avail—$K_{3,3}$ remains nonplanar.

It follows from (9.2) that if $\delta(G) \geq 6$, then G is nonplanar. This is so because $\sum \deg(v) = 2q$ and $\delta(G) \geq 6$ implies that $2q \geq 6n$, which in turn implies that $q \geq 3n > 3n - 6$. So G is nonplanar.

Formally stated, we have the following theorem and corollary.

Theorem 9.3 If G is a planar graph with n vertices ($n \geq 3$) and q edges, then $q \leq 3n - 6$. ∎

Corollary 9.3a If G is a planar graph, then $\delta(G) \leq 5$. ∎

Using a similar technique to that just given, it is not difficult to prove the following result. The proof will be left as an exercise. Recall that the girth of a graph is the length of a shortest cycle.

Theorem 9.4 If G is a planar graph with n vertices, q edges, and finite girth g, then $q \leq \frac{g(n-2)}{g-2}$. ∎

Corollary 9.4a If G is a planar graph with n vertices, q edges, and no triangles, then $q \leq 2n - 4$. ∎

A Characterization of Planar Graphs

Given a graph G, a **subdivision** of G is a graph that can be obtained by inserting any number of vertices of degree 2 along edges of G. This is equivalent to replacing certain edges by paths. Note that not every edge of G need be subdivided. Furthermore, distinct edges can be replaced by paths of different lengths. Graph G has many different subdivisions, depending on how many edges we subdivide and the path lengths that we use. Figure 9.7 exhibits a graph G and two nonisomorphic subdivisions H and J. A graph is considered to be a (trivial) subdivision of itself. So Figure 9.7 actually shows three subdivisions of G.

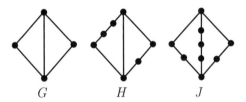

Figure 9.7 Three nonisomorphic subdivisions of G.

The next theorem, due to Kuratowski [7], reduces the problem of determining planarity of a graph to a search for two forbidden subdivision subgraphs! It also shows that K_5 and $K_{3,3}$ are special when it comes to planarity. We shall omit the lengthy proof.

Theorem 9.5 A graph G is planar if and only if it contains no subgraph that is a subdivision of K_5 or $K_{3,3}$. ∎

From our description of subdivisions, we see that not only are K_5 and $K_{3,3}$ not permitted as subgraphs of a nonplanar graph, but also any graph that could be obtained from these two graphs by subdividing edges is a forbidden subgraph. It is important to note that the offending subgraph need *not* be induced. If it appears at all, it would make G nonplanar.

Example 9.2 Prove that the Petersen graph is nonplanar.

9.1 Planarity

Solution

We begin by trying to use Theorem 9.3. The Petersen graph has $q = 15$ edges and $n = 10$ vertices. Testing the inequality of Theorem 9.3, we get $15 \leq 3(10) - 6 = 24$. Unfortunately, the theorem did not help. Satisfying the inequality does not guarantee planarity. But if the Petersen graph had failed to satisfy the inequality, we could have concluded that it was nonplanar. So we move on to another tool, Theorem 9.5.

Since any subdivision of K_5 has vertices of degree 4, and the Petersen graph is 3-regular, we instead search for subdivisions of $K_{3,3}$. A labeled Petersen graph is displayed in Figure 9.8(a). We display a subgraph that is a subdivision of $K_{3,3}$ in Figure 9.8(b) and redraw it in Figure 9.8(c) to clarify that it is indeed a subdivision of $K_{3,3}$. Thus, by Theorem 9.5, the Petersen graph is nonplanar. ◇

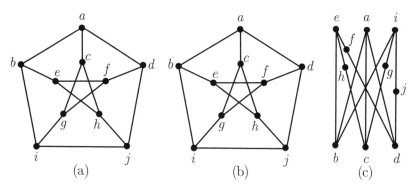

Figure 9.8 A subgraph of the Petersen graph is a subdivision of $K_{3,3}$.

Exercises 9.1

1. Find a drawing of $K_{3,4}$ in which there are only two edge crossings.

2. Verify Euler's formula $n - q + f = 2$ for a tetrahedron, a cube, and an octahedron (see Figure 9.9 for graphs of these **Platonic solids**).

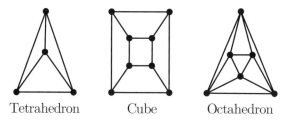

Tetrahedron   Cube   Octahedron

Figure 9.9

3. Verify Euler's formula for the dodecahedron (see Figure 5.11).

4. Verify Euler's formula for the icosahedron (see Figure 9.10).

5. Given a graph G, we obtain **the subdivision graph** of G, denoted $S(G)$, by subdividing each edge of G exactly once. Show that for every graph G, its subdivision graph $S(G)$ is bipartite.

6. If G is a connected unicyclic graph, show that $S(G)$ is equitable.

7. Show that $\bar{C}_n$ is nonplanar when $n \geq 8$, using Corollary 9.3a.

8. Show that although $\bar{C}_7$ satisfies the inequality in Theorem 9.3, $\bar{C}_7$ is nevertheless nonplanar. Explain why this does not contradict the statement of the theorem.

Icosahedron

Figure 9.10

9. Graph G in Figure 9.11 is planar. Redraw G with no crossing edges.

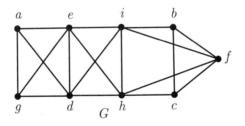

Figure 9.11

10. Consider graph H in Figure 9.12.

   (a) Verify Euler's formula for graph H.

9.1 Planarity

(b) A graph is **maximal planar** if no additional edges can be added to G while keeping the graph planar. By Theorem 9.3, a maximal planar graph has $q = 3n - 6$ edges. Redraw H from Figure 9.12 and then add additional edges to H so that the resulting graph becomes maximal planar.

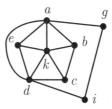

Figure 9.12 A plane graph that is not yet maximal planar.

11. The graph in Figure 9.13 has a subgraph that is a subdivision of $K_{3,3}$. Find that subgraph and draw it.

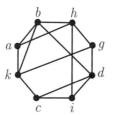

Figure 9.13 A nonplanar graph.

12. For $m \geq 1$ and $n \geq 3$, the **generalized wheel** $W_{m,n}$ is defined as follows. $V(W_{m,n}) = V(C_n) \cup V(\bar{K}_m)$ and $E(W_{m,n})$ consists of the edges of C_n together with an edge from each of the m isolated vertices to every vertex of the cycle C_n. Prove that $W_{m,n}$ is planar if and only if $m < 3$.

13. Draw a connected plane graph that has seven vertices and five faces.

14. Prove that there does not exist a connected plane graph that has seven faces and all vertices of degree 4 (use n vertices, where n is an arbitrary positive integer). Note that the integer is arbitrary; this does *not* allow you to choose it. *Hint*: Use Euler's formula.

15. Suppose that a graph G has 10 vertices of degree 4, 8 vertices of degree 5, and all other vertices have degree 7. Determine the maximum number of vertices of degree seven that G could have and still remain planar.

16. Use induction on the number of components k to prove Corollary 9.2a.

17. Use Euler's Formula to prove Theorem 9.4.

18. Use Theorem 9.4 to prove that the Petersen graph is nonplanar.

9.2 Planar Graphs, Graph Coloring, and Embedding

Two centuries ago, mathematicians wondered how few colors were needed to color planar graphs. This question was also of interest to mapmakers for the following reason. A map featuring countries with shared boundaries is easier to read if "adjacent" countries—that is, countries that share a common segment of their boundaries, are colored differently. However, since color printing was expensive, it was desirable to minimize the number of colors used in coloring the maps.

Graphs and Maps

There is a strong connection between maps and planar graphs. First, let's be specific about what we mean by a map. A **map** consists of a set of contiguous countries (no isolated nations) where each country is all in one piece (unlike Russia, for example, which has the Kaliningrad enclave on the Baltic Sea, which is bounded by Poland and Lithuania, but separated from the rest of Russia). Draw the capitol of each country and draw a road between two capitals whenever their countries are adjacent. This road can be drawn to lie entirely within the two countries. The resulting graph consisting of capitals and the roads connecting them is a planar graph! If this graph is k-colorable for some number k, then so is the map. A glance at Figure 9.14 might be useful at this point. It depicts a map of the Totalitarian Alliance and its five member nations Maoland, Stalinland, Leninland, Polpotland, and Caligulaland, with their capitals labeled M, S, L, P, and C, respectively. The roads are drawn as dotted lines. Thus the graph related to the map is $W_{1,4}$ with one spoke of the wheel removed.

This transformation from a map to the corresponding graph is closely related to the dual of a plane graph, which we will discuss in Section 9.3. The transformation converts the problem from one of coloring regions on a map to one of coloring the vertices of a plane graph. The chromatic number $\chi(G)$ is a graph invariant, so its value is unchanged for any of the isomorphic drawings of G—that is, whether or not the planar graph G is actually drawn as a plane graph.

9.2 Planarity, Coloring, and Embedding

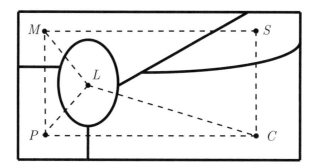

Figure 9.14 A map of the dreaded Totalitarian Alliance.

The four-color problem began in the 1850s when Francis Guthrie, a student at University College, London, found that he could color the counties of England using just four colors so that adjacent counties could be distinguished from one another. He then tried to prove that the same could be done for any map. His brother Frederick was a student of Augustus De Morgan, famous for De Morgan's laws of sets and logic. Although De Morgan publicized the problem, there was no real progress on it until the late 1870s. At that time, Arthur Cayley expressed interest in the problem, which was apparently enough of an endorsement to get many graph theorists hooked on the problem.

Although the four-color problem is so easy to describe, it took over one hundred years to solve. When the proof was finally published in 1977, it created a fair amount of controversy since it was the first proof that made extensive use of a computer. The amazing feat of Appel and Haken (with Koch) [1] is the following.

Theorem 9.6 *Every planar graph is 4-colorable.* ▌

The proof extended over 100 pages with dozens of additional pages of microfiche containing figures. When Appel and Haken and Koch [2] filled in further details, the proof extended to over 700 pages.

The first major step in determining an upper bound for the chromatic number of planar graphs was the following theorem of Heawood [6].

Theorem 9.7 *All planar graphs are 5-colorable.*

Proof. The proof uses induction on n, the order of the graph. When $n \leq 5$, it is obvious that five or fewer colors suffice. Assume, then, that all planar graphs with order $n = k$ or less are 5-colorable and let G be a planar graph on $n = k+1$ vertices. By Corollary 9.3a, a planar graph must have a vertex of degree at most 5. So let $v \in V(G)$ be such a vertex. By the inductive hypothesis, $G-v$ is 5-colorable since it has k vertices. Let's see what happens when we put v back into $G-v$, which has a coloring using the five colors c_1, c_2, c_3, c_4, and c_5.

If $\deg_G(v) \leq 4$, then v has at most four neighbors. This means that some color, c_i, is not used by any of the neighbors of v. Then use color c_i for v, thereby coloring the vertices of G with five colors, and we are done.

So assume that $\deg_G(v) = 5$. Now if the five neighbors of v do not use one of the colors, we use this omitted color on v, and we have a coloring of the vertices of G with five colors, and we are done once again.

This brings us to the final situation in which $\deg_G(v) = 5$, and the five neighbors of v use all five colors. This is depicted in Figure 9.15, in which case, without loss of generality, color c_i is given to vertex x_i.

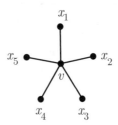

Figure 9.15 Vertex v is surrounded! Its five neighbors use all five colors.

In graph G, for each pair of numbers i and j between 1 and 5, let $G(i,j)$ be the induced subgraph on all vertices colored with either of the colors c_i and c_j. These $\binom{5}{2} = 10$ induced subgraphs may or may not be connected. Some of them may consist of several components. Consider $G(1,3)$ in particular. Obviously, $x_1, x_3 \in G(1,3)$. (The notation $a, b \in A$ means that $a \in A$ and $b \in A$.) We must now resort to two cases.

Case 1: Vertices x_1 and x_3 belong to different components of $G(1,3)$.

In the component containing x_1, switch the colors c_1 and c_3. It is not hard to see that we will still have a valid coloring of G—that is, no adjacent vertices will have the same color. Vertex x_1 will now have the color c_3 while x_3 still has the color c_3, thereby freeing up the color c_1 for v. No neighbor of v will have the color c_1. Then color v using c_1 and we have a coloring of G with five colors!

Case 2: Vertices x_1 and x_3 belong to the same component of $G(1,3)$.

Since G is planar, the subgraph $G(1,3) \cup \{vx_1, vx_3\}$ has a cycle that "surrounds" vertex x_2, effectively cutting it off from x_4 in $G(2,4)$. Then x_2 and x_4 belong to different components of $G(2,4)$, enabling us to switch the colors c_2 and c_4 in the component of $G(2,4)$ containing x_2. Now proceed in a manner similar to what we did in Case 1. Color v with the color c_2, since it is not used by any of its neighbors after we pull the "switch." This yields a valid coloring of the vertices of G using five colors and we are done. ∎

This proof was given by Heawood in 1890. In 1977, when Appel and Haken's computer-aided proof that planar graphs are 4-colorable was pub-

Embeddings

When a graph is drawn with no crossing edges on a particular surface, we say that the graph is **embedded** in the surface. Thus, a plane graph is an embedding of a planar graph in the plane. Although a nonplanar graph G cannot be embedded in the plane, we can embed G in more complex surfaces. For example, we saw that K_5 is nonplanar and has crossing number 1. If we use a torus (shaped like a donut, bagel, or inner tube), we can achieve an embedding. (If you leave a bagel out for a couple of days, it becomes so hard that you can actually draw on it.) An embedding of K_5 on a torus is shown in Figure 9.16. The dashed edge ad goes behind the torus (through the hole), while edge ce goes around the hole.

In the same paper containing the 5-color theorem, Heawood proved that any graph that can be drawn on a torus without any pair of edges intersecting is 7-colorable. A graph that can be embedded on a torus is called **toroidal**. It is interesting that the k-colorability question was settled for the torus before it was for the plane.

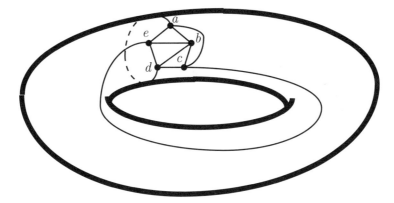

Figure 9.16 K_5 embedded on a torus.

A torus is often referred to as a sphere with one hole. Imagine flattening the sphere a bit and then punching a hole in it as a baker might when it's time to bake the donuts. Why stop with one hole? A doughnut with k holes is said to have **genus** k. A torus with four holes is shown in Figure 9.17.

Heawood [6] had thought he had proved the following theorem about the maximum possible chromatic number of a graph G that can be embedded on a surface of genus n. He had only established an upper bound in his paper in 1890. It took until 1968 for equality to be proven, which was achieved by Ringel and Youngs [9].

Theorem 9.8 The maximum chromatic number of all graphs G embedded on a surface of genus n is given by

$$\chi(G) = \left\lfloor \frac{7 + \sqrt{1 + 48n}}{2} \right\rfloor \qquad (9.3)$$

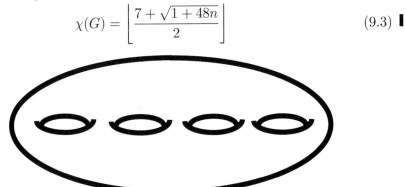

Figure 9.17 A torus with four holes—its genus is 4.

Note that if Theorem 9.8 had also included the case $n = 0$, it would have given the four-color theorem for the plane. This is because a graph can be embedded in the plane if and only if it can be embedded on a sphere.

When $n = 1$, Theorem 9.8 says that toroidal graphs are 7-colorable.

The **genus of a graph** G is the minimum genus of surfaces on which G can be embedded. Theorem 9.8 follows (see Harary [5, p. 136]) from Ringel and Youngs' [9] result that the genus of K_n is $\left\lceil \frac{(n-3)(n-4)}{12} \right\rceil$. Ringel [8] wrote a whole book devoted to the map coloring theorem.

Exercises 9.2

1. Show that a graph can be embedded in the plane if and only if it can be embedded on a sphere.

Theorem 9.8 implies that a toroidal graph is 7-colorable. However, a toroidal graph may have chromatic number much smaller than seven. Exercises 2–5 illustrate this idea.

2. Draw the 4-cube Q_4.

3. Find a subgraph of Q_4 that is a subdivision of $K_{3,3}$, thereby establishing by Kuratowski's theorem that Q_4 is nonplanar.

4. Display an embedding of Q_4 on the torus.

9.3 Duals and Planar Graph Applications

5. Determine $\chi(Q_4)$, thereby showing that a toroidal graph can have chromatic number much smaller than 7.

Ringel [8] showed that for $K_{m,n}$ the genus is $\left\lceil \frac{(m-2)(n-2)}{4} \right\rceil$ for $m, n \geq 2$. Exercises 6–10 relate to that result.

6. Draw an embedding of $K_{3,3}$ on the torus. How many regions (faces) are there in your embedding?

7. Draw an embedding of $K_{3,4}$ on the torus.

8. What is the smallest n for which $K_{3,n}$ is *not* toroidal?

9. What is the smallest n for which $K_{4,n}$ is *not* toroidal?

10. Use Ringel's result to show that there are bipartite graphs of arbitrarily large genus.

11. In a **maximal plane graph** G, every face is a triangle. Thus no additional edges can be added to G while keeping G planar. Prove that every planar graph is 4-colorable if and only if every maximal plane graph is 4-colorable.

12. Prove that the states within a map of the United States are not 3-colorable. *Hint*: Look for an odd wheel $W_{1,2n+1}$; then explain.

9.3 Graph Duals and a Planar Graph Application

In Section 9.2, we saw how to go from a map to a related dual graph. We now examine the concept of duality for plane graphs in general, not just those corresponding to maps. We then examine a facility layout problem that involves planarity.

Duality

A planar graph G divides the plane into regions called **faces**. The **degree of a face** is the number of edges incident with it. Notice that this is similar to the definition of the degree of a vertex. The faces f_1, f_2, f_3, and (the outer face) f_4 in the planar graph of Figure 9.18 have degrees 3, 4, 4, and 5, respectively. Note that two faces can share more than one edge. The faces f_2 and f_3, in this example, share two edges, while the outer face f_4 shares two edges with each of f_1 and f_3.

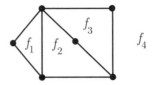

Figure 9.18 A planar graph G with faces of varying degrees.

Given a plane graph G with faces $f_1, f_2, \ldots, f_k$, the **dual** of G, denoted G^*, is defined as follows. Graph G^* has vertex set $\{f_1, f_2, \ldots, f_k\}$; that is, the vertices of G^* are the faces of G. The edge set of G^* is obtained by placing an edge between vertices f_i and f_j for every edge shared by faces f_i and f_j in G. Note that the dual of the graph of Figure 9.18 is a multigraph. It is drawn in Figure 9.19.

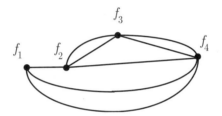

Figure 9.19 G^*, which is the dual of G.

The dual of a plane graph is a plane multigraph. Note that there is a one-to-one correspondence between the edges of a plane graph and the edges of its dual. Also, the degree of each vertex $f_i \in V(G^*)$ equals the degree of its corresponding face f_i in G.

The graph of G^* in Figure 9.19 has four vertices since G has four faces. You might have noticed that G^* has six faces (regions between parallel edges are also considered faces) while G has six vertices. The next theorem says that this is not a coincidence.

Theorem 9.9 Let G be a planar graph of order n. Then the dual graph of G has n faces.

Proof. Let f and g denote the number of faces of G and its dual G^*, respectively. Then G^* has order f. From the definition of the dual of a graph, $E(G)$ and $E(G^*)$ have the same size, say q. Then, using Euler's formula on G and G^* (see Theorem 9.2), we obtain $n - q + f = 2$ for G and $f - q + g = 2$. Taking the difference of these two equations yields $n - g = 0$, or $g = n$. ∎

Second Proof. If a drawing of the dual G^* of G is superimposed on a drawing of G so that corresponding edges of G and G^* intersect, then each vertex

9.3 Duals and Planar Graph Applications

of G is surrounded by a cycle of edges of G^* corresponding to their incident edges in G. This implies that to each vertex v of G, there corresponds a unique face of G^*, namely, the face inside of which v is located. ∎

The method of this second proof enables us to see that the degree of the vertex $f_i \in V(G)$ equals the degree of its corresponding face f_i in G^*, as noted previously. Furthermore, the very same diagram that superimposes a drawing of G^* on a drawing of G implies that the dual of G^* is G! This justifies the use of the word "dual." We then have the following theorem.

Theorem 9.10 For any plane graph G, we have $(G^*)^* \cong G$. ∎

The planar graph G of Figure 9.20 has a bridge, xy. Unlike the other edges of G, each of which belongs to a cycle, edge xy is not shared by two faces. Then in the dual graph G^*, the vertex corresponding to the outer face, f_3, will require an edge not shared with any other vertex! This edge in G^* is incident with the vertex f_3 twice and is called a **loop**. The resulting structure, called a **pseudograph**, is depicted in Figure 9.21.

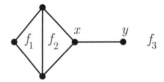

Figure 9.20 A planar graph G with a bridge.

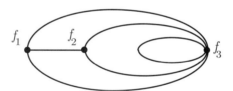

Figure 9.21 G^* has a loop, so it is a pseudograph.

Example 9.3 Given the plane graph G in Figure 9.22, draw its dual G^*. Then list the number of vertices, edges, and faces of both G and G^*.

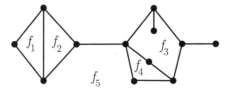

Figure 9.22

Solution

First note that G^* will have five vertices, which correspond to the five faces of G. Furthermore, there will be two loops at vertex f_5 of G^* and one at vertex f_3. The dual G^* is displayed in Figure 9.23. Graph G has 12 vertices, 15 edges, and 5 faces. Graph G^* has 5 vertices, 15 edges, and 12 faces. ◊

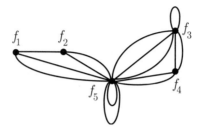

Figure 9.23

Facility Layout

When designing the layout of facilities such as libraries, hospitals, shopping malls, or department stores, we may wish to make certain regions adjacent to certain others. For example, in a shopping mall, it is usual to locate restaurants near the theater. To make this problem specific, suppose there are eight departments, labeled a through h, planned for within a new department store, and the desired adjacencies are as described in Table 9.1.

Department	Desired Adjacencies
a	b, f, h
b	a, c, f
c	b, d, g, h
d	c, e, f
e	d, f, h
f	a, b, d, e, g
g	c, f, h
h	a, c, e, g

Table 9.1 Desired adjacencies for a department store

Now draw the graph with vertices corresponding to the departments and each edge corresponding to an adjacency in Table 9.1. Graph G is displayed in Figure 9.24. If the resulting graph is planar, then it is possible to satisfy all of the desired adjacencies. The difficulty lies in rearranging the crossing edges within the figure.

9.3 Duals and Planar Graph Applications

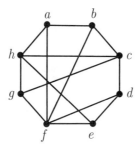

Figure 9.24 Graph G for desired adjacencies.

Graph G is hamiltonian. This will not always be the case, but the more adjacency conditions that are desired, the more likely it will be that the resulting graph is hamiltonian. Suppose we label the edges within the hamiltonian cycle C arbitrarily and form the **crossing graph** H. Each internal edge joining nonadjacent vertices on the cycle C of G becomes a vertex in the new graph H, and two vertices of H are adjacent if the corresponding edges cross one another in G. Figure 9.25 displays graph G with the internal edges labeled and then displays the crossing graph H.

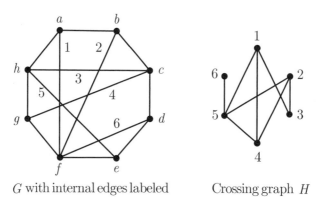

G with internal edges labeled   Crossing graph H

Figure 9.25

It turns out that graph G is planar if and only if its crossing graph H has $\chi(H) = 2$. To see why, remember that the vertices of H represent the internal edges of G, so a 2-coloring of H corresponds to 2-coloring of the internal edges such that edges of the same color do not cross in G. This means that we could move the edges of one color to the outside of the cycle C and keep those of the second color inside and achieve a plane embedding of G.

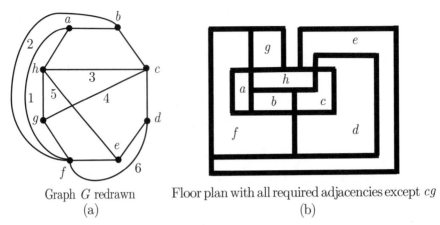

Graph G redrawn
(a)

Floor plan with all required adjacencies except cg
(b)

Figure 9.26

For our example, however, a 2-coloring of H is not possible (H contains a triangle). A best possible coloring would be, for example, coloring 1, 2, and 6 red, 3 and 5 blue, and 4 green. This means that G is not planar, so not all of the desired adjacency conditions can be achieved. In Figure 9.26(a), we display a drawing of G with the red edges 1, 2, and 6 outside, blue edges 3 and 5 inside, and edge 4, which necessarily must then cross one of the others. Then in Figure 9.26(b), we display one possible floor plan based on the graph in Figure 9.26(a). *Note*: All desired adjacencies described in Table 9.1 are satisfied by this layout except the adjacency between departments c and g.

We can adjust the arrangement of the layout in various ways to achieve other desired conditions, such as the relative size of the departments. The layout given is simply a starting point for the designer. Weighted graphs could be used to help model the additional desired conditions, such as relative size. We should mention that there is a more general area of facility layout studied in operations research, where rather than specifying adjacency between departments, we specify distance between them. Typically, we want to minimize the sum of all distances in the final layout while at the same time trying to satisfy as many of the distance constraints as possible.

Exercises 9.3

1. Show that $K_4* = K_4$; that is, that K_4 is self-dual.
2. Draw the dual of the hypercube Q_3.
3. Show that the dual of Q_3 is an octahedron.
4. Show that the 2-mesh $M(a-1, b-1)$ is an induced subgraph of the dual of $M(a,b)$.

9.3 Duals and Planar Graph Applications

5. Find the dual of the wheel $W_{1,n}$.

6. Verify that $(G^*)^* \cong G$ for the plane graph in Figure 9.12.

7. Show that the dual of the icosahedron (see Figure 9.10) is the dodecahedron (see Figure 5.11).

8. Explain why the result of Exercise 7 automatically implies that the dual of the dodecahedron is the icosahedron.

9. Consider the graphs of G and H in Figure 9.27.

    (a) Show that G and H are isomorphic.
    (b) Draw G^* and H^*.
    (c) Show that G^* and H^* are *not* isomorphic.

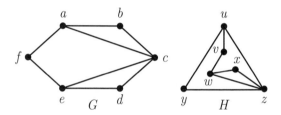

Figure 9.27

10. Prove that if a nontrivial plane graph G is eulerian, then its dual is bipartite.

11. An interesting problem of avoiding edge crossings is Dudeney's puzzle. Consider the subgraph H displayed in Figure 9.28 of a graph G, where G has the additional edges Aa, Bb, Cc, and Dd. Do not move any of the vertices or edges of H, and draw in the missing edges so that they are completely contained within the cycle A, B, C, D, b, A and do not cross one another.

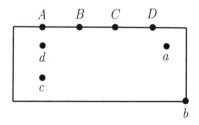

Figure 9.28

12. If $G \cong G^*$, then G is **self-dual**. Show that if G is self-dual, then $2n = q + 2$. Then find a plane graph H for which $2n = q + 2$ and H is not self-dual.

13. Consider a facility layout problem for a small shopping mall with nine stores. The desired adjacencies are displayed in Table 9.2.

	Desired Adjacencies
a	b, d, i
b	a, c, e
c	b, d, e, f, h
d	a, c, g, h
e	b, c, f, i
f	c, e, g
g	d, f, h, i
h	c, d, g, i
i	a, e, g, h

Table 9.2

(a) Draw the related graph G.
(b) Graph G is hamiltonian. Find a hamiltonian cycle C for G and redraw G so that all edges not on C are inside the cycle.
(c) Label the internal edges and draw the crossing graph H.
(d) Find an optimal coloring of the vertices of H and use it to find a redrawing of G with as few crossings edges as possible.
(e) Draw an optional arrangement of the stores of the shopping mall.

14. Consider a facility layout problem for a small shopping mall with nine stores. The desired adjacencies are displayed in Table 9.3.

	Desired Adjacencies
a	b, d, e
b	a, c, e
c	b, e, f, h
d	a, g, h
e	a, b, c, f, i
f	c, e, g
g	d, f, h
h	c, d, g, i
i	e, h

Table 9.3

(a) Draw the related graph G.

(b) Graph G is hamiltonian. Find a hamiltonian cycle C for G and redraw G so that all edges not on C are inside the cycle.

(c) Label the internal edges and draw the crossing graph H.

(d) Find an optimal coloring of the vertices of H and use it to find a redrawing of G with as few crossings edges as possible.

(e) Draw an optional arrangement of the stores of the shopping mall.

References for Chapter 9

1. Appel, K., and W. Haken, Every planar map is four colorable. *Illinois J. Math* 21 (1977) 429–567.

2. Appel, K., and W. Haken (with J. Koch), Every planar map is four colorable. *Contemporary Mathematics*, 98, *AMS*, Providence, RI (1989).

3. Fáry, I., On the straight line representation of planar graphs. *Acta. Sci. Math.* 11 (1948) 229–233.

4. Heawood, P. J., Map coloring theorem. *Quarterly J. Math.* 24 (1890) 332–229.

5. Fritsch, R., and G. Fritsch, *The Four Color Theorem*. Springer, New York (1998).

6. Harary, F., *Graph Theory*. Addison-Wesley, Reading, MA (1969).

7. Kuratowski, K., Sur le problèmedes courbes gauches en topologie. *Fund. Math.* 15 (1930) 271–283.

8. Ringel, G., *Map Color Theorem*. Springer-Verlag, Berlin (1974).

9. Ringel, G., and J. W. T. Youngs, Solution of the Heawood map coloring problem. *Proc. Nat Aca. Sci. U.S.A.* 60 (1968) 438–445.

10. Wagner, K., Bemerkung zum Vierfarbenproblem. *Jber. Deutch. Math. Verein.* 46 (1936) 21–22.

Additional Readings

11. Cipia, B. A., Advances in map coloring: complexity and simplicity. *SIAM News* 29:10 (Dec. 1996) 20.

12. Frederickson, G.N., Geometric dissections now swing and twist. *Mathematical Intelligencer* 23:3 (Summer 2001) 9–20.

13. Johnson, S., and H. Walser, Pop-up polyhedra. *Mathematical Gazette* 81 (1997) 364–380.

14. Polster, B., A. E. Schroth, and H. Van Maldeghem, Generalized flatland. *Mathematical Intelligencer* 23:4 (Fall 2001) 33–47.

15. Stewart, I., Mathematical recreations: empires and electronics. *Scientific American* 277:3 (September 1997) 92–94.

16. Samelson, H., In defense of Euler. *L'Enseignment Mathématique* 42: (1996) 377–382.

17. Temple, D. W., Colored polygon triangulations. *College Mathematics Journal* 29 (1998) 43–47.

Chapter 10

Digraphs and Networks

We have seen several instances where a structure other than a simple graph is needed to model real-life problems. For example, in some problems a multigraph or a weighted graph is more appropriate. In this chapter, we consider some additional structures. Digraphs are similar to graphs except that there are directions on the edges. Digraphs are used to model problems where the direction of flow of some quantity (information, traffic, liquid, electrons, and so on) is of importance. When there are limits placed on how much of that quantity can flow through a particular directed edge, we obtain a network. A special type of digraph having no directed cycles, called an activity digraph, has weights on the directed edges indicating the duration of a given activity. These digraphs are used to aid in scheduling individual activities that compose a complex project. In this chapter, we will explore various types of digraphs and their properties, will study network flows and present an algorithm to maximize total flow, and will discuss activity digraphs and their uses.

10.1 Directed Graphs

A **directed graph**, abbreviated **digraph**, D, consists of a finite nonempty set of vertices $V(D)$ together with a set of **ordered** pairs of distinct vertices called **arcs**. An arc xy goes from x to y and is represented by an arrow from x toward y, as shown in Figure 10.1.

Figure 10.1 A digraph with two vertices and arc xy.

When speaking about adjacencies in a digraph, we must keep track of the direction on an arc. Thus in the digraph of Figure 10.1, x is **adjacent** to y while y is **adjacent from** x. Note that y is *not* adjacent *to* x. We may think of an arc as a one-way street. When arcs uv and vu are both present, they are **symmetric arcs**. Given a digraph, the graph with each arc replaced by an edge is called the **underlying (multi-) graph**. For the digraph of Figure 10.1, the underlying graph is K_2. The **outdegree od(v)** of vertex v is the number of vertices that v is adjacent to and the **indegree id(v)** is the number of vertices that v is adjacent from. Equivalently, od(v) is the number of arcs that point away from v and id(v) is the number of arcs that point toward v.

Example 10.1 Draw the digraph D with $V(D) = \{a,b,c,d,e\}$ and $E(D) = \{ab, ac, ae, bc, bd, cb, ce, ed\}$. Then state the indegree and outdegree of each vertex. Which arcs are symmetric?

Solution
The digraph D is displayed in Figure 10.2. We have id(a) = 0, od(a) = 3, id(b) = 2, od(b) = 2, id(c) = 2, od(c) = 2, id(d) = 2, od(d) = 0, id(e) = 2, od(e) = 1. The arcs bc and cb are symmetric. ◇

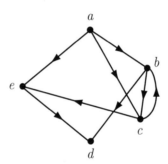

Figure 10.2

In Figure 10.2 vertex a has indegree 0. Such a vertex is referred to as a **transmitter** because in a telecommunications context, it sends information but does not receive any. Similarly, a vertex with outdegree zero is called a **receiver**. In the digraph of Figure 10.2, vertex d is a receiver. It receives information but does not send any.

For any digraph D, we have the following simple result that is analogous to Theorem 2.1 for graphs.

Theorem 10.1 If D is a digraph with vertex set $V(D) = \{v_1, v_2, \ldots, v_n\}$ and having q arcs, then $\sum_{i=1}^{n} \text{id}(v_i) = \sum_{i=1}^{n} \text{od}(v_i) = q$.

Proof. Each arc contributes precisely one to the indegree of the vertex it points toward and one to the outdegree of the vertex it points from. ∎

Notice that the digraph of Figure 10.3 has a disturbing property. It looks connected, right? It's easy to find a directed path from y to x. Try, however,

10.1 Directed Graphs

to find a directed path from x to y. There is none. Nevertheless, since the underlying graph is connected, we call the digraph **weakly connected**. If for every pair of vertices u, v in a digraph D there is either a directed u–v path or a directed v–u path, then D is called **unilateral**. On the other hand, a digraph with the property that any pair of vertices u and v have both a directed u–v path and a directed v–u path is called **strongly connected**, or simply, **strong**.

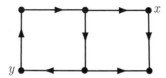

Figure 10.3 A weakly connected digraph.

Example 10.2 Draw a digraph on four vertices that is
a. weakly connected but not unilateral
b. unilateral but not strong
c. strong

Solution
There are various possibilities, but examples of the desired digraphs are shown in Figure 10.4. In (a), there is no directed a–d path nor directed d–a path. In (b), there is a directed path from b, c, and d to each other vertex, but there is no a–d path, for example. In (c), we can travel between any two vertices in either direction. Note that although (c) is strong, it has no directed hamiltonian cycle. ◇

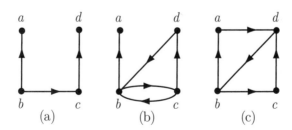

Figure 10.4

Given a connected graph G, we describe the act of assigning a direction to each edge as **orienting** the graph. If the resulting graph is strongly connected, we call the orientation **strong**. If we reverse the direction of the lower right horizontal edge of the digraph of Figure 10.3, we obtain a strong orientation. The problem of obtaining a strong orientation for a connected graph is a very practical one. It is useful in streamlining traffic flow in a busy city by making all streets one way.

Achieving a Strong Orientation

It should be clear that a graph with a bridge cannot be strongly oriented. Since the deletion of a bridge of a connected graph G disconnects it into two components, say H and J, as soon as we orient the bridge, say from H to J, there is no way to travel from J to H. On the other hand, if G has no bridges, it is always possible to obtain a strong orientation, as indicated in the following theorem due to Robbins [13].

Theorem 10.2 *A connected graph G has a strong orientation if and only if it contains no bridges; that is, every edge is contained in some cycle.*

To show how to obtain a strong orientation in a connected bridgeless graph, we need to employ the **depth-first search** (DFS) algorithm (see Algorithm 8.2) for finding a spanning tree. The following simple algorithm can be used to obtain a strong orientation.

Algorithm 10.1 (Orienting a Connected Bridgeless Graph)
1. Obtain a DFS tree T using Algorithm 8.2.
2. Orient each edge of T toward the vertex with higher number.
3. Orient each of the remaining edges of G—that is, those not in T—toward the vertex with lower number.

You will have to spend a bit of time convincing yourself that this works. Part of the work is obvious. The root can reach every vertex. Now show that every vertex can reach the root. Bear in mind that G has no bridges.

Figure 10.5 depicts a strong orientation for $M(3,3)$ obtained by Algorithm 10.1. Note that when the DFS algorithm brings us to vertex 5 in Figure 10.5, there is no new unlabeled neighbor and we must backtrack, eventually to vertex 2 before we can move forward again. The new neighbor of vertex 2 is then numbered 6, as Algorithm 8.2 indicates. From vertex 6, we continue to vertices 7 and 8 and we are done.

Acyclic Digraphs and Partial Orders

A digraph that has no directed cycles is called **acyclic**. Of course, if the underlying graph of a digraph D is a forest, then D is acyclic. That condition is not necessary, however, as the two digraphs of Figure 10.6 indicate.

10.1 Directed Graphs

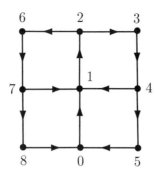

Figure 10.5 A strong orientation for $M(3,3)$.

A **partial order** $\sim$ on a set is a relation that is **reflexive** ($a \sim a$, "a is related to a"), **antisymmetric** ($a \sim b$ and $b \sim a$ implies $a = b$), and **transitive** ($a \sim b$ and $b \sim c$ implies $a \sim c$). Two common partial orders you are familiar with are $\leq$ on a set of numbers and $\subseteq$ on a collection of sets. A set together with an associated partial order is called a **partially ordered set** (**poset**, for short). A poset is often modeled by using an acyclic digraph, even though, technically, there should be a loop at each vertex because of reflexivity. Thus, the loops are "understood."

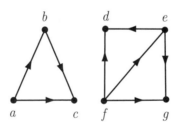

Figure 10.6 Two acyclic digraphs.

Example 10.3 Consider the set $A = \{2, 3, 5, 6, 10, 12, 15, 30\}$ with the relation $|$ (read "divides").
a. Show that the relation "$|$" is a partial order.
b. Draw the poset $(A, |)$ as an acyclic digraph (omit the loops).
Solution
a. We must show that "$|$" is a reflexive, antisymmetric and transitive. First $a \,|\, a$—that is, for any $a \in A$, a divides evenly into itself. Thus "$|$" is reflexive. Next, if $a \,|\, b$—that is, a divides evenly into b—then a cannot exceed b, so $a \leq b$. Similarly, if $b \,|\, a$, then $b \leq a$, so $a = b$. Hence, "$|$" is antisymmetric. Finally, if $a \,|\, b$, then $b = ka$, where $k \in Z$. If $b \,|\, c$, then $c = tb$, where $t \in Z$. So $c = tb = t(ka) = (tk)a$. Since $t \in Z$ and

$k \in Z$, $tk \in Z$. Thus $c = (tk)a$, where $tk \in Z$. But this means that $a \mid c$. Therefore, we have shown that $a \mid b$ and $b \mid c$ implies that $a \mid c$—that is, "$\mid$" is transitive. Hence "$\mid$" is a partial order relation.

b. The acyclic digraph (with loops omitted) representing the poset $(A, \mid)$ is depicted in Figure 10.7. It is easy to see that the digraph is acyclic because all arcs point upward. ◇

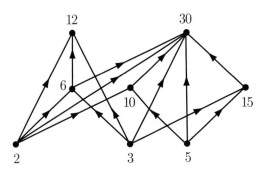

Figure 10.7

A very simple but useful theorem about acyclic digraphs is the following result.

Theorem 10.3 Every acyclic digraph has at least one vertex of outdegree zero and at least one vertex of indegree zero.

Proof. Consider the last vertex v in any longest path in the digraph. Vertex v has no vertices adjacent from it or there would be a cycle. Thus v has outdegree zero. Now consider the first vertex u of a longest path P in the digraph. There can be no arc wu pointing toward u, or we could either extend P to a longer path, a contradiction, or we would get a cycle, also a contradiction. Thus u has indegree zero. ∎

Example 10.4 Identify all acyclic digraphs that have appeared in figures in this section. In each such digraph, find the vertices of outdegree zero and indegree zero guaranteed by Theorem 10.3.

Solution
The digraph in Figure 10.1 is acyclic; y has outdegree zero and x has indegree zero. The digraph in Figure 10.4(a) is acyclic; a and d have outdegree zero and b has indegree zero. The digraphs in Figure 10.6 are acyclic; in the first, c has outdegree zero and a has indegree zero, whereas in the second, d and g have outdegree zero and f has indegree zero. Finally, the digraph of Figure 10.7 is acyclic; vertices 12 and 30 have outdegree zero and vertices 2, 3, and 5 have indegree zero. ◇

We should mention that researchers who work on posets usually represent them in an even simpler manner than as in Figure 10.7. Note that we had omitted the loops. Some researchers additionally omit all the transitive arcs, such as $(2, 12)$ or $(2, 30)$, treating them as implied [see Figure 10.8(a)]. Even more such researchers use a layered graph, called a Hasse diagram to represent the poset. The **Hasse diagram** is formed as follows. In the acyclic digraph D representing the poset, determine the set X_1 of vertices of indegree zero and line them up horizontally at level 1. Next, consider the acyclic digraph $D_1 = D - X_1$. In this digraph, determine the set X_2 of vertices of indegree zero. Line these vertices up horizontally at level 2 above level 1. In general, consider the acyclic digraph $D_i = D_{i-1} - X_i$ and locate the set X_{i+1} of vertices of indegree zero. Place those vertices horizontally at level $i + 1$ above the vertices at level i. Finally, for each arc joining a vertex u at level i to a vertex v at a higher level in the simplified (transitive arcs removed) digraph, we put in edge uv in the Hasse diagram [see Figure 10.8(b)].

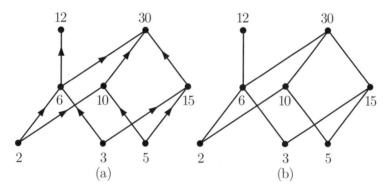

Figure 10.8 The simplified acyclic digraph and Hasse diagram for a poset.

The Hasse diagram may seem somewhat artificial to you because our digraphs were already drawn in a layout format. However, the layout format is not always given. The procedure for generating the Hasse diagram tells how to create the layout. Note that an edge may skip between layers in a Hasse diagram. For example, if we consider the poset $B = \{5, 8, 10, 40\}$ ordered by divisibility, then 5 and 8 would be at level 1, 10 at level 2, and 40 at level 3. There would be edges $(5, 10)$, $(10, 40)$, and $(8, 40)$. The last edge joins a vertex at level 1 to a vertex at level 3. To learn more about posets, see Grimaldi [8] or Kolman, Busby, and Ross [9].

Tournaments

When sports teams play to determine who is the best team, there are two major ways the games are organized. In an **elimination tournament**, once a team loses a game, it is out of the competition. In a **round-robin tournament**, each team plays each other team exactly once. In each game, there is a winner and a loser; no ties are allowed. Although both types of competitions can be modeled using digraphs, we shall focus our attention on round-robin tournaments.

A **tournament** is a directed graph, where for each pair of vertices, u and v, precisely one of the arcs uv or vu is present. Thus, a tournament on n vertices is an orientation of K_n. When there are no labels on the vertices, there is only one tournament on one vertex and one on two vertices. There are two unlabeled tournaments on three vertices, a cyclic one and a acyclic one. There are four nonisomorphic unlabeled tournaments on four vertices. They are displayed in Figure 10.9. An arc from u to v indicates that vertex u defeated vertex v.

Figure 10.9 The tournaments of order 4.

After a round-robin tournament is over, we would like to be able to rank the players from best to worst. Unfortunately, this is not always possible. For example, it should be easy for you to construct a tournament of order 5 for which each vertex has indegree 2 and outdegree 2. This means that each player in the corresponding round-robin tournament has two wins and two losses. Thus you could not declare a best player based on wins alone. This accounts for part of the popularity of elimination tournaments. However, elimination tournaments suffer from a different flaw. The winning team in those tournaments may not be the best team, which may have been eliminated earlier in the competition. Round-robin tournaments do have some nice properties. Here is one: In any round-robin tournament, it is always possible to arrange the players on a list $p_1, p_2, \ldots, p_n$ so that player i beat player $i + 1$, for $1 \leq i < n$. This result is implied in the following theorem of Rédei [12].

Theorem 10.4 Every tournament contains a directed hamiltonian path.

Proof. The proof is by induction on n. For $n \leq 3$, we can easily check that the statement is true. Assume that every tournament with $n = k$ vertices has a directed hamiltonian path, and let T be a tournament on $k+1$ vertices. For any vertex v of T, consider $T - v$, which has k vertices. By the inductive hypothesis, $T - v$ has a directed hamiltonian path $P = v_1, v_2, \ldots, v_k$. Either vv_1 or v_1v is in T. If $vv_1 \in T$, then $v, v_1, v_2, \ldots, v_k$ is a directed hamiltonian path in T and we are done. If $vv_1 \notin T$, then let v_i be the first vertex for which $vv_i \in T$. If there is no such vertex v_i, then $v_1, v_2, \ldots, v_k, v$ is a

10.1 Directed Graphs

directed hamiltonian path. If there is such a vertex, v_i, then $v_{i-1}v \in T$. So $v_1, v_2, \ldots, v_{i-1}, v, v_i, v_{i+1}, \ldots, v_k$ is a directed hamiltonian path. ∎

Another interesting theorem about tournaments is due to Landau [10] and was discovered during an empirical study of pecking orders among hens.

Theorem 10.5 *In any tournament, the distance from a vertex of maximum outdegree to any other vertex is at most two.*

Proof. Suppose to the contrary that $d(u,v) \geq 3$ for some vertex v, where u has maximum outdegree. Then there are arcs from v to all neighbors of u and to u itself. So the outdegree of v exceeds that of u, a contradiction. ∎

Theorem 10.5 is commonly called the **king-chicken theorem**. It says that a king chicken u (a vertex of maximum outdegree) either pecks a given chicken v or pecks a chicken w, who in turn pecks v. Anyone who knows the difference between a chicken and a rooster realizes that the theorem was somewhat misnamed.

Exercises 10.1

1. Use a depth-first search algorithm to produce a strongly connected digraph whose underlying graph is Q_3. Do the same for Q_4.

2. Repeat Exercise 1, but for the mesh $M(3,3,3)$.

3. Given a digraph with underlying graph K_3, show that there is a directed spanning path.

4. Draw the digraph D for which $V(D) = \{a, b, c, d, e, f\}$ and $E(D) = \{af, bc, bd, bf, cb, cf, db, dc, df\}$ and then state the indegree and outdegree of each vertex. Verify Theorem 10.1 for D.

5. Is digraph D of Exercise 4 weakly connected, unilateral, or strong?

6. Identify any transmitters and any receivers in the digraph of Figure 10.6.

7. Consider the set $A = \{\emptyset, \{1,3\}, \{2,4\}, \{3,5\}, \{3,6\}, \{1,2,4\}, \{1,3,4\}, \{1,2,3,4\}, \{1,3,4,6\}\}$ with the partial order $\subseteq$.

    (a) Explain why $\subseteq$ is indeed a partial order on A.

    (b) Draw the acyclic digraph (loops omitted) corresponding to $(A, \subseteq)$.

    (c) Draw the Hasse diagram for $(A, \subseteq)$ using the procedure to find successive levels described in this section.

8. Draw all nonisomorphic unlabeled digraphs that have four vertices and four arcs.

9. Explain why the number of nonisomorphic unlabeled digraphs that have four vertices and four arcs is equal to the number of nonisomorphic unlabeled digraphs that have four vertices and eight arcs.

10. A digraph D is **r-regular** if $\text{id}(v) = \text{od}(v) = r$ for each vertex v of D. Find conditions on n and r that guarantee the existence of an r-regular digraph on n vertices.

11. Prove that if $\text{od}(v) > 0$ for all v in D, then D has a directed cycle.

12. A matrix is **upper triangular** if its only nonzero entries appear above the diagonal—that is, in positions whose address is (i, j), where $i < j$. Prove that a digraph is acyclic if and only if it is possible to label the vertices so that the adjacency matrix $A(D)$ is upper triangular.

13. The **converse digraph** D' of D has $V(D') = V(D)$ and arc uv is in D' if and only if vu is in D. Thus the directions on all arcs of D are reversed. Give an example of a digraph that is not a directed cycle and is isomorphic to its converse.

14. A tournament is **transitive** if $uv \in T$ and $vw \in T$ implies that $uw \in T$. Prove that a tournament T is transitive if and only if T is acyclic.

15. Prove that in a transitive tournament there is a unique vertex that should be declared the winner in the related round-robin tournament.

16. Use Exercise 15 and mathematical induction to show that there is a unique unlabeled tournament on n vertices that has outdegree sequence $n-1, n-2, n-3, \ldots, 0, 1$.

17. Determine whether the relation $\sim$ on the set Z is a partial order if $a \sim b$ means $a = 3b$.

18. Determine whether the relation $\sim$ on the set Z is a partial order if $a \sim b$ means $a = bk$ for some $k \in N$.

19. You are traffic manager for Graphland Valley. Due to traffic congestion caused by a new shopping mall, you have determined that making all streets in town one-way streets will make traffic flow more smoothly. Find an orientation of the edges of the graph in Figure 10.10, representing the Graphland Valley street map, using Algorithm 10.1, thereby obtaining a strong digraph to achieve a good traffic flow.

20. Determine whether the relation $\sim$ on the set Z is a partial order if $a \sim b$ means $a + 2b = 0$.

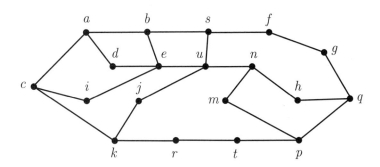

Figure 10.10 The road map of Graphland Valley.

21. You have done such a good job as traffic manager for Graphland Valley that you have been called in as a consultant to help ease traffic congestion in a neighboring area Networkville. Again you have determined that making all streets into one-way streets will make traffic flow more smoothly. Find an orientation of the edges of the graph in Figure 5.10 from Section 5.1, representing the Networkville area map, using Algorithm 10.1, thereby obtaining a strong digraph to achieve good traffic flow.

22. Explain why in a tournament on n vertices, it is impossible to have two vertices with outdegree $n-1$.

23. Find a condition on the outdegrees of a tournament on n vertices that guarantees that some vertex has outdegree zero.

10.2 Networks

When analyzing various transportation, telecommunication, or operations research problems, digraphs are use in which additional information is needed besides the direction on the arcs. For example, the required information might be how much traffic a particular arc (road or cable) can carry and how much traffic it is currently carrying. In such situations, a slightly more complex structure called a network is used to model and analyze the problem. In this section, we discuss the methods used in those problems.

Distance in Networks

We begin this section with a brief discussion of distance, a subject never far from our thoughts. Consider a weighted digraph, where the weight $w(uv)$ on arc uv denotes the distance from u to v traveling along that arc. (Note that here distance takes the weights on the arcs into account and is *not* simply the number of arcs on a shortest path.) Of course, the actual distance from u to v may in fact be less than $w(uv)$. For example, uv may represent a winding road, whereas, by taking road ux followed by xv, we may get from u to v in a shorter distance. Sometimes the weights represent travel times. The same idea applies there. Arc uv may represent a very congested road through a city, whereas, by taking road ux followed by xv, we may get from u to v much quicker. How do we find a best route from one vertex to another? Fortunately, there is an algorithm due to Dijkstra [4] designed explicitly to solve this important problem. This algorithm shares some similarities with BFS and Prim's algorithm; however, here a vertex v receives a temporary label denoting the distance of v from the root r. That distance is updated when a shorter path from r to v is found. The **target vertex** is the vertex whose distance from the root we are trying to determine. $N(v)$ denotes the adjacency list of vertex v.

Algorithm 10.2 (Dijkstra's Algorithm)
1. Input weighted digraph W with all adjacency lists, the root r, and the target t.
2. $d(r) = 0$; for $v \neq r$, set $d(v) = \infty$ [initialize all distances]
3. $T = V(W)$; $u = r$.
4. While $u \neq t$ do
   begin
      for each $v \in N(u)$ do
        if $v \in T$ and $d(v) > d(u) + w(uv)$ [a shorter path to v has been found] then
          $d(v) = d(u) + w(uv)$ [redefine the distance to v]
      $T = T - u$
      let u be a vertex in T for which $d(u)$ is minimum
   end while
   output $d(t)$.

Dijkstra's algorithm applies to both directed and undirected graphs. In either case there are generally weights on the edges; however, if there are no such weights, we initially define them to be one. The main difference for directed weighted graphs is that the adjacency list $N(v)$ for each vertex v corresponds just to the vertices that v is adjacent *to*, not those that v is adjacent *from*. Thus, we might think of $N(v)$ as the *out-neighborhood* of v.

Example 10.5 Use Dijkstra's algorithm to find the distance from r to t in the weighted digraph of Figure 10.11.

10.2 Networks

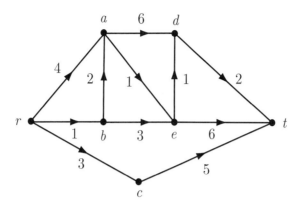

Figure 10.11

Solution
First, we assign the distance 0 to r and ∞ to all other vertices. All vertices are placed in T and we begin our search from $u = r$. The out-neighborhood of r is $\{a, b, c\}$. Since the current distances to those vertices are ∞, step 4 redefines the distances as follows: $d(a) = 4$, $d(b) = 1$, $d(c) = 3$. Next, we remove r from T (we have already searched from r) and since $d(b) = 1$ is the minimum distance assigned, we let $u = b$. Beginning our search from b, we assign the distances $d(a) = 1 + 2 = 3$ and $d(e) = 1 + 3 = 4$. Vertex b is removed from T and we proceed to search from a (which is tied for lowest distance with c among vertices in T). Vertex d gets assigned distance 9, and a is removed from T. Next searching from c, vertex t gets assigned distance 8, and c is removed from T. Only d, e, and t remain in T. Their current distances are 9, 4, and 8, respectively, so we search from e next. Vertex d gets assigned the new distance $4 + 1 = 5$, and e is removed from T. Now we search from d and assign the distance $5 + 2 = 7$ to t, and remove d from T. Finally, since we are now ready to search from vertex t, the while loop in step 4 causes us to stop and output the distance $d(t) = 7$. ◇

It is important to note that although we were finding the distance from s to t in the last example, Dijkstra's algorithm actually finds the distance from the root vertex (s in our case) to *all* other vertices in the network. By successively modifying the root, we can find the distance between all pairs of vertices using Dijkstra's algorithm.

Network Flows

A **two-terminal network** N consists of a weighted connected digraph that has two distinguished vertices s and t, called **terminals**, together with a nonnegative real-valued function c defined on the arcs of N. Vertex s is called the **source** and has indegree zero, so it is a transmitter in the underlying digraph. Vertex t is called the **sink** and has outdegree zero, so it is a receiver in the underlying digraph. The remaining vertices are called **intermediate vertices** of the network. In general, a network need not have any terminals, and some may have several, but we will focus on the two-terminal problem, with one source and one sink. For each arc e the weight $c(e)$ assigned is called the **capacity** of that arc. A network is displayed in Figure 10.12.

In a transport problem, vertices a, b, c, and d represent points at which items will be transferred from one shipping mechanism to another. You may think of the items as being transferred from a truck to a railway or ship. In a telecommunications context, you may think of the vertices as points where information is transferred from (for example, a local network to a wide area network or a trunk line).

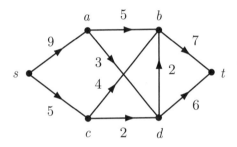

Figure 10.12

A (legal) **flow** f in a network N is an assignment of a nonnegative real number $f(e)$ to each arc e of N so that the following two conditions hold:

$$\text{for each arc } e, 0 \leq f(e) \leq c(e) \quad (10.1)$$

$$\text{for each vertex } v \text{ other than } s \text{ and } t, \sum_{uv \in N} f(uv) - \sum_{vx \in N} f(vx) = 0 \quad (10.2)$$

Condition (10.1) guarantees that the capacity of a given arc is not exceeded. Condition (10.2), which is called **conservation of flow**, ensures that the flow into an intermediate vertex equals the flow out of it. It sometimes helps to think of the flow as water passing through pipes (arcs) on the way from the source, s, to the sink, t. In that context, (10.1) says that no more water will pass through a given pipe than that pipe can bear, and (10.2) says that there are no leaks at the pipe joints (vertices). The general problem of interest in network problems is to maximize the flow of goods from the source s to the sink t with legal flows along the arcs of the network N. The **total flow** F in a network N is defined to be the flow out of the source s, or, equivalently, the flow into the sink t; that is,

10.2 Networks

$$F = \sum_{su \in N} f(su) = \sum_{xt \in N} f(xt) \tag{10.3}$$

Thus we want to maximize F subject to conditions (10.1) and (10.2).

Rather than presenting an algorithm right away, it will be worthwhile first to explore some of the concepts and situations that will be of concern while trying to achieve a maximum flow. We begin with all flows at zero and label each arc e with the ordered pair $(c(e), f(e))$. The **slack** sl(e) of arc e is the amount by which the capacity exceeds the flow, so sl$(e) = c(e) - f(e)$. Thus sl(e) represents the remaining capacity available in arc e. If $sl(e) = 0$, then arc e is **saturated**. Clearly, if every arc into the sink t is saturated, then we have achieved a maximum flow. Similarly, if every arc out of source s is saturated, then we have achieved a maximum flow. However, neither of those two situations is necessary as we shall soon see.

Example 10.6 Attempt to achieve a maximum flow for the network in Figure 10.12.

Solution

We begin by assigning zero flow to each arc to yield the network in Figure 10.13(a) with each arc e labeled by $(c(e), f(e))$. Our initial flow F is zero. We will start by successively finding paths from s to t along which we can increase the flow. For such a path, we will be able to increase the flow by the minimum of the slack values for the arcs of that path. Although many path choices are possible, we shall choose our paths to illustrate a certain situation that can occur during this flow augmentation process.

We start with path s, c, b, t. See Figure 10.13(a). Arc sc has slack value $5 - 0 = 5$, arc cb has slack value $4 - 0 = 4$, and arc bt has slack value $7 - 0 = 7$. The minimum slack value is 4, so we send a flow of 4 along the path. This increases $f(e)$ by 4 for each arc e of the path. See Figure 10.13(b).

Next we use path s, c, d, t, for which the minimum slack value is on arc sc, namely, $5 - 4 = 1$. Thus we increase the flow along arcs sc, cd, and dt by 1. Then using path s, a, d, t, we find minimum slack value 3 on arc ad. So we increase the flow along arcs sa, ad, and dt by 3. The result of these two augmentations is shown in Figure 10.14(a). Now arcs sc, cd, and ad are saturated.

Next we find path s, a, b, t, for which the minimum slack value is 3 on arc bt. We increase the flow by three along arcs sa, ab, and bt to obtain the network flow in Figure 10.14(b). If you check carefully, you will see that there is no path from s to t that can be augmented. So do we have a maximum flow? The surprising answer is no! With the choice of augmenting paths, we have sort of painted ourselves into a corner. How then can we increase the flow and get out of that corner? By using *semipaths*, where we do not need to follow the directions on the arcs. The difference is that if we traverse an arc e in the wrong direction, we will decrease the flow on that arc. This has the effect of diverting some of the prior flow on that arc. How much additional flow can we move along a semipath? We can move the minimum of the slack values for the forward arcs and the flow values for the reverse arcs.

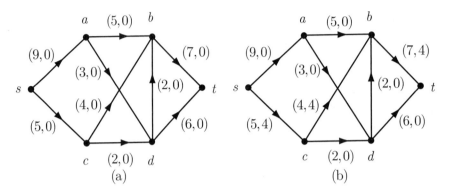

Figure 10.13

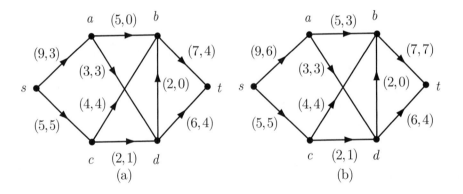

Figure 10.14

Consider the semipath s, a, b, c, d, t [see Figure 10.14(b)]. The minimum slack on the forward arcs is one and the flow on the one reverse arc bc is 4. Thus we can augment the flow by one in the forward arcs and decrease the flow by one in the reverse arc. The effect is to divert one unit of flow from arc cb and send it through cd instead. The resulting network is shown in Figure 10.15.

The total flow is 12 (measure it either at s or t) and it is maximum. We shall soon see why. ◇

10.2 Networks

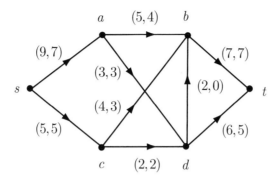

Figure 10.15

Minimal Cuts and Minimum Cuts

We will be able to confirm that we have achieved a maximum flow in a network by using an important and powerful theorem due to Ford and Fulkerson [6] called the max-flow min-cut theorem. It uses the concept of a minimum cut. For a network N with source s and sink t, a **cut** is a set S of arcs such that every directed path from s to t passes through at least one of the arcs of S. Another way of saying this is that if the arcs of S were removed from N, there would be no directed path from s to t.

The **value of the cut** or **cut value** is the sum of the capacities of the arcs in the cut. A **minimum cut** for a network N is a cut that has the smallest possible cut value. A **minimal cut** S is a cut such that for all $e \in S$, $S - e$ is not a cut. Thus if any edge of S is removed, we no longer have a cut. We may say that there are no **redundant arcs** in a minimal cut. It should be clear that a minimum cut is necessarily minimal, but a minimal cut is not necessarily a minimum cut.

Example 10.7 Find all minimal cuts for the network in Figure 10.15. Which of these cuts are also minimum cuts?

Solution
We must find minimal sets of arcs whose removal would cut off any possibility of flow from s to t. Clearly, $\{sa, sc\}$ and $\{bt, dt\}$ are minimal cuts. Since sa cuts flow to a, a minimal cut that includes sa would never contain ab or ad—they would be redundant. Similarly, a minimal cut that includes sc would not contain cb or cd; a minimal cut that includes bt would not contain ab, cb, or db; and a minimal cut that contains dt would not contain ad or cd. Thus, besides the minimal cuts $\{sa, sc\}$ and $\{bt, dt\}$, we also have $\{sa, cb, cd\}$, $\{sa, cb, db, dt\}$, $\{sa, cd, bt\}$, $\{sc, ab, ad\}$, $\{sc, ab, db, dt\}$, $\{sc, ad, bt\}$, $\{ab, ad, cb, cd\}$, $\{ab, cb, db, dt\}$, and $\{ad, cd, bt\}$. To get the cut values, we add the capacities of the arcs in a given cut. The respective cut

values are 14, 13, 15, 21, 18, 13, 18, 15, 14, 17, and 12. The only minimum cut is $\{ad, cd, bt\}$, which has cut value 12, the same value as the total flow we found in Example 10.6. ◇

Max-Flow Min-Cut Theorem

The fact that the maximum flow for the network in Example 10.6 has the same numerical value as the minimum cut value determined in Example 10.7 is no coincidence. We now examine why that is the case. First note that any cut S partitions the vertex set of N into two sets X and Y, where $s \in X$ and $t \in Y$. The set X consists of all vertices that s can reach in $N - S$, and the set Y consists of all other vertices. Thus each arc of S is incident with a vertex in X and a vertex in in Y. Let $C(S)$ denote the cut value of S.

Theorem 10.6 If F is the total flow in a network N and S is any cut, then $F \leq C(S)$.

Proof. For $v \in X$, consider the value $z(v) = \sum f(v,x) - \sum f(u,v)$—that is, the total flow out of v minus the flow into v. For $v = s$, $z(v) = F$. For $v \neq s$, $z(v) = 0$ because of (10.2) conservation of flow. Thus

$$F = \sum_{v \in X} z(v) = \sum_{v \in X} \left[\sum f(v,x) - \sum f(u,v) \right] \tag{10.4}$$

For an arc ab with $a \in X$ and $b \in X$, $f(a,b)$ appears once with a positive sign (when a plays the role of v in the outer summation) in (10.4) and once with a negative sign (when b plays the role of v). So these terms cancel. If $a \in X$ but $b \in Y$, then $f(a,b)$ appears just once in (10.4), with a positive sign. If $b \in X$ and $a \in Y$, then $f(a,b)$ appears just once in (10.4), with a negative sign. So (10.4) simplifies to

$$F = \sum_{a \in X, y \in Y} f(a,y) - \sum_{b \in Y, v \in X} f(b,v) \tag{10.5}$$

The first sum in (10.5) is at most $C(S)$ since each flow $f(a,y)$ is bounded by the capacity $c(ay)$. The second sum is nonnegative because each flow $f(b,v)$ is nonnegative. So the right side of (10.5) is at most $C(S)$; that is $F \leq C(S)$. ∎

Since the total flow F is bounded from above by the cut value of any cut, F can be no larger than the cut value of a minimum cut. So we have the following.

Corollary 10.5a For any network N, the maximum flow is less than or equal to the minimum cut value. ◇

Ford and Fulkerson [6] showed that not only is Corollary 10.5a true, we actually get equality.

10.2 Networks

Theorem 10.7 (Max-Flow Min-Cut Theorem) For any network N, the value of a maximum flow in N equals the cut value of a minimum cut.

Proof. (sketch) For a minimum cut S, Corollary 10.5a establishes that the maximum flow value F satisfies $F \leq C(S)$.

To complete the proof, we would need to show that the upper bound can always be achieved. This can be done by developing an algorithm that finds semipaths along which we can increase the flow (as was done in Example 10.6). There are two main items to contend with. The first is to show that the algorithm will terminate in a finite number of steps. This is fairly straightforward when the capacities are all integer valued, since each augmenting flow will increase the total flow by a positive integer and F is bounded by $C(S)$. It is somewhat more difficult when the capacities are not all integer valued. The second main item to contend with to complete the proof is to show that when the algorithm terminates, there will be a cut, all of whose edges are saturated, and at that point, we will have achieved a maximum flow and located a minimum cut. ∎

The interested reader may find a more detailed proof in Chartrand and Oellermann [2].

Finding a Flow-Augmenting Semipath

We now describe an algorithm that involves labeling the vertices. On each pass through the algorithm, each vertex gets labeled at most once. If the sink t gets labeled, then a semipath has been found along which the flow can be augmented to increase the total flow in N. Such a semipath is called a **flow-augmenting semipath** or simply an **augmenting path**. We will use the latter term because it is the standard term that has been used over the years, but keep in mind that it is actually a semipath in N. It is only a path in the underlying graph, where there are edges rather than directed arcs to traverse. If the sink t does not get labeled, the algorithm terminates and a maximum flow has been found.

When we say **scan** a vertex v, we mean do a BFS at that vertex; that is, look at all the neighbors of v. This, of course, would generally be done with the aid of adjacency lists.

Ford and Fulkerson's original algorithm handled networks with integer or even rational capacities without any problems. However, they showed that if the capacities were irrational—that is, in $\Re - Q$ (see Section 1.1, if necessary)—then their algorithm might take an infinite number of steps. No one wants to wait that long, so Edmonds and Karp [5] fixed the shortcoming by using a breadth-first search BFS (see Chapter 8) in the following process. They showed that by doing so, the number of steps until termination is a linear function of $|V|^2 |E|$.

Algorithm 10.3 (Maximum Flow Algorithm)
1. Label s by *. Make all initial flows 0 [$f(e) = 0$ for all arcs e].
2. Find a labeled but unscanned vertex u [initially s is such a vertex]. For each arc uv where v is unlabeled and $f(uv) < c(uv)$, label v with u^+ [this means you get to v on a forward arc from u]. For each reverse arc uw where w is unlabeled and $f(wu) > 0$, label w with u^- [you get to w on a reverse arc uw from u].
3. If t has been labeled, stop; an augmenting path exists. Backtrack from t to s using the labels on the vertices to generate the augmenting path P. Calculate the augmenting flow Δ as follows. For each forward arc $e_i = uv$ on P, let $\Delta_i = c(uv) - f(uv)$ [available additional capacity]; and for each reverse arc $e_i = uw$ on P, let $\Delta_i = f(wu)$ [the flow on uw available for rerouting]. Let $\Delta = \min_{e_i \in P} \{\Delta_i\}$. Increase the flow by Δ on each forward arc uv and decrease the flow by Δ on each reverse arc uw of P. Delete all vertex labels except * on s and repeat step 2.
4. If t has not been labeled, then
   a. If all labeled vertices have been scanned, stop; we have achieved a maximum flow.
   b. If not all labeled vertices have been scanned, return to step 2.

Example 10.8 Use Algorithm 10.3 to determine a maximum flow for the network in Figure 10.16.

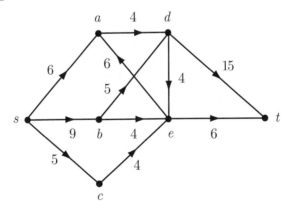

Figure 10.16

Solution
We label s by *, assign initial flows of 0, then scan s. Vertices a, b, and c get labeled s^+, and then we scan a. Next, d gets labeled a^+. (*Note*: b is already labeled, so it does not get labeled again; e does not get labeled a^- because there is no flow in the reverse arc ae.) Next, e gets labeled b^+. Finally, t

gets labeled d^+. We have reached t, so we backtrack to find the augmenting path. See Figure 10.17.

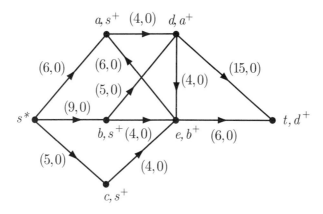

Figure 10.17

The vertex labels tell us that we got to t from d on a forward arc. Thus the augmenting path is s, a, d, t, and $\Delta = \min\{6 - 0, 4 - 0, 15 - 0\}$. We increase the flow along arcs sa, ad, and dt by 4. We delete all vertex labels except $*$ on s and repeat step 2 to search for another augmenting path.

We label a, b, and c by s^+. Arc ad is saturated, so d does not get labeled by a^+ but by b^+ instead. Vertex e also gets labeled b^+, then t gets labeled by d^+. See Figure 10.18. We backtrack to find the augmenting path s, b, d, t and find $\Delta = 5$. We increase flow along arcs sb, bd, and dt by 5. In the process, bd becomes saturated.

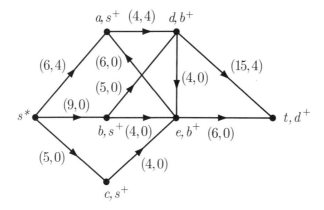

Figure 10.18

We delete all labels except $*$ on s. Then a, b, and c get labeled s^+, e gets labeled by b^+, and t gets labeled e^+. We get $\Delta = 4$ and increment the flow on arcs sb, be, and et by 4. After this step and deleting all labelings except * on s, the network appears as in Figure 10.19.

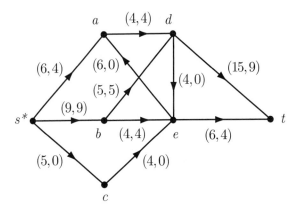

Figure 10.19

Next, a, and c get labeled s^+ (arc sb is saturated, so b gets no label at this point), e gets labeled c^+, and t gets labeled e^+. We find $\Delta = 2$ and increase the flow in each of sc, ce, and et. Arc et is saturated. We delete all labels except * on s, and continue. Label a, and c by s^+. Then e gets labeled c^+ and then b gets label e^- because of the flow in the reverse arc eb. Then no further labeling is possible. See Figure 10.20.

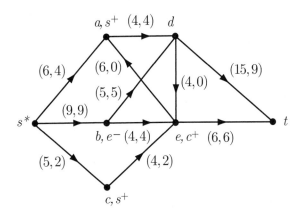

Figure 10.20

The sink t is unlabeled but all labeled vertices have been scanned, so, according to step 4a of Algorithm 10.2, we have achieved a maximum flow. The maximum flow is 15. To confirm that this is indeed a maximum flow, we should be able find a minimum cut with cut value 15. This is fairly easy to do because the arcs of the cut must be saturated and the minimum cut must also be minimal. The saturated arcs are sb, be, bd, ad, and et. The combination of those that form a minimum cut is ad, bd, and et. The sum of their capacities is $4 + 5 + 6 = 15$. ◇

Numerous varieties of networks are more general than the ones we have considered. For example, we may have multiple sources and/or multiple sinks in N. These are handled easily. For example, when there are multiple sources, we may put in a single super source A and have arcs from A to each of the multiple sources. Give each of those arcs huge capacity—for example, some number exceeding the sum of all the capacities in N. Then proceed with Algorithm 10.2 on this new network.

Networks, Matchings, and Connectivity

Network flows can be used to easily solve bipartite matching problems. Given a bipartite graph G with parts A and B, suppose we want to achieve a maximum matching. We can do so as follows. Create a network N from G by introducing two new vertices s and t. Put arcs from s to each vertex in A, direct all edges from A to B, and put in arcs from each vertex in B toward t. See Figure 10.21. Next, put capacity 1 on every arc. Now apply Algorithm 10.3. When a maximum flow is found, the saturated arcs from A to B correspond to a maximum matching in G.

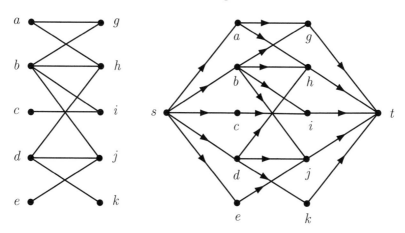

Figure 10.21 A bipartite graph and related network.

Networks are also used in certain connectivity problems. For example, Dantzig and Fulkerson [3] showed that the max-flow min-cut theorem can be used to prove Menger's theorem (see Theorem 4.7). Gould [7] gives a very readable account of how this is done.

Exercises 10.2

1. Apply Algorithm 10.3 to the network of Figure 10.12 to obtain a maximum flow.

2. In Figure 10.20, explain why $\{dt, et\}$ is not a minimum cut.

3. Solve the job assignment problem, Exercise 15 from Section 3.4, using network flows.

4. Solve the job assignment problem, Exercise 16 from Section 3.4, using network flows.

5. Consider the weighted graph in Figure 5.20 and direct all edges toward the right. Then find a maximum flow from a to d.

6. Consider the digraph in Figure 10.2. Suppose you make $c(x) = 1$ and $f(x) = 1$ for every arc x in the digraph. Is f a (legal) flow? Explain.

7. Assign flow values to each arc in the digraph of Figure 10.4(c) using only the values 1 and 2, where the capacity of each arc is 2, in order to produce a legal flow.

8. Apply Algorithm 10.3 to the network of Figure 10.22 to find a maximum flow from s to t.

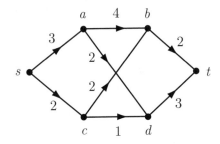

Figure 10.22

9. Find all minimal cuts for the network in Figure 10.22 with source s and sink t. Which of them are minimum cuts?

10. In some pipe systems, it is desirable not only to achieve flows that meet conditions (10.2) but also a modified version of (10.1), namely,

$$\text{for each arc } e, \ m(e) \leq f(e) \leq c(e) \qquad (10.1')$$

10.2 Networks

Here, $m(e)$ is a minimum flow allowed in arc e. Such minimums may help prevent sediments from accumulating or help deter frozen pipes in winter. Suppose you had a network N that had maximum total flow and you wanted to decrease that to minimum total flow suggested by condition (10.1′) on each arc. Explain how you might alter Algorithm 10.3 to accomplish this.

11. Find a maximum flow from s to t in the network of Figure 10.23.

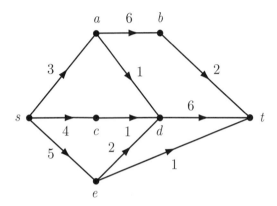

Figure 10.23

12. Find a minimum cut for the network in Figure 10.23.

13. Find a maximum flow from s to t for the network in Figure 10.24(a). Then find a minimum cut.

14. Find a maximum flow from s to t for the network in Figure 10.24(b). Then find a minimum cut.

15. Prove that if N is a network where each capacity is an integer, then N has a maximum flow where the flow on each arc is an integer.

16. Use Dijkstra's algorithm to find the distance from s to t in the network of Figure 10.22.

17. Use Dijkstra's algorithm to find the distance from s to t in the network of Figure 10.23.

18. Use Dijkstra's algorithm to find the distance from s to t in the network of Figure 10.24(a).

19. Use Dijkstra's algorithm to find the distance from s to every vertex in the network of Figure 10.23.

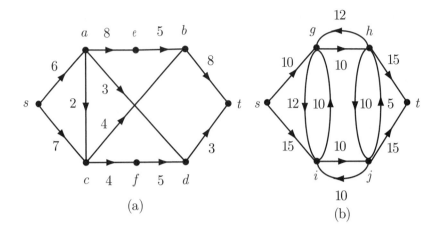

Figure 10.24

20. Rewrite Algorithm 10.2, introducing **predecessor labels** so that when the algorithm is complete a shortest path from the root r to any other vertex v can be determined by backtracking from v using the predecessor labels.

21. Suppose that the weights on the arcs for the network in Figure 10.11 correspond to capacities. Find a maximum flow from r to t for the network in Figure 10.11. Then find a minimum cut.

10.3 The Critical Path Method

Every large project consists of numerous activities that must be scheduled to occur in a certain order for prompt and successful completion of the project. The project manager makes estimates of the duration of each individual activity and determines a precedence relation for the activities. For example, on a construction project we would generally install the electrical wiring before installing the wall board. The activity durations and precedence relations among the activities can be modeled using a certain weighted acyclic digraph. Using that digraph, the manager can determine the appropriate schedule for the activities. We discuss this process in this section.

Activity Digraphs

In order to describe the various concepts that are needed we shall consider an example. Suppose that you plan to open a self-serve laundromat. The

10.3 The Critical Path Method

various activities that you must complete during the course of the project are described in Table 10.1.

The **activity digraph** is constructed as follows. Each activity corresponds to a vertex. There is an arc from vertex i to vertex j if activity i is an **immediate predecessor** of activity j. In that case, vertex j is called a **successor** of vertex i. Note that *start* has no immediate predecessors and *finish* has no successors. An arc from i to j has weight equal to the duration of activity i. We should mention that an activity digraph is actually a *network*, but the term activity *digraph* has been the standard terminology.

Example 10.9 Draw the activity digraph for the project described in Table 10.1.

Solution
Using the precedence relations described in Table 10.1, we see that there must be an arc from *start* to A with weight 0. Then there is an arc from A to B with weight 3 and from A to C with weight 3. Each arc from B has weight 3 and B has arcs toward D, E, and F. C also has arcs toward D, E, and F, but they each have weight 2. Continuing in this way, we obtain the activity digraph in Figure 10.25. ◇

Activity	Description	Immediate Predecessors	Duration in Days
Start	—	—	0
A	meet with real estate agent	*Start*	3
B	meet with lawyer	A	3
C	meet with accountant	A	2
D	obtain mortgage from bank	B, C	4
E	obtain washers and dryers	B, C	3
F	obtain vending machines	B, C	2
G	hook up electricity	D	1
H	hook up gas	D	1
I	get city license	D	1
J	set up equipment	E, F, G, H	2
Finish	—	I, J	0

Table 10.1

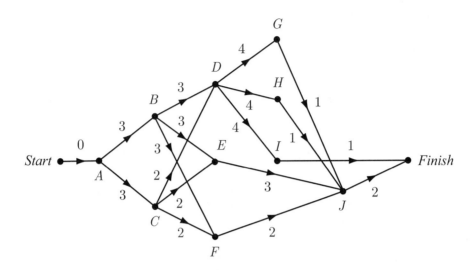

Figure 10.25

For the "duration in days" column of Table 10.1, the numbers sum to 22, which means that the project could definitely be completed in 22 days. However, since certain activities can be handled simultaneously, the project can be completed quite a bit faster than that. To complete the project as quickly as possible, we need to know the earliest possible time that each activity can start and can finish. For flexibility in scheduling, it is also useful to know the latest that a given activity can start without delaying the project. Some activities are generally critical in the sense that if they start late, the completion of the project will be delayed.

Critical Path Method

In 1957, E. I. Dupont deNemours and Company developed a technique to control the duration of construction projects. That technique is called the **critical path method (CPM)**. Here is an outline of the **CPM**.

> **Note 10.1 (Outline of the Critical Path Method)**
> 1. Identify the activities, their immediate predecessors, and their duration.
> 2. Draw the activity digraph.
> 3. Use the activity digraph to determine the earliest start time and the earliest finish time for each activity.
> 4. Compute the slack time for each activity.
> 5. Identify all critical activities and critical paths.

10.3 The Critical Path Method

Certain items in Note 10.1 may need some further explanation. The **earliest start time (EST)** of an activity is the earliest time that activity can begin, assuming all preceding activities were completed as soon as possible. The **earliest finish time (EFT)** of an activity is the EST of that activity plus its duration. We move forward through the activity digraph to compute these values as follows.

Note 10.2 (Computing EST and EFT)
1. EST($start$) = 0
2. For $v \neq start$,
   EST(v) = max{EFT(w) : w is an immediate predecessor of v}.
3. EFT(u) = EST(u)+ duration(u) for all u.
4. EFT(finish) = **completion time** of the project.

The **latest start time (LST)** and **latest finish time (LFT)** of an activity are the latest that activity can start and finish, respectively, without delaying the project completion time. We move backward through the activity digraph to compute these values as follows.

Note 10.3 (Computing LST and LFT)
1. Determine project completion time using Note 10.2.
2. LFT(finish) = project completion time.
3. For $v \neq finish$,
   LFT(v) = min{LST(w) : w is the successor of v}.
4. LST(u) = LFT(u)− duration(u) for all u.

The **slack time** of an activity is the amount by which we can delay beginning that activity without affecting the project completion time. It is easily calculated as follows: slack-time(v) = LST(v)− EST(v).

Example 10.10 Set up a table listing each activity, its duration, EST, EFT, LST, LFT, and slack time for the activity digraph of Figure 10.25 which we constructed in Example 10.9.

Solution

Using Notes 10.2 and 10.3, we compute each value. They are shown in Table 10.2. As we move forward through the activity digraph, we can complete the EST and EFT columns from top to bottom. As we move backward through the digraph, we complete columns LFT and LST from bottom to top. Then we complete the slack time column for each activity v by calculating LST(v)− EST(v).

Activity	Duration	EST	EFT	LST	LFT	Slack Time
Start	0	0	0	0	0	0
A	3	0	3	0	3	0
B	3	3	6	3	6	0
C	2	3	5	4	6	1
D	4	6	10	6	10	0
E	3	6	9	8	11	2
F	2	6	8	9	11	3
G	1	10	11	10	11	0
H	1	10	11	10	11	0
I	1	10	11	12	13	2
J	2	11	13	11	13	0
Finish	0	13	13	13	13	0

Table 10.2

The project completion time is 13 days. ◇

A **critical activity** is an activity whose slack time is zero. Any delay in starting such an activity will delay completion of the project. On the other hand, a project with positive slack time is not critical since its start time is somewhat flexible. For example, activity E has slack time 2, which means there is a two-day leeway in when activity E can begin without delaying the project. A **critical path** in an activity digraph is a directed path from start to finish, all of whose activities are critical (have slack-time $= 0$). An activity digraph always has at least one critical path, but it may have several.

Example 10.11 Use Table 10.2 to find all critical activities for the activity digraph in Figure 10.25. Then describe all critical paths.

Solution
The critical activities have slack time equal to zero, so they are $Start$, A, B, D, G, H, J, and $Finish$. There are two critical paths, namely, $Start, A, B, D, G, J, Finish$ and $Start, A, B, D, H, J, Finish$. ◇

The critical path method assumes that an activity's duration can be predicted with relative certainty. This is often the case in projects such as construction, training, and maintenance. Certain projects, however, such as computer software development, drug development, and research projects, often have activities for which the duration may be less certain. For those problems, a technique called **PERT (program evaluation and review technique)** that is related to **CPM** may be more appropriate. In those problems, the durations are considered to be random variables, with a range of times rather that a specific time. For example, activity T might have duration between 11 and 15 weeks. Based on probability considerations, estimates of actual duration are made. After that, the procedure for PERT is

10.3 The Critical Path Method

rather similar to CPM. An activity digraph is constructed, EST, EFT, LST, LFT, and slack times are determined. Then probability calculations are used to estimate project completion time. PERT was developed independently from CPM by the U.S. Navy in 1958 to manage its huge Polaris submarine construction project. For further information on PERT, see Ragsdale [11].

Activity	Immediate Predecessors	Duration (in weeks)
Start	—	0
A	Start	3
B	A	2
C	A	4
D	B	8
E	C, D	2
F	C, D	1
G	E, F	1
H	E	2
I	F	1
Finish	G, H, I	0

Table 10.3

Exercises 10.3

1. Using the data in Table 10.3,

   (a) Construct the activity digraph

   (b) For each activity, determine EST, EFT, LST, LFT, and slack time.

   (c) Determine all critical activities

   (d) Find all critical paths in the activity digraph.

2. Given the activity digraph shown in Figure 10.26, find all critical paths and the completion time for the project.

3. Explain why the digraph in Figure 10.27 cannot be an activity digraph.

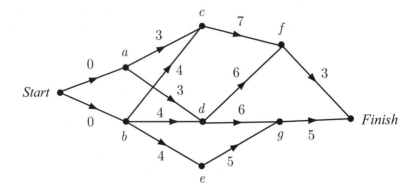

Figure 10.26

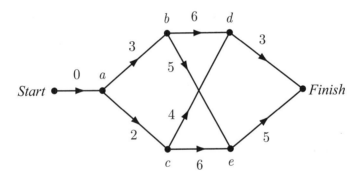

Figure 10.27

4. Using the data in Table 10.4,

   (a) Construct the activity digraph.

   (b) For each activity, determine EST, EFT, LST, LFT, and slack time.

   (c) Determine all critical activities.

   (d) Find all critical paths in the activity digraph.

5. Explain why an activity digraph must be acyclic.

10.3 The Critical Path Method

Activity	Immediate Predecessors	Duration (in weeks)
Start	—	0
A	Start	4
B	Start	5
C	A	2
D	A, B	3
E	C, D	6
F	B, E	5
G	C, D	7
Finish	F, G	0

Table 10.4

6. Suppose that an acyclic digraph D has a $Start$ and $Finish$ vertex, where $Start$ has only out degree and $Finish$ has only indegree. Additionally, there are positive weights on every arc except those from $Start$ and those to $Finish$. Explain why this is not sufficient to guarantee that D is an activity digraph. Draw an example of a digraph that has all those properties and is still not an activity digraph. Start an additional property that is necessary.

7. In preparing an annual report that will be sent to all shareholders, a corporation performs the activities described in Table 10.5. Using the data in Table 10.5,

    (a) Construct the activity digraph

    (b) For each activity, determine EST, EFT, LST, LFT, and slack time.

    (c) Determine all critical activities.

    (d) Find all critical paths in the activity digraph.

8. Explain why every activity except $Finish$ must appear in the immediate predecessors column for the table of an digraph.

Activity	Description	Immediate Predecessors	Duration in Weeks
Start	—	—	0
A	Decide on a theme	Start	4
B	Have articles written	A	5
C	Obtain artwork	A	2
D	Layout report	B, C	3
E	Prepare a mailing list	D	6
F	Proofread	D	2
G	Make all changes and have printed	F	5
H	Retrieve report from printer	E, G	1
I	Mail out report	H	1
Finish	—	I	0

Table 10.5

9. Given the activity digraph shown in Figure 10.28, find all critical paths and the completion time for the project.

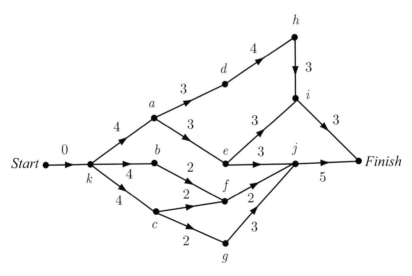

Figure 10.28

References for Chapter 10

1. Bang-Jensen, J., and G. Gutin, *Digraphs: Theory, Algorithms and Applications*, Springer, London (2001).

2. Chartrand, G., and O. R. Oellermann, *Applied and Algorithmic Graph Theory*, McGraw-Hill, New York (1993).

3. Dantzig, G. B., and D. R. Fulkerson, On the max-flow min-cut theorem of networks: linear inequalities and related systems. *Annals of Math.* 38, Princeton University Press (1956) 215–221.

4. Dijkstra, E. W., A note on two problems in connection with graphs. *Numeriske Math* 1 (1959) 269–271.

5. Edmonds, J., and R. M. Karp, Theoretic improvements in algorithmic efficiency for network flow problems. *Journal of the Association for Computing Machinery* 19 (1972) 248–264.

6. Ford, L. R., and D. R. Fulkerson, Maximal flow through a network. *Canadian Journal of Mathematics* 8 (1956) 399–404.

7. Gould, R., *Graph Theory*, Benjamin Cummings, Menlo Park, CA (1988).

8. Grimaldi, R. P., *Discrete and Combinatorial Mathematics: An Applied Introduction*, 4th ed., Addison-Wesley, Reading, MA (1999).

9. Kolman, B., R. C. Busby, and S. Ross, *Discrete Mathematical Structures*, 3rd ed., Prentice-Hall, Upper Saddle River, NJ (1996).

10. Landau, H. G., On dominance relations and the structure of animal societies, III: the conditions for a score situation. *Bulletin of Mathematical Biophysics* 15 (1953) 143–148.

11. Ragsdale, C. T., *Spreadsheet Modeling and Decision Analysis*, 3rd ed., South-Western College Publishing (2001).

12. Rédei, L., Ein kombinatorischer Satz, *Acat Litt. Szeged* 7 (1934) 39–43.

13. Robbins, H. E., A theorem on graphs, with an application to a problem of traffic control. *American Mathematical Monthly* 46 (1939) 281–283.

Additional Readings

14. Benson, C., and F. Lowenthal, The marriage lemma for polygamists. *Mathematics Magazine* 66 (1993) 238–242.

15. Wallis, W. D., One factorizations of graphs: tournament applications. *College Mathematics Journal* 18 (1987) 116–123.

Chapter 11

Special Topics

We now consider two additional topics, Ramsey theory and graph domination. The first is related to edge colorings in graphs and the second is related to both distance and independence. The one thing that these last two topics have in common is that they are lots of fun to work on.

11.1 Ramsey Theory

How large a group of people must there be to guarantee that there are three mutual acquaintances or three mutual strangers in the group? The answer turns out to be six, as we shall see. Some readers may be thinking, "Half of six is three, so that makes sense." Unfortunately, that logic is flawed. How large a group would be needed to guarantee that there are four mutual acquaintances or four mutual strangers? The surprising answer is eighteen! Even more surprising is that if we want to guarantee five mutual acquaintances or five mutual strangers, the answer is unknown!

Ramsey's Theorem

Frank Ramsey worked on set theory in the early part of the twentieth century. His brother, the Archbishop of Canterbury in the 1960s, was incredulous when informed that a whole area of graph theory was named for his brother. First, Frank Ramsey had not actually worked on graph theory but set theory instead, and second, Frank Ramsey had died at the young age of 26. In his short, busy life, he also established fundamental results in philosophy and in economics.

Ramsey's theorem [8] is a generalization of a counting principle called the **pigeonhole principle**, which, stated informally, says that if there are lots of pigeons flying into not too many pigeonholes, then some pigeonholes will contain many pigeons. Ramsey's theorem is far more profound. To see his statement in its general form, see Ryser [9], where a proof is also given. Here we state the graph theoretic version.

Theorem 11.1 (Ramsey's Theorem) For any two positive integers m and n, there exists a least positive integer $r(m, n)$, such that if the edges of the complete graph on $r(m, n)$ vertices are colored with two colors, say red and blue, then there is a complete subgraph K_m all of whose edges are red or there is a complete subgraph K_n all of whose edges are blue. ▮

Note that Ramsey's theorem guarantees the *existence* of the number $r(m, n)$. It neither tells what that number is nor gives an algorithm for finding it. The number $r(m, n)$ is called a **ramsey number**. It is not difficult to show that $r(m, n) = r(n, m)$ (see Exercise 1). Let's look at some small ramsey numbers.

It is easy to see that $r(m, 1) = 1$. This is trivially true. If we color the edges of K_1 (it has no edges) red and blue, then we will find either a K_m in red or a K_1 all of whose edges (it has none) are in blue. A similar argument shows that $r(1, n) = 1$. Equally easy is the fact that $r(2, 2) = 2$. If we color the *only* edge of K_2 either red or blue, then we are guaranteed to find either a red K_2 or a blue K_2. It is also easy to show that $r(2, n) = n$. Here is the first nontrivial case.

Theorem 11.2 $r(3, 3) = 6$.

Proof. We must show two things. First we must show that it is possible to 2-color the edges of K_5 and avoid both a red K_3 and a blue K_3. Then we must show that no matter how we 2-color the edges of K_6, we must get a monochromatic K_3. Consider K_5 drawn in normal position with a pentagon on the outside and a pentagram (five-pointed star) on the inside. Color the edges of the pentagon red and the edges of the pentagram blue and you have avoided a red K_3 and a blue K_3. Thus $r(3, 3) > 5$. Now suppose $V(K_6) = \{v_1, v_2, \ldots, v_6\}$. Then at least three of the edges incident with v_1 have the same color. Without loss of generality, suppose that v_1v_2, v_1v_3, and v_1v_4 are all blue. Then if none of the edges v_2v_3, v_3v_4, and v_2v_4 is blue, they must all be red, so we have a red K_3, namely v_2, v_3, v_4, v_2. On the other hand, if one of them, say v_3v_4, is blue, we have a blue K_3, namely v_1, v_3, v_4, v_1. Thus any 2-coloring of the edges of K_6 forces a monochromatic K_3 to occur. Hence $r(3, 3) = 6$. ▮

If we interpret a red K_3 as three mutually acquainted people and a blue K_3 as three mutual strangers, we now see why a group of at least six people is required to guarantee three mutual acquaintances or three mutual strangers. Proving that $r(4, 4) = 18$ is far more challenging. A useful approach to examining ramsey numbers is by using some known bounds. These bounds at least give a starting point for trying to establish the actual value of $r(m, n)$.

11.1 Ramsey Theory

Theorem 11.3 For integers $m \geq 2$ and $n \geq 2$, we have $r(m,n) \leq r(m-1,n) + r(m, n-1)$.

Proof. Consider a 2-coloring of the edges of the complete graph G with $r(m-1, n) + r(m, n-1)$ vertices. We will show that G contains a red K_m or a blue K_n. Consider any vertex v of G and suppose that at least $r(m-1, n)$ red edges are incident with v. Then its "red neighborhood," $RN(v)$, has the property that the induced subgraph $\langle RN(v) \rangle$ contains a red K_{m-1} or a blue K_n. If $\langle RN(v) \rangle$ contains a blue K_n, then G does as well and we are done. Otherwise, $\langle RN(v) \rangle$ contains a red K_{m-1}. But all $m-1$ edges from v to the vertices of the red K_{m-1} are also red, so we have a red K_m.

We must consider a second case—namely, when v does not have at least $r(m-1,n)$ red incident edges. Then since $|V(G)| = r(m-1,n) + r(m, n-1)$, v must have at least $r(m, n-1)$ blue incident edges. An analogous argument shows that G must contain a red K_m or a blue K_n. So $r(m,n) \leq r(m-1,n) + r(m,n-1)$. ∎

Another upper bound for $r(m,n)$ is given by the following result of Erdős and Szekeres [3].

Corollary 11.3a For every two positive integers m and n, the ramsey number $r(m,n) \leq \binom{m+n-2}{m-1}$.

Proof. The proof is by induction on $m+n$. For $m = 1$ or 2 with n arbitrary, the result is straightforward. In fact, $r(1,n) = \binom{1+n-2}{1-1} = \binom{n-1}{0} = 1$ and $r(2,n) = \binom{2+n-2}{2-1} = \binom{n}{1} = n$. So assume that m and n are each at least 3. Assume that for a and b, where $a + b < m + n$ and $m + n \geq 6$, that

$$r(a,b) \leq \binom{a+b-2}{a-1} \tag{11.1}$$

Inequality (11.1) is our inductive hypothesis. So by (11.1), we have $r(m-1,n) \leq \binom{m-1+n-2}{m-1-1} = \binom{m+n-3}{m-2}$. Similarly, $r(m, n-1) \leq \binom{m+n-3}{m-1}$. So, using Theorem 11.3 and the recursion, formula (1.6), from Chapter 1, we get

$$r(m,n) \leq r(m-1,n) + r(m,n-1)$$
$$\leq \binom{m+n-3}{m-2} + \binom{m+n-3}{m-1} = \binom{m+n-2}{m-1}. \quad \blacksquare$$

It is important to note that Theorem 11.3 generally gives a better upper bound than Corollary 11.3a. However, it requires us to know the numerical value of $r(m-1,n)$ and $r(m, n-1)$ to get a numerical upper bound for $r(m,n)$. Corollary 11.3a simply requires the values of m and n to obtain such a bound, but an important part of Corollary 11.3a is that it confirms that the numbers $r(m,n)$ are indeed finite. In Table 11.1, we list the ramsey numbers for $m, n \geq 3$ that were known at the beginning of 2002.

$m \setminus n$	3	4	5	6	7	8	9
3	6	9	14	18	23	28	36
4	9	18	25				
5	14	25					
6	18						
7	23						
8	28						
9	36						

Table 11.1

It is amazing how few ramsey numbers are known, especially considering the fact that ramsey theory has been an extremely active area of research for several decades. In addition to known values, there are some bounds for various missing numbers in Table 11.1. These are as follows: $40 \leq r(3,10) \leq 43$, $35 \leq r(4,6) \leq 41$, $43 \leq r(5,5) \leq 49$, $58 \leq r(5,6) \leq 87$, $102 \leq r(6,6) \leq 165$, and $205 \leq r(7,7) \leq 540$. Since $r(m,n) = r(n,m)$, by symmetry, we have bounds for three other missing numbers in Table 11.1.

Generalized Ramsey Numbers

There are many ways in which we can generalize the study of ramsey numbers. For example, we could vary the number of colors, we could alter the subgraphs being searched for, or we could look for multiple copies of particular subgraphs. We consider 2-colorings where we will be searching for subgraphs that need not be K_n.

For graphs G and H, the **ramsey number $r(G, H)$** is the smallest positive integer so that if we 2-color the edges of the complete graph having $r(G, H)$ vertices, then there is a copy of G, all of whose edges are red, or there is a copy of H, all of whose edges are blue. To simplify terminology, we call these a red G and a blue H, respectively. Also, we call an edge coloring where we are trying to avoid a red G and a blue H a (red, blue)-coloring. In this notation, $r(K_m, K_n) = r(m, n)$.

Example 11.1 Show that $r(P_3, K_3) = 5$.

Solution
First we must show that in a (red, blue)-coloring of K_4, we can avoid a red P_3 and a blue K_3. To do this, find a perfect matching in K_4 and color those edges red. The remaining edges induce the subgraph C_4. Color those edges blue. We have avoided a red P_3 and a blue K_3. Next we must show that in any (red, blue)-coloring of K_5, we cannot avoid both a red P_3 and a blue K_3. That is, at least one of them must occur. Consider any vertex v in K_5. At most one edge incident with v is red, or we would get a red P_3, which is not permitted. Thus there are at least three blue edges vx, vy, vz. If any of

xy, yz, or xz is blue, we get a blue K_3. If none of them is blue, we get a red K_3, and therefore a red P_3. So in either case we are forced to get one of the forbidden subgraphs. Thus $r(P_3, K_3) = 5$. ◇

Generally, as the graphs get larger, so do the challenges. For our next example, we use a sequential join $K_1 + K_1 + K_2$, which consists of a K_3 with a pendant edge attached.

Example 11.2 Show that $r(P_4, K_1 + K_1 + K_2) = 7$.

Solution
First we show that in a (red, blue)-coloring of K_6, we can avoid a red P_4 and simultaneously avoid a blue $K_1 + K_1 + K_2$. In K_6, find two disjoint K_3's and color them red. Color all remaining edges blue. The blue edges induce $K_{3,3}$, which has no triangles, so there is no blue $K_1 + K_1 + K_2$. The red $2K_3$ contains no P_4. Thus $r(P_4, K_1 + K_1 + K_2) > 6$.

Next, we must show that any (red, blue)-coloring of K_7 must produce a red P_4 or a blue $K_1 + K_1 + K_2$. Suppose the vertices of K_7 are labeled a, b, c, d, e, f, g. We consider two cases: (1) At least three edges, say, ab, ac, ad, incident with a are red, or (2) at least three edges, say, ab, ac, ad, incident with a are blue.

Case 1: To avoid a red P_4, all other edges incident with b, c, and d must be blue [see Figure 11.1(a), where red edges are shown]. But then edges bc, bd, cd and de induce a blue $K_1 + K_1 + K_2$.

Case 2: In this case, edges bc, bd, and cd must all be red to avoid a blue $K_1 + K_1 + K_2$. But then to avoid a red P_4, all other edges incident with b, c, and d must be blue. [See Figure 11.1(b), where red edges are thicker.] Then if any of the edges ef, fg, ga is blue, we would get a blue $K_1 + K_1 + K_2$. To avoid that, all of ef, fg, and ga must be red. But then those edges induce a red P_4. Hence $r(P_4, K_1 + K_1 + K_2) = 7$. ◇

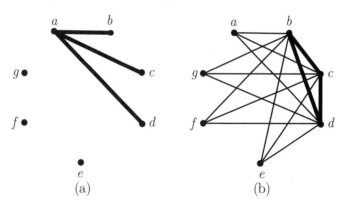

Figure 11.1 Partial (red, blue)-colorings of K_7.

Now we consider 3-colorings of the edges of K_n. The ramsey number $r(F, G, H)$ is the smallest positive integer such that any (red, blue, green)-coloring of the edges of the complete graph on $r(F, G, H)$ vertices contains a red F, a blue G, or a green H.

Example 11.3 Determine $r(P_3, P_4, K_3)$.

Solution

First we show that $r(P_3, P_4, K_3) > 6$ by exhibiting a (red, blue, green)-coloring of K_6 that avoids a red P_3, a blue P_4, and a green K_3. Let the vertices of K_6 be labeled 1, 2, 3, 4, 5, 6. Join all odd vertices by blue edges and join all even vertices by blue edges, thereby making two blue K_3's. Color by red the edges $(1, 2)$, $(3, 4)$, and $(5, 6)$. Color all remaining edges of K_6 by green. The red edges induce $3K_2$; that is, a perfect matching in red. The green edges induce a green C_6. So there is no red P_3, no blue P_4, and no green K_3.

Next we show that any (red, blue, green)-coloring of K_7 produces a red P_3, a blue P_4, or a green K_3. Label the vertices of K_7 by a, b, c, d, e, f, g. Again, at most one edge incident with a is red or we would get a forbidden red P_3. Thus at least three edges incident with a are blue (case 1), or at least three edges incident with a are green (case 2).

Case 1: Without loss of generality, suppose ab, ac, and ad are blue (dotted in Figure 11.2). None of the edges bc, bd, or cd can be blue or we would get a forbidden blue P_4. Furthermore, exactly one of these edges, say bc, must be red (thick in Figure 11.2) or we would get a red P_3 or a green K_3. So bd and cd are both green (thin and solid in Figure 11.2). To avoid a red P_3 and to avoid a blue P_4, be, bf, bg, ce, cf, and cg must all be green. Then to avoid a green K_3 and a blue P_4, edge dg must be red. We then find it is impossible to color edge de and avoid a red P_3, a blue P_4, or a green K_3. See Figure 11.2.

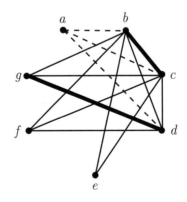

Figure 11.2

11.1 Ramsey Theory

Case 2: Suppose that ab, ac, and ad are green. By case 1, we may conclude that no three blue edges meet at a given vertex. This means that not only must we avoid a blue P_4, we must also now avoid a blue $K_{1,3}$. To avoid a green K_3, none of bc, bd, or cd can be green. Furthermore, to avoid a red P_3, at least two of those edges, say bc and cd, must be blue. Focus on vertex a for a second. If some additional edge incident with a, say ag, is green, then to avoid a green K_3 and a blue P_4, both bg and dg must be red. But this would produce a forbidden red P_3. So no additional edges incident with a are green. To avoid a red P_3 and a blue $K_{1,3}$, exactly two of the remaining edges incident with a, say ae and af, are blue, and the other, ag, is red. See Figure 11.3(a).

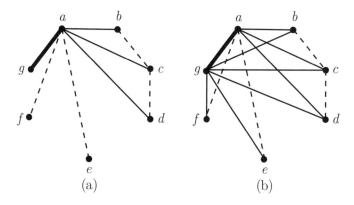

Figure 11.3

Now certain edges are forced. Edges bg, cg, and dg must be green (do you see why?). Also, ge and gf must be green (why?). Our graph is shown in Figure 11.3(b). Finally, be must be red (why?) and de must be red (why?). But this produces a red P_3. Hence $r(P_3, P_4, K_3) = 7$. ◇

Ramsey theory is a very extensive subject. We have only touched on a small portion of it. There is a whole book by Graham, Rothschild, and Spencer [4] on the subject. Be forewarned, it is quite challenging. But it is loads of fun.

Exercises 11.1

1. Explain why $r(m, n) = r(n, m)$ for all positive integers m and n.

2. Show that Theorem 11.3 gives a better bound for $r(4, 4)$ than Corollary 11.3a does. *Hint*: Table 11.1 will be needed.

3. Use Theorem 11.3 and Corollary 11.3a to obtain bounds for $r(5, 5)$. Then compare those bounds with the upper bound given in the text.

4. Prove that $r(2,n) = n$ for all positive integers n.

5. Prove that for all $n \in N$, $r(3,n) \leq (n^2+n)/2$.

6. Suppose that G and H have order m and n, respectively. Explain why $r(G,H) \leq r(m,n)$.

7. Prove that $r(a,b) \leq r(m,n)$ when $a \leq m$ and $b \leq m$.

8. Prove that $r(P_3, P_3, P_3) = 5$.

9. Prove that $r(P_4, P_4) = 5$.

10. Prove that $r(2K_2, K_3) = 5$.

11. Find $r(K_2, K_2, K_3)$.

12. Find $r(K_3, K_3, K_2)$.

13. Show that $r(C_4, C_4) = 6$.

14. Prove that $r(K_{1,m}, K_{1,n}) = \begin{cases} m+n-1 & \text{if } m \text{ and } n \text{ are even.} \\ m+n & \text{otherwise} \end{cases}$

15. Prove that $r(P_4, K_1 + K_2 + K_1) = 7$.

16. Prove that $r(P_5, K_{1,3}) = 5$.

17. Another generalization of ramsey numbers is to 2-color the edges of some graph other than K_n. Show a (red, blue)-coloring of $K_{4,4}$ that avoids $K_{2,2}$ in both red and blue.

18. Show that any (red, blue)-coloring of $K_{5,5}$ will contain a red $K_{2,2}$ or a blue $K_{2,2}$.

19. Determine $r(K_{1,4}, C_4)$.

11.2 Domination in Graphs

As our final topic of this text, we consider graph domination. Spurred on mainly by its wide variety of interesting applications, research in graph domination has experienced a truly explosive growth in the last few decades. So much so that the fascinating book [8] by Haynes, Hedetniemi, and Slater, devoted exclusively to that topic, contains over 1200 references. Some of the many applications discussed in [8] are job assignments, school bus routing,

11.2 Domination in Graphs

communication networks, placement of radio stations, social network theory, and land surveying. In a recent interesting paper, Haynes, Hedetniemi, Hedetniemi, and Henning [7] discussed power grid domination.

Domination Concepts

The first instance of domination appearing in mathematics was in about 1850 in what is often called the **n queens problem**. In the game of chess, a queen Q at position x on the chessboard is permitted to move horizontally, vertically, or diagonally. Q is said to **attack** a position y if Q could capture a chess piece at position y. That is, position y is in the direct line of sight from Q's position x either horizontally, vertically, or diagonally. In the n queens problem, the challenge is to place n queens on a previously empty chessboard so each of the 64 squares is either occupied by a queen or being attacked by at least one queen. The queens are said to dominate all the squares they either occupy or attack, and the set of queens is a dominating set for the board. The problem is to determine the minimum number of queens in such a dominating set. The answer is 5; one solution is displayed in Figure 11.4. There is also a solution where all five queens are in the same row, and there is a solution where all the queens are on the main diagonal. The solution in Figure 11.4 is quite special because it has the additional property that no two queens are attacking one another.

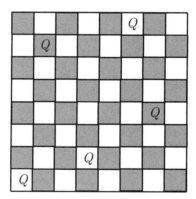

Figure 11.4 Five dominating queens, no two attacking one another.

The concept of a dominating set for a graph was not formally defined until 1958, when Berge [1] wrote the second graph theory book (it was in French); the first graph theory book was written by König [7] in 1936 (it was in German). A set of vertices S in graph G is a **dominating set** in G if every vertex of G is either in S or adjacent to a vertex in S. Of course, $V(G)$ is a dominating set for G, but we are generally interested in finding a much smaller set. A **minimal dominating set** S is a dominating set for which no proper subset is a dominating set. That is, if S' is a dominating set and $S' \subseteq S$, then $S' = S$. A **minimum dominating set** is a dominating set of minimum cardinality. If the vertices of dominating set S are mutually nonadjacent, then S is an **independent dominating set**. For the graph of Figure 11.5, $\{c, d, g, j\}$ and $\{c, f, i\}$ are each dominating sets, but of those two only $\{c, f, i\}$ is an independent dominating set. Set $\{c, h\}$ is a minimum size independent dominating set. An independent dominating set is necessarily minimal, but not necessarily minimum.

Example 11.4 For the graph G in Figure 11.5, find
a. all minimal dominating sets
b. all independent dominating sets
c. all minimum dominating sets

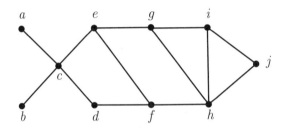

Figure 11.5

Solution
Since there are many sets, it helps to organize them alphabetically.
a. We find the following minimal dominating sets: $\{a, b, d, e, h\}$, $\{a, b, d, e, i\}$, $\{a, b, d, e, j\}$, $\{a, b, d, g, i\}$, $\{a, b, d, g, j\}$, $\{a, b, e, f, j\}$, $\{a, b, f, g, j\}$, $\{a, b, f, h\}$, $\{a, b, f, i\}$, $\{c, d, e, j\}$, $\{c, d, g, j\}$, $\{c, e, i\}$, $\{c, f, g, j\}$, $\{c, f, i\}$, and $\{c, h\}$. It seems rather surprising that there are so many. You should verify that none of these sets is a subset of another.
b. Of the dominating sets found in part (a), the independent sets are $\{a, b, d, e, h\}$, $\{a, b, d, e, i\}$, $\{a, b, d, e, j\}$, $\{a, b, d, g, j\}$, $\{a, b, f, g, j\}$, $\{a, b, f, i\}$, $\{c, f, g, j\}$, $\{c, f, i\}$, and $\{c, h\}$.
c. The only minimum dominating set is $\{c, h\}$. ◇

Example 11.5 Suppose that a county contains eight cities connected by highways as indicated in Figure 11.6. The county supervisor wants to upgrade their medical facilities so that each city has a major burn unit within their hospital system or is adjacent to a city that does. Due to budgetary constraints only a minimum number of units an be upgraded.
a. In which cities should the burn units be placed if there must be one at c?
b. In which cities should the burn units be placed if there is already one at b, the largest city in the county?

11.2 Domination in Graphs

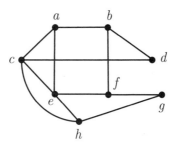

Figure 11.6

Solution

Here we need a minimum dominating set.

a. The burn center at c will dominate c, a, d, e, and h. That is, each of those cities will have burn center service nearby. To dominate the remaining cities b, f, and g, we place the burn center at f. Thus there will be burn centers at c and f.

b. The burn center at b serves b, a, d, and f. To service the remaining cities, build just one additional burn center at h. So the burn centers will be at b and h. This is a second minimum dominating set. ◇

The **domination number**, **$\gamma(G)$**, of a graph G is the cardinality of a minimum dominating set. For the graph G in Figure 11.6, we have $\gamma(G) = 2$. Determining $\gamma(G)$ for some graphs is rather easy. For example, it is easy to show that $\gamma(K_n) = \gamma(W_{1,n}) = 1$. Also, $\gamma(W_{m,n}) = \min\{m, n\}$. For graphs in general, however, calculating $\gamma(G)$ can be extremely difficult. There is no known efficient algorithm for calculating $\gamma(G)$.

Example 11.6 The city planners of Mesh City have determined that not only is their city a mesh, it is also a mess. To correct the problem, they have decided to place waste receptacles at various intersections. Determine the fewest number of receptacles that they can use so that for each intersection, there is either a receptacle or there is one at an intersection one block away. Their street grid is $M(4, 5)$ and is shown in Figure 11.7.

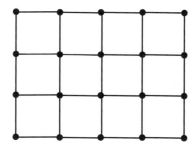

Figure 11.7 The street grid of Mesh City.

Solution
The problem is equivalent to finding $\gamma(G)$ for graph G in Figure 11.7. First note that there are 20 vertices and a vertex dominates its neighborhood and itself. Since $\Delta(G) = 4$, each vertex can dominate at most five vertices, so $\gamma(G) \geq 20/5 = 4$. However, that lower bound cannot be achieved. Here's why. To dominate a corner vertex such as $v = (1, 4)$, we must include either v or a neighbor of v in the dominating set. Vertex v dominates just three vertices and its neighbors dominate only four. So already the bound creeps up to $\gamma(G) \geq 5$ if we use neighbors each time [$20 - 4(4) = 4$, so at least one additional vertex is needed to dominate the four remaining vertices]. To achieve $\gamma(G) = 5$, we could not use more than one corner (which dominates only three vertices). For example, if we use two corners and two neighbors of corners, they could dominate at most $2(3) + 2(4) = 14$ vertices. At least two additional vertices would be needed in a dominating set to dominate the remaining six vertices.

Now we show that $\gamma(G) > 5$. Since there must be at least three neighbors of corners in a minimum dominating set, and having them adjacent would waste a possible vertex domination, there must be two such vertices in consecutive columns of $M(4,5)$. Without loss of generality, we may assume that $(2, 1)$ and $(1, 3)$ are in a minimum dominating set S. Then all vertices in the first two columns are dominated except $(2, 4)$. Vertex $(3, 1)$ is also dominated. To dominate $(2, 4)$, we must use $(3, 4)$, because using any other vertex would dominate at most two new vertices, which would be equivalent to using two corners (which is not allowed). Thus $(3, 4) \in S$. Now $(3, 2)$ is undominated. The only vertex that dominates $(3, 2)$ and more than one other new vertex is $(4, 2)$, which dominates five new vertices. So $(4, 2) \in S$. Now there are three undominated vertices: $(5, 1)$, $(5, 3)$, and $(5, 4)$. At least two vertices are needed to dominate them. So $\gamma(G) > 5$. Since $S = \{(1,3), (2,1), (3,4), (4,2), (5,1), (5,3)\}$ is a dominating set for $M(4, 5)$, we see that $\gamma(G) = 6$. Thus, the minimum number of trash receptacles that are required is six. Now if the city managers could just convince the residents to *use* the receptacles, the suburban Graphland Valley folks might come visit. ◊

Covers, Domination, and Independent Sets

We saw in Chapter 6 that $\beta(G)$ denotes the maximum cardinality of an independent set of vertices in a graph. We also saw that certain bounds on the chromatic number can be obtained in terms of the order, n, and $\beta(G)$. The domination number is also related to $\beta(G)$.

Theorem 11.4 For any graph G, $\gamma(G) \leq \beta(G)$.

Proof. Let I be an independent set of vertices in G, where $|I| = \beta(G)$. So I is an independent set of maximum cardinality. For $v \in V(G) - I$,

11.2 Domination in Graphs

v must be adjacent to some vertex of I. Otherwise, $I \cup \{v\}$ would be a larger independent set, a contradiction. So I dominates all of $V(G)$. Thus $\gamma(G) \leq \beta(G)$. ∎

A concept related to domination is vertex coverings. A **vertex cover** is a set S of vertices such that every edge of G is incident with at least one vertex of S. The minimum cardinality of a vertex cover of G is denoted $\boldsymbol{\alpha(G)}$. It is easy to prove that $\alpha(K_n) = n - 1$. Simply use induction on n. For complete bipartite graphs, we have $\alpha(K_{m,n}) = \min\{m, n\}$.

A simple application involving vertex covers is the placement of guards in a museum. Because of the valuable paintings, the museum will place a guard at various intersections of hallways. Assuming a guard can watch each hallway incident with where he is posted, what is the fewest number of guards that are needed? This is equivalent to finding $\alpha(G)$ for the graph modeling the museum's hallway floor plan.

Incidentally, some people confuse $\alpha(G)$ with $\gamma(G)$. With $\alpha(G)$, we are covering edges with vertices, whereas with $\gamma(G)$, we are dominating vertices with vertices. Suppose, for example, that the hallway floor plan of the museum is modeled by C_5. Then $\gamma(C_5) = 2$ while $\alpha(C_5) = 3$. If only two guards were in place, one hallway would be unwatched.

An interesting connection between vertex covers and independent sets is the following.

Theorem 11.5 If S is any vertex cover of G, then $V(G) - S$ is an independent set.

Proof. Suppose to the contrary that $V(G) - S$ is not independent. Then some pair of vertices x and y of $V(G) - S$ is adjacent. Then S does not cover the edge xy, contradicting the fact that S is a vertex cover. ∎

Corollary 11.5a For any graph G on n vertices, $\alpha(G) + \beta(G) = n$.

Proof. Let S be a minimum cardinality vertex cover. Then $|S| = \alpha(G)$. By Theorem 11.5, $V(G) - S$ is an independent set. So $\beta(G) \geq n - \alpha(G)$, which means that $\alpha(G) + \beta(G) \geq n$. On the other hand, if I is an independent set such that $|I| = \beta(G)$, then $V(G) - I$ must be a cover. To see this, note that the only edges in G are edges incident with $V(G) - I$, so $V(G) - I$ covers all those edges. This implies that $\alpha(G) \leq n - \beta(G)$ or, equivalently, $\alpha(G) + \beta(G) \leq n$. So we have $\alpha(G) + \beta(G) \geq n$ and $\alpha(G) + \beta(G) \leq n$. Thus $\alpha(G) + \beta(G) = n$. ∎

Exercises 11.2

1. Find a dominating set of five queens, all in the same row for an 8×8 chessboard.

2. Find another nonisomorphic (that is, rotating and/or flipping the board from Figure 11.4 is not allowed) placement of five queens on an 8×8

chessboard such that every position is dominated. A queen being attacked is permitted, but not all queens are in the same row, column or diagonal.

3. Find a dominating set of five queens, all on the main diagonal for an 8 × 8 chessboard.

4. Prove that an independent dominating set must be a minimal dominating set.

5. Draw a graph for which a minimum dominating set is not an independent set.

6. For graph G in Figure 11.8, find

    (a) all minimal dominating sets
    (b) all independent dominating sets
    (c) all minimum dominating sets

7. Determine $\alpha(G)$, $\beta(G)$, and $\gamma(G)$ for graph G in Figure 11.8.

8. Determine $\alpha(G)$, $\beta(G)$, and $\gamma(G)$ for the graph in Figure 11.5.

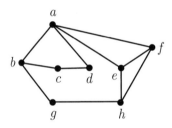

Figure 11.8

9. Determine $\alpha(G)$, $\beta(G)$, and $\gamma(G)$ for the graph in Figure 11.6.

10. Find α, β, and γ for the 2-mesh $M(4,3)$.

11. Find $\gamma(M(2,n))$.

12. Show that $\gamma(P_6) = 2$ and $\gamma(P_9) = 3$.

13. Show that $\gamma(P_{3k}) = k$.

14. Show, with the help of the previous exercise, that $\gamma(P_n) = \lceil \frac{n}{3} \rceil$.

15. Find α, β, and γ for C_n.

References for Chapter 11

1. Berge, C., *Théorie des Graphes et ses Applications*, Dunod, Paris (1958).

2. Bóna, M., and G. Tóth, A Ramsey-type problem on right-angled triangles in space. *Discrete Mathematics* 150 (1995) 61–67.

3. Erdős, P., and G. Szekeres, A combinatorial problem in geometry. *Compositio Math.* 2 (1935) 463–470.

4. Graham, R. L., B. L. Rothschild, and J. E. Spencer, *Ramsey Theory*. Wiley, New York (1990).

5. Haynes, T. W., S. M. Hedetniemi, S. T. Hedetniemi and M. A. Henning, Power domination in graphs applied to electrical power networks, *SIAM J. Discrete Mathematics* (to appear).

6. Haynes, T. W., S. T. Hedetniemi, and P. J. Slater, *Fundamentals of Domination in Graphs*, Marcel Dekker Inc., New York (1998).

7. König, D., *Theorie der endlichen und unendlichen Graphen*, BSB B. G. Teubner Verlagsgesellschaft, Leipzig (1936).

8. Ramsey, F. P., On a problem of formal logic. *Proceedings of the London Mathematical Soc.* 2nd Ser., 30 (1930) 264–286.

9. Ryser, H. J., *Combinatorial Mathematics*. Carus Mathematical Monographs, no. 14, The Mathematical Association of America, Washington, DC (1963).

Additional Readings

10. Albertson, M. O., People who know people. *Mathematics Magazine* 67 (1994) 278–281.

Answers to Exercises

Chapter 1

Exercises 1.1 (page 12)

1. $3\lfloor x \rfloor \leq \lfloor 3x \rfloor$, while $3\lceil x \rceil \geq \lceil 3x \rceil$.
2. $n\lfloor x \rfloor \leq \lfloor nx \rfloor$, while $n\lceil x \rceil \geq \lceil nx \rceil$.
3. Let $x = n + a$ and let $y = m + b$, where n and m are integers, and a and b satisfy $0 \leq a < 1$ and $0 \leq b < 1$. Then $\lfloor x + y \rfloor = \lfloor n + m + a + b \rfloor \geq n + m = \lfloor x \rfloor + \lfloor y \rfloor$. If we let $x = y$, we get inequality (1.1).
4. $f(n) = \lceil n/3 \rceil$.   5. The maximum value is 9.
6. $(6k+1)(6j+1) = 36kj + 6k + 6j + 1 = 6(6kj + k + j) + 1$. The sum of n awesome numbers has the form $n(6t) + n = 6tn + n$. This is an awesome number precisely when $n = 6j + 1$, for some integer j. When $j = 1$, notice that $n = 7$.
7. The collection of tall people, the collection of difficult subjects, and the collection of expensive cars.
8. a. The collection of all people aged 65 or older.
   b. The collection of all cars costing over $35,000.
   c. The collection of all students who scored at least 1200 on their SATs.
9. $\{0, 3, -3, 6, -6, 7, -7, 9, -9, 12, -12, 14, -14, 15, -15, 18, -18, 21, -21, 24, -24, 27, -27, 28, -28, 30, -30, 33, -33, \ldots\}$.
10. $\{1, 2, 3, 4, 5, 6, 7, 8\}$ 11. $\{12, 13, 14, 15, 16, \ldots, 90\}$
12. $\{36, 41, 46, 51, 56, 61\}$ 13. $\{x : x = 2n, n \in N, 1 \leq n \leq 45\}$
14. $\{x : x \in N, 3 \leq x \leq 10\}$ 15. $\emptyset$
16. $\{\text{Uruguay}\}$ 17. $\{1, -1, 3, -3, 5, -5, \ldots\}$
18. $\{x : x = 20n, n \in N\}$
19. Sets in Exercises 14, 15, and 16 are finite; sets in Exercises 17 and 18 are infinite.
20. Statements (a) and (c) are true.
21. $Y \subset F$, $Y \subset W$, $F \subseteq W$, $W \subseteq F$, $D \subset Y$, $D \subset F$, $D \subset W$ (note that $D = \emptyset$). Sets W and F are equal.
22. The first set contains the element 6, but the second does not.
23. Yes. No; for example, $\{1, 3\}$ is a finite subset of the infinite set Z.
24. a. $\{6, 12\}$ b. $\{2, 4, 10\}$ c. $\{4, 10\}$ d. $\{4, 10\}$
30. $B \in A$, and $B \subseteq A$ are true. The other relations are false.
31. $S = \{\emptyset, 1, \{1\}\}$ 32. $S = \{\emptyset, \{\emptyset\}, \{\{\emptyset\}\}, \{\{\{\emptyset\}\}\}, \ldots\}$
33. a. $P(A) = \{\emptyset, \{a\}, \{b\}, \{c\}, \{a,b\}, \{a,c\}, \{b,c\}, \{a,b,c\}\}$
    b. $\{(a,c), (a,d), (b,c), (b,d), (c,c), (c,d)\}$
    c. $\{(c,a), (c,b), (c,c), (d,a), (d,b), (d,c)\}$
    d. $\{(a,a), (a,b), (a,c), (b,a), (b,b), (b,c), (c,a), (c,b), (c,c)\}$

Exercises 1.2 (page 23)

1. Note that $\sum_{i=1}^{1}(2i-1) = 1^2$ is true. Now assume that $\sum_{i=1}^{k}(2i-1) = k^2$ is true. We have $\sum_{i=1}^{k+1}(2i-1) = \sum_{i=1}^{k}(2i-1) + 2(k+1) - 1 = \sum_{i=1}^{k}(2i-1) + 2k + 1 = k^2 + 2k + 1 = (k+1)^2$.

2. For $n = 0$: $\sum_{i=0}^{0} 2^i = 2^0 = 1 = 2^{0+1} - 1$. Assume for $n = k$: $\sum_{i=0}^{k} 2^i = 2^{k+1} - 1$. Now show for $n = k+1$: $\sum_{i=0}^{k+1} 2^i = \sum_{i=0}^{k} 2^i + (2^{k+1}) = 2^{k+1} - 1 + (2^{k+1}) = 2(2^{k+1}) - 1 = 2^{k+2} - 1$.

6. Write the sum as $\sum_{i=1}^{n}(4i-1)$. When $n = 1$, we have $\sum_{i=1}^{1}(4i-1) = 3 = 1(2(1)+1)$. Now assume that $\sum_{i=1}^{k}(4i-1) = k(2(k)+1)$, and observe that $\sum_{i=1}^{k+1}(4i-1) = \sum_{i=1}^{k}(4i-1) + (4(k+1) - 1) = k(2k+1) + (4k+3) = 2k^2 + 5k + 3 = (k+1)(2k+3)$.

7. Note that $\sum_{i=1}^{1} i^3 = 1^3 = 1 = \frac{1^2(1+1)^2}{4} = \frac{4}{4} = 1$. Now assume that $\sum_{i=1}^{k} i^3 = \frac{k^2(k+1)^2}{4}$. Finally, observe that $\sum_{i=1}^{k+1} i^3 = \sum_{i=1}^{k} i^3 + (k+1)^3 = \frac{k^2(k+1)^2}{4} + (k+1)^3 = \frac{k^2(k+1)^2}{4} + \frac{4(k+1)^3}{4} = \frac{(k+1)^2(k^2+4k+4)}{4} = \frac{(k+1)^2(k+2)^2}{4}$.

18. When $n = 1$, we obtain $1^3 + 2(1) = 3$, which is divisible by 3. Now assume that $k^3 + 2k$ is divisible by 3 and show that $(k+1)^3 + 2(k+1)$ is divisible by 3. $(k+1)^3 + 2(k+1) = k^3 + 3k^2 + 3k + 1 + 2(k+1) = k^3 + 3k^2 + 3k + 1 + 2k + 2 = k^3 + 2k + 3(k^2 + k + 1)$. By the inductive hypothesis, $k^3 + 2k$ is divisible by 3. Since $3(k^2 + k + 1)$ is divisible by 3, so is the sum.

Exercises 1.3 (page 31)

1. $9! = 362,880$ 2. $(n+2)(n+1) = n^2 + 3n + 2$

8. Each route is an ordered sequence of m U's (for "up") and n R's (for "right"). There are $m + n$ places in each sequence. A sequence is fully determined by choosing the m places to be occupied by U's.

9. $20(7)(2) = 280$ 10. $6(3)(3)(2) = 108$
11. $26^2 \cdot 10^4 = 6,760,000$ 12. $26^3 \cdot 10^3 = 17,576,000$
13. $26^4 \cdot 10^2 = 45,697,600$ 14. $26(25)(10)(9)(8) = 468,000$
15. $27 \cdot 26 \cdot 10^3 = 702,000$ 21. $5 \cdot 9 \cdot 10^6 = 45,000,000$
22. $7(5) = 35$

Answers/Solutions to Selected Exercises   315

46. The factors of $n = p^a q^b r^c s^d$ are of the form $p^w q^x r^y s^z$, where $0 \le w \le a$, $0 \le x \le b$, $0 \le y \le c$, and $0 \le z \le d$, where $w, x, y, z \in N \cup \{0\}$. Then there are $a + 1$ choices for w, $b + 1$ choices for x, $c + 1$ choices for y, and $d + 1$ choices for z. By the multiplication principle, we obtain $(a+1)(b+1)(c+1)(d+1)$ factors of n. To generalize this, write $n = a_1^{e_1} a_2^{e_2} \cdots a_r^{e_r}$, where e_1 is the number of times the prime factor a_1 appears as a factor of n.

47. $5!/2 = 60$. Let one of the repeated letters be A'. Then there are $5!$ ordered sequences of A, A', B, C, D. Now removing the prime from the A' cuts the number in half, as, for example, there is no distinction between $ABDA'C$ and $A'BDAC$. If there are n symbols with k repetitions, we get $n!/k!$ ordered sequences.

49. $(n-1)!$. Each of the $n!$ configurations may be rotated into n positions without affecting the seating arrangement. Hence we divide $n!$ by n.

Exercises 1.4 (page 44)

1. $[1 \cdot 3 \cdot 5 \cdots (2n-1)] \cdot [2 \cdot 4 \cdot 6 \cdots (2n)] = (2n)!$, $2 \cdot 4 \cdot 6 \cdots (2n) = (2 \cdot 1) \cdot (2 \cdot 2) \cdot (2 \cdot 3) \cdots (2 \cdot n) = 2^n \cdot n!$

2. Use the recursion formula to show that the initial and final 1's in row 2^k change the parities of their neighbors in row $2^k + 1$. These in turn change the parities of their inner neighbors in row $2^k + 2$, and so on. Subsequent outer 1's have a similar effect.

4. For $n = 1$, $\sum_{i=1}^{1} i(i!) = (1+1)! - 1 = 1$. Assume $\sum_{i=1}^{k} i(i!) = (k+1)! - 1$, and show that $\sum_{i=1}^{k+1} i(i!) = (k+2)! - 1$. We have $\sum_{i=1}^{k+1} i(i!) = \sum_{i=1}^{k} i(i!) + (k+1)(k+1)! = (k+1)! - 1 + (k+1)(k+1)! = (k+1)!(k+2) - 1 = (k+2)! - 1$.

6. $1, 9, 36, 84, 126, 126, 84, 36, 9, 1$

7. The left side counts the number of r-element subsets of an n-element set A. Let x and y be two element of A. Separate the $\binom{n}{r}$ r-element subsets of A into three groups: those that contain neither x nor y, those that contain just one of x and y, and those that contain both x and y. The right side counts each of those groups.

8. $r\binom{n}{r} = r \cdot \frac{n!}{r!(n-r)!} = \frac{n!}{(r-1)!(n-r)!} = \frac{n(n-1)!}{(r-1)![(n-1)-(r-1)]!} = n \cdot \frac{(n-1)!}{(r-1)![(n-1)-(r-1)]!} = n\binom{n-1}{r-1}$

9. $\sum_{r=1}^{n} r\binom{n}{r} = \sum_{r=1}^{n} n\binom{n-1}{r-1} = n \sum_{r=1}^{n} \binom{n-1}{r-1} = n \sum_{j=0}^{n-1} \binom{n-1}{j} = n \cdot 2^{n-1}$

17. We want to choose three x's and six y^2's from the nine parentheses. The coefficient is $\binom{9}{3} = 84$.

18. $\binom{5}{3}(-2)^3 = -80$

Chapter 2

Exercises 2.1 (page 55)

1. a.

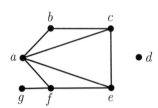

  b. $\{a, b, c\}$ and $\{a, e, f\}$ **c.** $4, 2, 3, 0, 3, 3, 1$

  d. $4 + 2 + 3 + 0 + 3 + 3 + 1 = 16 = 2(8)$ and there are eight edges in G.

2. a. The sum of the degrees is odd, contradicting Corollary 2.1a.

  b. The first vertex has degree 10, which means that it must be adjacent to 10 other vertices. But the length of the degree sequence indicates that there are only seven other vertices.

3. a. The vertices correspond to cities and an edge corresponds to a road between a given pair of cities.

  b. The vertices correspond to atoms, and an edge indicates a bond between a given pair of atoms.

  c. The vertices correspond to stations (subway stops), and an edge indicates that a subway travels directly from one given station to another.

  d. The vertices correspond to people in a given clan, and an edge indicates that one of the people is a child (son or daughter) of the other person linked by that edge.

  e. The vertices correspond to jobs and applicants, and an edge joins an applicant to a job that that applicant is qualified for.

7.

8. If there were such a 3-regular graph, then the sum of the degrees would be odd, which is impossible.

Answers/Solutions to Selected Exercises 317

Exercises 2.2 (page 62)

1. Cycles
2. Each vertex is adjacent to all $n-1$ of the remaining vertices.
4. mn
5. a. 91   b. 54   c. 24   d. 80   e. 19
6. Draw two copies of K_4.
7. Graphs H and K are subgraphs of G. Graph J is not because J contains edge bd, which is not an edge of G.
8. We found in Exercise 7 that just graphs H and K are subgraphs of G. Of those two graph, H is an induced subgraph. For an induced subgraph, we must use the maximal subgraph with the given vertex set. For graph H, all necessary edges appear appear. Thus, H is induced. For graph K, edge ce would also need to appear since that edge is in G. Thus, K is not an induced subgraph of G.
9. The walks are a,b,a,b,f; a,b,e,d,f; a,b,f,b,f; a,b,f,d,f; a,d,e,b,f; a,d,f,b,f; a,d,f,d,f; a,e,a,b,f; a,e,a,d,f; and a,e,c,b,f
10. a,b,e,d,f; a,d,e,b,f; and a,e,c,b,f are paths. Each one uses all distinct vertices.
11. a,e,b,c,e,d,f is an a–f trail of length six. Note that although it reuses some vertices, which is allowed, it does not reuse any edges.
12. A longest path in a graph with six vertices has at most five edges. Thus an a–f path of length six is impossible.
13. b,a,d; b,e,d; and b,f,d are the b–d geodesics in graph G.
14. Graph K has $2^6 = 64$ spanning subgraphs. For each of the six edges we must decide whether to include it or not. The multiplication principle then yields the result.
18. Prove this by contradiction. Assume that G is a connected graph of order n such that $q < n-1$ and obtain a contradiction as follows. Show using Theorem 2.1 that G has an end vertex v, adjacent to another vertex u. Now delete v and edge uv, producing a graph H that is obviously connected. For H, the number of edges still lags behind the number of vertices by at least two. ($q < n-1$ is equivalent to $q \leq n-2$.) Now continue reducing the order of the graphs one vertex at a time until you obtain a supposedly connected graph with three vertices and at most one edge, which is absurd.
19. It has $(n-1)(n-2)/2$ edges. Consider a graph with two complete graph components, one of order k. It has $f(k) = (1/2)k(k-1) + (1/2)(n-k)(n-k-1) = k^2 - nk + (1/2)(n^2 - n)$ edges. The graph (not in the graph theory sense) of this function is an upright parabola. Hence the maximum in the range $1 \leq k \leq n-1$, is achieved by taking $k=1$ or $n-1$.

20. Since H spans G and L spans H, it follows that the three graphs have the same vertex set. Furthermore, $E(L) \subseteq E(H) \subseteq E(G)$, implying that $E(L) \subseteq E(G)$. Then it follows that L spans G.

Exercises 2.3 (page 72)

2. The graph is $P_3 \cup K_1$. Draw it.

4. The graphs are $C_5 \cup P_2$, $C_4 \cup P_3$, and $C_3 \cup P_4$. Draw them.

5. The vertex of degree three in the first graph is adjacent to only one end vertex.

6. If v and w are corresponding vertices in isomorphic graphs G and H, respectively, then since an isomorphism preserves adjacencies, there is a one-to-one correspondence between the neighbors of v in G and the neighbors of w in H. Hence v and w have the same degree.

8. $3,3,3,3,3,3,3,3 \Rightarrow 2,2,2,3,3,3,3 \Rightarrow 3,3,3,3,2,2,2 \Rightarrow 2,2,2,2,2,2,$ which is the degree sequence of C_6. Two nonisomorphic graphs with the original degree sequence $3,3,3,3,3,3,3,3$ are drawn below.

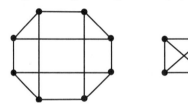

9. Do this the long way, or notice that it is the degree sequence of the wheel $W_{1,n}$.

10. $m+1$. A vertex of degree m requires m neighbors.

11. Let $V(G) = \{v_1, v_2, \ldots, v_k, v_{k+1}, \ldots, v_n\}$, where $\deg(v_i) = d_i$ for each $i = 1, 2, \ldots, n$. Now let $H = G \cup \{w\} \cup \{wv_1, wv_2, \ldots, wv_k\}$. Then $\deg_H(w) = k$, while $\deg_H(v_i) = d_i + 1$, for $i = 1, 2, \ldots, k$.

12. Start with the graphical sequence $1,1$. Now use the previous exercise repeatedly to obtain the graphical sequences $2,2,2$, then $3,3,3,3$, and so on.

13. n vertices. The number of edges is given by $(1/2)(d_1 + d_2 + \cdots + d_n)$. In particular $d_1 + d_2 + \cdots + d_n$ must be even. Since $5+4+4+4+4 = 21$, the sequence is not graphical. Easier—if $n = 5$, then $\Delta \leq 4$.

14. Label the consecutive vertices of C_4 by a, b, c, d, and label the vertices in one part of $K_{2,2}$ by A and C and those in the other part by B and D. Then show that with the function that relates a vertex with a given lowercase letter with the corresponding uppercase letter, adjacencies are preserved.

16.

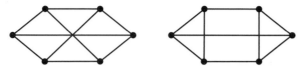

19. To get a subgraph isomorphic to K_5, we must choose any five vertices from the seven and then include all edges joining those five vertices. There are $\binom{7}{5} = 21$ ways we could select the five labeled vertices, and then just one way to include all the edges. Thus 21 of the subgraphs are isomorphic to K_5.

21. There are many characteristics that distinguish them. For example, graph G has just one triangle while H has two. Also, in graph H the vertices of degree two are adjacent while in graph G they are not.

Exercises 2.4 (page 82)

1.

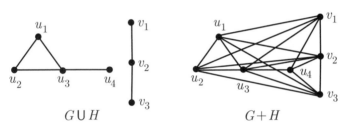

2. We showed in this section that Q_n has 2^n vertices. Each of those vertices has degree n. By Theorem 2.1, the number of edges is therefore $(1/2)n \cdot 2^n = n \cdot 2^{n-1}$.

3.

4. $2n$

5. $Q_3 = K_2 \times Q_2 = K_2 \times (K_2 \times Q_1) = K_2 \times K_2 \times K_2 \cong P_2 \times P_2 \times P_2 = M(2,2,2)$

6. Every vertex v of $\bar{G}$ is adjacent to each other vertex that v is not adjacent to in G. Since there are n vertices in G, there are $n-1$ vertices other than v. Vertex v is adjacent to k vertices in G, so it is not adjacent to $n-1-k$ of the other vertices. Since this is true for each vertex v in $\bar{G}$, graph $\bar{G}$ is $(n-1-k)$-regular.

7. Let the center of $W_{1,n}$ be v and let x and y be adjacent vertices. Now delete edge xy and delete all edges incident with v except for vx and vy.

8. Using the vertices a, b, c, r, s draw edges rs, rb, rc, sa, sc. This is isomorphic to the original graph.

10. For a fixed a, consider a layer of a 3-mesh $M(a, b, c)$. That layer is a 2-mesh $M(b, c)$ that has a spanning path that begins at a corner vertex $(1, 1)$ and goes vertically as far as possible, then horizontally one space, then vertically as far as possible, then horizontally one space, and so on. This path will end at (b, c) if b is odd, and at $(1, c)$ if b is even. We can use the same spanning path in each layer, and then easily link the end vertices of the paths. The resulting spanning tree can be layed out on one layer as a 2-mesh $M(ab, c)$. So that 2-mesh spans the 3-mesh $M(a, b, c)$.

11. Think of the 3-mesh as a stack of horizontal 2-meshes. Now delete all "vertical edges" except for columns of boundary edges alternating between the left and right sides of each horizontal 2-mesh. Alternatively, let P be a spanning path of the bottom 2-mesh. Now add all vertical edges.

12. $\deg(1, 1, \ldots, 1) = n$. The vertex $(2, 2, \ldots, 2)$ has degree $2n$, as does every "interior" vertex. Let the entries be 1's and 2's to achieve intermediate degrees. If $a_i = 2$, there are no interior vertices.

14. To form the complement of this graph, B and $\bar{B}$ are interchanged while A remains unchanged. The joins between them (which make the graph look like a letter "A") are removed and copies of A, B, and $\bar{B}$ that were not joined in the original graph are now joined, thereby restoring the "A."

15. $\bar{K}_m + \bar{K}_n$ consists of mutually nonadjacent vertex sets of orders m and n, respectively, and each vertex in one set is adjacent to each vertex in the other.

16. K_{m+n}

18. Isomorphism also preserves nonadjacencies.

19. Since G has six vertices, $\deg(u)$ in G and $\deg(u)$ in $\bar{G}$ add up to 5. Then one of them has to be at least 3. Now assume u is adjacent to x, y, and z in G. If x is adjacent to y, then x, y, and u form a K_3 in G, and so on for x and z and y and z. If no pair among x, y, and z is adjacent in G, then $\{x, y, z\}$ forms a K_3 in $\bar{G}$. Done. Since C_5 is self-complementary and contains no K_3, we need at least six vertices.

Answers/Solutions to Selected Exercises 321

Chapter 3
Exercises 3.1 (page 91)

2.

3.

4.

5. 49

7. K_1, K_2. When $n \geq 3$, a tree on n vertices has end vertices of degree 1 and has internal vertices of degree greater than 1.

10. If T is a tree, then by definition T is connected; that is, every pair of vertices is connected by a path. Since T contains no cycles, the path joining the pair of vertices is unique.

13. $1 + 2 + 2^2 + 2^3 + \cdots + 2^k = 2^{k+1} - 1$. It has 2^k end vertices.

17. $3,3,2,2,1,1,1,1 \Rightarrow 2,1,1,1,1,1,1 \Rightarrow 0,0,1,1,1,1 \Rightarrow 1,1,1,1,0,0$, which is the union of two K_2's and two K_1's. Here are the three nonisomorphic trees:

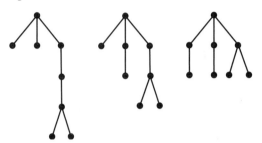

18. One component is the star $K_{1,3}$, and the other is $K_1 + K_1 + K_2$.

Exercises 3.2 (page 99)

1. **a.** The spanning trees are P_6, $K_1+K_1+K_1+K_1+\bar{K}_2$, $K_1+K_1+K_1+\bar{K}_3$, $\bar{K}_2+K_1+K_1+\bar{K}_2$, $K_{1,5}$, and the graph consisting of P_5 with a pendant edge attached at the middle vertex of the path.
   b. The spanning trees are P_6, $\bar{K}_2 + K_1 + K_1 + \bar{K}_2$, and the graph consisting of P_5 with a pendant edge attached at the middle vertex of the path.

2. Vertices a, d, e, f, n have deficiency zero, b, g, h, m have deficiency one, and c, i, j, k, l each have deficiency two. Thus, the sum of the deficiencies is fourteen. For the original graph in Figure 3.9, we have $n = 14$, and $q = 20$, so $2(q - n + 1) = 2(20 - 14 + 1) = 2(7) = 14$, consistent with Theorem 3.3.

6. Here are spanning trees with two, three and four degree-preserving vertices, in that order.

7. A spanning path will have two end vertices. Now pivot on an end vertex to produce a spanning tree with three end vertices, and so on.

8. By using Kruskal's algorithm, the unique edge with smallest weight is selected first, then the edge with next smallest weight, and so on. Thus

the edges with the $n-1$ distinct smallest weights will be precisely the edges of the spanning tree. Hence the spanning tree is unique.

9. Use a greedy algorithm such as Kruskal's, but select maximum weight edges at each stage rather than minimum weight ones.

12. **a.** af, cf, ab, ah, cg, and draw the tree.

   b. gc, cf, af, ab, ah, and draw the tree.

Exercises 3.3 (page 105)

1. The graphs are $4K_1$, $2K_1 \cup K_2$, $2K_2$, $K_1 \cup P_3$, $K_{1,3}$, P_4, and $K_{2,2}$. Draw them.

3. Suppose, without loss of generality, that G has an edge $e = uv$. Then each vertex of H is adjacent to both u and v in $G + H$. This implies that e together with any vertex of H induces a triangle in $G+H$. Since $G+H$ has a triangle, it is not bipartite.

5. If G is a subgraph of bipartite graph H, then H has no odd cycle. Then G also has no odd cycles, so by Theorem 3.4, G is bipartite.

7. If V_1 and V_2 are the partite sets of k-regular bipartite graph G, then each vertex of V_1 and V_2 has degree k. Thus there are $k|V_1| = k|V_2|$ edges between V_1 and V_2. But this implies that $|V_1| = |V_2|$.

8. If G is semiregular bipartite, where the vertices in V_1 have degree s and vertices in V_2 have degree t, then each edge is incident with precisely $s+t-2$ other edges. Since an edge becomes a vertex of the line graph, $L(G)$ is regular of degree $s+t-2$.

11. If G_1 has p_1 vertices and G_2 has p_2 vertices, then $G_1 \times G_2$ can be drawn with the vertices laid out in a $p_1 \times p_2$ rectangular grid. In the drawing, $G_1 \times G_2$ consists of p_2 horizontal copies of G_1 and p_1 vertical copies of G_2 with no additional edges. If G_1 and G_2 are both bipartite, then we can see that $G_1 \times G_2$ is also bipartite.

14. If $|X| = m$ and $|Y| = n$, then $\deg(x) = m$ for all $x \in X$ and $\deg(y) = n$ for all $y \in Y$. There are mr edges incident with vertices of X and ns edges incident with vertices of Y. Since each edge is incident with a vertex in each color set, $mr = ns$. Since $|Y| = n$, we have $r \leq n$. By the same logic, $s \leq m$. Note that $mr = ns$ implies that m divides ("goes into") ns. If the greatest common divisor of m and n is 1, then m divides s; that is, $s = km$, where $k \in N$. Then we have $mr = nkm$ or $r = nk$. This implies that $r \geq n$. But $r \leq n$, in which case $r = n$ and $k = 1$. Then $s = km = 1m = m$, and the graph must be $K_{m,n}$.

15. One such graph is shown here. (This graph is disconnected!) Another pair of values for r and s is $r = 8$ and $s = 6$.

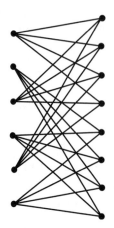

Exercises 3.4 (page 115)

1. Each of the three sets of (four) edges that divide Q_3 into copies of Q_2, and the four rotations of the matchings shown below (solid edges).

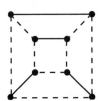

2. Use each edge that goes from the first copy of G to its corresponding vertex in the second copy of G. That will always produce a perfect matching.
3. Use $K_{1,n-1}$.
4. If n is even, then $W_{1,n}$ has odd order and so cannot have a perfect matching. If n is odd, however, label the vertices of the cycle $1, 2, 3, \ldots, n$. Then join each even labeled vertex k to vertex $k+1$, and join the center vertex of the wheel to vertex 1. That produces a perfect matching.
6. If T is a spanning tree of G, then each edge of T is an edge of G so each edge of a perfect matching of T is an edge of G. Since T spans G, the perfect matching includes all vertices of G and so is a perfect matching for G also. To show that the converse does not hold, use a wheel with a spanning star.
7. Use, for example, $K_{1,7}$, $K_{2,6}$, $K_{3,5}$, and $K_{1,9}$. There are infinitely many other choices.
8. One of them consists of the five rungs, four have three rungs, and three have one rung, making a total of eight perfect matchings, as expected.
9. The hint should suffice.

Answers/Solutions to Selected Exercises

10. There are $2n + 1$ ways to delete a vertex. The resulting graph is K_{2n}, which has, by the previous exercise, $1 \cdot 3 \cdot 5 \cdots (2n-1)$ perfect matchings. Then K_{2n+1} has $1 \cdot 3 \cdot 5 \cdots (2n-1) \cdot (2n+1)$ maximum matchings.

14.

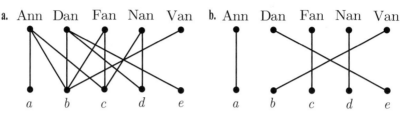

19. $(A_1, 5), (A_2, 1), (A_3, 7), (A_4, 2), (A_5, 3)$ is a system of distinct representatives for the sets.

21. $(A_1, 5), (A_2, 3), (A_3, 4), (A_4, 2), (A_5, 6), (A_6, 1)$ is a system of distinct representatives for the sets.

Chapter 4

Exercises 4.1 (page 127)

2. In C_n, $e(x) = \lfloor n/2 \rfloor$. When n is odd, x has two eccentric vertices.

3. $\text{rad}(C_n) = \text{diam}(C_n) = \lfloor n/2 \rfloor$. $\text{rad}(P_n) = \lfloor n/2 \rfloor$, $\text{diam}(P_n) = n - 1$. $|C(P_n)| = 2$ when n is even.

4. H is created by deleting edges from G. Then distances can either increase or stay the same.

5. Use the previous exercise and the fact that the eccentricity of a vertex is the maximum of its distances from the remaining vertices of the graph. P_n is a spanning subgraph of C_n, such that $\text{rad}(P_n) = \text{rad}(C_n)$, while $\text{diam}(P_n) > \text{diam}(C_n)$.

6. Use Exercise 2. Another infinite class is given by K_n.

7. Denote by x and y two randomly chosen vertices of K_n. Now remove edge xy and call this graph G_n. Then in G_n, $e(x) = e(y) = 2$, while all other vertices have eccentricity 1. It follows that $|C(G_n)| = n - 2$. Here's another example. Given the cycle C_n, with vertices $v_1, v_2, \ldots, v_n$, add vertices $w_1, w_2, \ldots, w_n$, and the n edges $v_1 w_1, v_2 w_2, \ldots, v_n w_n$. Call this graph H_n. Then $C(H_n) = \{v_1, v_2, \ldots, v_n\}$. The so-called **central ratio**, $|C(H_n)| / |V(H_n)| = 1/2$. In the first example in this exercise, $|C(G_n)| / |V(G_n)| = (n-2)/n$, which goes to 1 as n goes to infinity.

9. Given x and y in G, let $c \in C(G)$. Then $d(x, y) \leq d(x, c) + d(y, c) \leq 2 \, \text{rad}(G)$. Now $\text{diam}(G)$ is the maximum value of $d(x, y)$ for all pairs of vertices x and y in G, each of which obeys this inequality.

10. Given three vertices x, y, and z of a tree, such that x and y are adjacent, either y lies on the x–z path or x lies on the y–z path. In either case, one of x and y is one edge closer to z, implying that $|d(x, z) - d(y, z)| = 1$.

If a vertex v that is not an end vertex claims to be an eccentric vertex of a given vertex u, then enlarge the u–v path to a neighbor of v which is not on the path, showing that v is not a "farthest" vertex from u.

13. The star $K_{1,n-1}$ spans $W_{1,n-1}$ and has one center vertex. To get a spanning tree with two center vertices, let u be the center of $W_{1,n-1}$, and let v be any other vertex. Call the neighbors of v on the outer cycle x and y. Finally, include edges uv, vx, and vy and attach all of the remaining vertices to u. The center of this spanning tree consists of u and v.

14. Since edges are deleted in the construction of a spanning tree, distances either increase or stay the same. Hence for all $x \in V(G)$, $e_T(x) \geq e_G(x)$. Hence $\text{rad}(T) \geq \text{rad}(G)$ and $\text{diam}(T) \geq \text{diam}(G)$. Now if $\text{rad}(T) > \text{rad}(G)$, let $\text{rad}(G) = r$. Then $\text{rad}(T) \geq r + 1$. Then $\text{diam}(T) \geq 2(r+1) - 1 = 2r + 1 > 2r \geq \text{diam}(G)$; that is, $\text{diam}(T) > \text{diam}(G)$.

17. Start with P_{2m}; then divide the remaining $n - m$ black vertices into m sets of approximately the same size and attach each set of black vertices to a unique white vertex in P_{2m}.

18. Let L and M have common length k. Then $d(y,w)$ and $d(z,w)$ equal $k - d(x,w)$, from which $d(y,w) = d(z,w)$.

19. $x, y \in A \Rightarrow d(x, y) \leq 2$, since for any vertex $z \in B$, there is a path x, z, y, and so on. The diameter is three because a vertex in one copy of $\bar{B}$ is at distance three from any vertex in the other copy of $\bar{B}$.

20. $x \in G$, $y \in H \Rightarrow d(x, y) = 1$. $x, y \in G$ or $x, y \in H \Rightarrow d(x,y) \leq 2$. If G and H are complete, $diam(G + H) = 1$.

21. $x \in G$, $y \in J \Rightarrow d(x,y) = 2$, and so on.

22. Here are the weights.

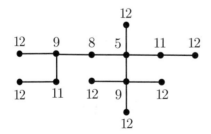

23.

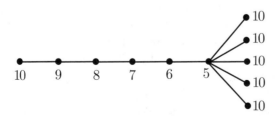

24. $f(i) = \max\{n-i, i-1\}$

25. Let T have $\Delta = k-j+1$ and let v be the unique vertex of degree $k-j+1$. Let all of the other vertices of T have degrees 1 or 2. Furthermore, let the $k-j+1$ components of $T-v$ be one P_j and $k-j$ isolated vertices.

27. Generalize the second part of the previous exercise. $G-v$ has order $n-1$.

Exercises 4.2 (page 135)

1. The cut vertices are the top center rightmost vertex, the second, third and fourth vertices in the row near "G" and the next to last vertex on the right. The bridges are the two edges incident with top center rightmost vertex, the leftmost edge, and the rightmost edge.

2. The cut vertices are a, e, g. The bridges are ag and ak.

3. $\{ab, eb, ib, fb, cb\}$ is a minimal cutset of size 5; $\{ab, eb, ib, fa, fe, fi\}$ and $\{hc, hd, hg, jc, jd, jg\}$ are minimal cutsets of size 6.

10. a. Since $\delta = 4$, we know that the connectivity $\kappa \leq 4$. So we need only show that $\kappa > 3$; that is, removal of any three vertices is not sufficient to disconnect G. Label the vertices around the cycle C_4, a, b, c, d, and label the two vertices of K_2 by u and v. By the symmetry of G, there are four cases to consider: removal of $\{a, u, v\}$, $\{a, b, u\}$, $\{a, c, u\}$, or $\{a, b, c\}$. $G - \{a, b, u\}$ is isomorphic to K_3; whereas $G - \{a, u, v\}$, $G - \{a, c, u\}$, and $G - \{a, b, c\}$ are all isomorphic to P_3. Thus four vertices are required to disconnect G.

Exercises 4.3 (page 141)

1. Start with K_k and then add a pendant edge to each vertex. When $k = 2$, this is P_4.

2. N_1 is the set of vertices of degree 3 and N_2 is the set of vertices of degree 2.

3. Yes. $f(a) + f(b)$, where $f(n) = \lfloor n/2 \rfloor$ if n is odd, and $f(n) = n/2 - 1$ if n is even.

4. No. Consider N_2 of $M(3,3)$.

5. Note that $\text{rad}(G \times H) = \text{rad}(G) + \text{rad}(H)$, but there are center vertices v and w of $G \times H$ such that $d(v, w) = \text{rad}(G) < \text{rad}(G \times H)$.

6. Here's an F-graph of radius 2 with three center vertices.

Chapter 5

Exercises 5.1 (page 149)

1. The degree sum is 28 and it has 14 edges.
2. All the degrees are even. Start like this, then add (in order) edges ab, bc, cd, da, ax.

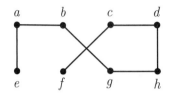

3. Since all degrees in G and H are even, it follows that all degrees in W are, too. (Degrees in W are sums of degrees in G and H.)
4. Degrees in W are sums of degrees in G and H.
5. Q_n is eulerian for all even n.
12. ($\Rightarrow$) Suppose that G is eulerian. Then we know that G is connected and every vertex has even degree. For any cut vertex $v \in G$, let the blocks at v be $B_1, B_2, \ldots, B_k$. There are edges from v into each block B_i. Since an eulerian circuit must begin and end at the same vertex, each time the circuit enters a given block at v, it must eventually return to v via a different edge within that block. Thus we see that not only is the degree of every vertex in G even for each cut vertex v, the number of edges spanning out from v into a given block B_i is also even. This means that as a subgraph, each block of G is connected and all its vertices have even degree. Thus, by Theorem 5.1, each block of G is eulerian.
($\Leftarrow$) If G is connected and each block of G is eulerian, then the vertices within each block all have even degree. This implies that G is connected and all its vertices have even degree, so Theorem 5.1 implies that G is eulerian.

Exercises 5.2 (page 157)

8. Case 1: x and y belong to one copy of G. Call the other copy G'. Let P be a spanning x–y path for G. Take w and z adjacent on P. Let P_x be the x–w subpath of P and let P_y be the y–z subpath of P. (It is assumed that these paths have no vertex in common.) Let their corresponding vertices in G' be w' and z'. Let M be a w'-z' spanning path for G'. The desired x–y spanning path for $G \times K_2$ is $P_x \cup \{ww'\} \cup M \cup \{z'z\} \cup P_y$. Case 2 is done similarly.
9. Line up a sequence $G_1, G_2, \ldots, G_n$ of G copies, corresponding to a spanning path for H. Now find a spanning path for G_1, say from vertex s_1 to vertex t_1. Start with an s_1–t_1 path in G_1; then traverse an H

edge from t_1 to its corresponding vertex t_2 in G_2. Then traverse a t_2–s_2 path in G_2, H edge $s_2 s_3$, and an s_3–t_3 path in G_3, and so on.
10. The outer 8-cycle is a spanning cycle. No spanning a-b path exists.
16. Use Exercise 10 from Section 2.4 and the fact that even 2-meshes are hamiltonian.
17. Let x and y be vertices incident with an interior vertical edge (called a rung!).
18. Let x and y be the first two vertices of the middle row (i.e., the second row).
19. Same as the previous exercise.
24. It looks like this. (The graph is hamiltonian.)

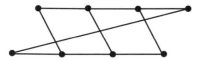

25. This graph is not hamiltonian.

Exercises 5.3 (page 163)

1. The vertices of odd degree are a, d, e, f. The shortest (weighted) paths joining pairs of these vertices have lengths as follows: $d(a, d) = 7$, $d(a, e) = 5$, $d(a, f) = 2$, $d(d, e) = 7$, $d(d, f) = 8$, and $d(e, f) = 3$. So our best choice of pairings to minimize distance sums is a, f and d, e for a total added distance of $2 + 7 = 9$. In each case the shortest paths happen to be single edges. We add in the additional (parallel) edges af and de so that all edges in the resulting multigraph have even degree. Then find an eulerian tour such as $a, b, c, d, e, b, d, e, f, a, c, f, a$ in the multigraph.

3. a. The vertices of odd degree in the graph of Figure 5.21(a) are b, c, d, f. The shortest (weighted) paths joining pairs of these vertices have lengths as follows: $d(b, c) = 5$, $d(b, d) = 5$, $d(b, f) = 8$, $d(c, d) = 2$, $d(c, f) = 4$, and $d(d, f) = 6$. So our best choice of pairings to minimize distance sums is b, d and c, f for a total added distance of $5 + 4 = 9$.

   b. The vertices of odd degree are in the graph of Figure 5.21(b) are l and j. The shortest (weighted) path joining these vertices has length: $d(b, c) = 5$, $d(b, d) = 5$, $d(b, f) = 8$, $d(c, d) = 2$, $d(c, f) = 4$, and $d(d, f) = 6$. So our best choice of pairings to minimize distance sums is b, d and c, f for a total added distance of $5 + 4 = 9$.

7. Suppose that v is a cut vertex in graph G. Suppose that a path P starts at a vertex v_1 in a component C_1 of $G - v$. Let v_2 be a vertex in different component C_2 of $G - v$. Then once P enters C_2, it cannot get back to v_1 in order to complete a cycle without revisiting the cut vertex v first. Thus G has no hamiltonian cycle.

8. We will use the labeling given in Figure 5.22. If the graph were hamiltonian, it would not matter where we start, so let's begin at vertex a. By symmetry, all vertices adjacent to a are equivalent, so go to d next. Again, the symmetry allows us to move, without loss of generality, to j and then to i next. So far our path is a, d, j, i. Next we can move either to vertex b (case 1) or vertex g (case 2). These choices are not equivalent since b is adjacent to a, while g is not.

Case 1 After moving to vertex b, we must move to vertex e or else we would close off the cycle too soon. Our path so far is a, d, j, i, b, e. We now have a choice of moving to either vertex f (case 1a) or h (case 1b) next. If we move to f next, then to avoid closing off the cycle too soon, we must follow that by g, c, h, yielding the path $a, d, j, i, b, e, f, g, c, h$. Since h is not adjacent to a and we have already visited all vertices, we cannot complete the cycle. In case 1b, we move to vertex h after e. Then to avoid closing off the cycle too soon, we must follow that by c, g, f, yielding the path $a, d, j, i, b, e, h, c, g, f$. Since f is not adjacent to a and we have already visited all vertices, we cannot complete the cycle.

Case 2 If we move to g after vertex i, then we have a choice of going to either c (case 2a) or f (case 2b) next. These choices are not equivalent since c is adjacent to a while f is not. If we go to c, then we must follow that by h, e. Now we have a dilemma. If we move to b, our only choice would be to return to a, closing off the cycle without visiting f or g. On the other hand, if we move from e to f, then that must be followed by g and we are at a dead end—we cannot get directly to a nor visit b without revisiting a vertex. Thus, in case 2a, we cannot complete a hamiltonian cycle. In case 2b, we go to vertex f after g. Our path thus far is, therefore, a, d, j, i, g, f. We must move to e next. Again we have a dilemma. If we move to b after e, we can only return to a, and we will miss c and h. On the other hand, if we go to h after e, then our only choice is c and then back to a, thereby missing vertex b. Thus, in case 2 there is no way to complete a hamiltonian cycle. Hence, the Petersen graph is not hamiltonian.

9. We can model this problem as follows. Let a different vertex correspond to each distinct room, and let an additional vertex represent the outside of the house. Put an edge between two vertices if there is a door joining the corresponding rooms or region, and two edges for the case where there are two doors. The resulting structure is a multigraph. The trip through the building corresponds to an eulerian circuit. However, the degree sequence of the resulting multigraph is $4, 4, 3, 3, 2, 2, 2, 2, 2$. Since there are vertices of odd degree not corresponding to doors, there is no eulerian circuit, so the desired tour through the building is impossible.

Chapter 6

Exercises 6.1 (page 171)

3. Each cycle is 3-colorable while the interconnecting tree structure is 2-colorable.
4. If H requires k colors, it still requires them when we view it as a subgraph of G. Hence G requires at least k colors.
10. $\lfloor n/2 \rfloor$
11. Permute the colors in the second copy of G, but use the same partition.
12. The removal of a vertex of $V(G)\backslash V(H)$ does not lower χ.
13. Let one of the copies of G be the subgraph of $G \times K_2$ of the previous exercise and note that $\chi(G) = \chi(G \times K_2)$.
14. $K_n - v \cong K_{n-1}$, for which $\chi = n - 1$.
15. Odd cycles. Deleting a vertex leaves a path, thereby lowering χ from 3 to 2.
16. $\chi(K_1) = 1$, while all nontrivial, connected bipartite graphs satisfy $\chi = 2$.
17. K_4
18. The other components can certainly be colored with $\chi(H)$ colors.
19. The removal of a vertex of a component other than H (of the previous exercise) does not lower χ.
20. If the triangles are edge disjoint, deleting any one vertex in the graph leaves at least one triangle intact, thereby preserving χ.
21. The larger color set is independent. The smaller color set forms a vertex cover.
22. One color set must have exactly one vertex. There are five ways to choose this vertex. Then there is only one way of dividing the remaining four vertices into two color sets, making a total of five ways.
23. The cardinalities of the three color sets can either be (a) 3, 3, and 1 or (b) 3, 2, and 2. There are seven ways to choose the vertex of the color set of cardinality 1 in case (a). The partition of the remaining six vertices can be done only one way. In case (b), the color set of three vertices can be selected in seven ways—this is the number of ways we can select a pair of adjacent vertices not in the set. (Any configuration of three independent vertices in C_7 has this property.) The partition of the remaining four vertices can be done in two ways. This yields 14 ways for case (b), bringing the total to 21.
24. The center of a wheel must have its own color. In the case of $W_{1,4}$, this leaves the uniquely colorable 4-cycle. For $W_{1,5}$, however, we then have a 5-cycle, which by a previous exercise has five coloring schemes.
25. Any vertex of G is joined to all of the vertices of H and hence cannot belong to any of the color sets of H in $G + H$. Thus each graph requires its own color sets.

26. Permute the colors of the sets of G in the second copy in such a way that no corresponding color sets have the same color. This is the same as finding permutations of A, B, and C such that A is not in the first place, B is not in the second place, and C is not in the third place. (This is called a **derangement**.) There are two ways to do this—BCA and CAB. Hence, while $\chi(G \times K_2) = 3$, the graph is not uniquely colorable.

27. Same idea as the previous exercise. $\chi(G \times K_2) = k$.

Exercises 6.2 (page 180)

1. Number the vertices K_6 by $1, 2, 3, 4, 5, 6$. Color all edges that join an odd-labeled vertex to an even-labeled vertex by red. Color edges that join two vertices whose labels have the same parity by blue. The resulting edge coloring has two blue triangles and no red triangles.

3. This is easiest to do using the line graph. Since $\chi(L(G)) = \chi_1(G)$, we need only determine $\chi(L(C_{2k+1}))$. $L(C_{2k+1}) = C_{2k+1}$, so $\chi_1(C_{2k+1}) = \chi(L(C_{2k+1})) = \chi(C_{2k+1}) = 3$.

5. For convenience, let the vertices of the outer cycle be labeled clockwise a, b, c, d, e, and let their respective neighbors on the inner cycle be A, B, C, D, E. Then, without loss of generality, label the edges of the outer cycle beginning at vertex a and going clockwise by red, blue, red, blue, green. To obtain a 3-edge coloring, the edges aA, bB, cC, dD, eE would then have to be blue, green, green, green, red. Now color the edges on the inner cycle. Edge AC is incident with a blue edge and a green edge, so AC must be red. Similarly, AD is incident with a blue edge and a green edge, so AD must be red. But then AC and AD are incident at A and have the same color, which is not allowed. Thus $\chi_1(\text{Petersen}) \geq 4$. We can achieve a 4-coloring by continuing our coloring with AC, AD, DB, BE, EC colored red, white, red, white, blue, respectively. Thus, $\chi_1(\text{Petersen}) = 4$.

7. Suppose that a is the inner vertex and the outer vertices are labeled consecutively b, c, d, e. Then since a has degree 4, its incident edges must use four distinct colors. There are $4! = 24$ ways to assign those colors. Now consider an outer edge such as bc. This edge cannot receive the same color as ab or ac, so it must receive one of the other two colors available. But then as we travel around the outer cycle, we find that there is always just one color available. Thus, by the multiplication principle, there are $24 \cdot 2 \cdot 1 \cdot 1 \cdot 1 = 48$ ways to 4-color the edges of $W_{1,4}$.

9. Theorem 6.8 says that for every graph, $\chi_1(G)$ is either $\Delta(G)$ or $\Delta(G)+1$. So for a k-regular graph, $\chi_1(G)$ is either k or $k+1$. If G is k-regular and $\chi_1(G) = k$, then there is precisely one edge at each vertex having one of the k colors. Since incident edges must have distinct colors, this

means that each color class induces a perfect matching. But this is only possible if the order of G is even, since a perfect matching requires an even number of vertices. Thus, when G is k-regular of odd order, $\chi_1(G) = k+1$.

13. When $n = 1$, $Q_1 = K_2$, and $\chi_1(Q_1) = \chi_1(K_2) = 1$. Assume that the result holds for $n = k$. Then since $Q_{k+1} = K_2 + Q_k$, and Q_{k+1} is $(k+1)$-regular, we know by Theorem 6.5 that $\chi_1(Q_{k+1}) \geq k+1$. A $(k+1)$-edge coloring of Q_{k+1} can be achieved as follows. Obtain a k-coloring of the edges of the first copy of Q_k in $Q_{k+1} = K_2 + Q_k$ (which is possible by the inductive hypothesis). Then obtain a k-coloring for the second copy, using the same k colors. Finally, use color $k+1$ for each edge matching a vertex in the first copy to a vertex in the second copy. Thus, $\chi_1(Q_{k+1}) = k+1$.

15. First note that $\chi_1(H) \geq 3$. Color the edges xy, vy, zy by red, blue, green, respectively. Then to achieve a coloring using just three colors, we must have vz red, then vu green, ux blue, xt green, and tz blue. So H is uniquely 3-edge colorable. Now consider graph G. Here also, $\chi_1 \geq 3$, so we aim for a 3-edge coloring. Without loss of generality, we color ac, af, ad by red, blue, green, respectively. Then, we would find that cf must be green, then ce blue, and de red. But then df is adjacent to a red, a blue, and a green, so df requires a fourth color, say white. Thus $\chi_1(G) = 4$, not three. In the 4-edge coloring just given, our edge partition is $\{ac, de\}, \{af, ce\}, \{ad, cf\}, \{df\}$. A different edge partition in a 4-edge coloring will show that G is not uniquely 4-edge colorable. Coloring ac and df red, ad and ce blue, af and de white, and cf green produces the partition $\{ac, df\}, \{ad, ce\}, \{af, de\}, \{cf\}$. Thus, G is not uniquely 4-edge colorable.

16. Simplify $\frac{(2k+1)(2k)(2k-1)}{6} - \frac{1}{2}(2k+1)k^2$.

17. The number of triangles in K_{2k} is $\binom{2k}{3} = \frac{2k(2k-1)(2k-2)}{6}$. By Theorem 6.11, at least $\frac{k(k-1)(k-2)}{3}$ of those triangles are monochromatic. Thus the probability that a randomly selected triangle is monochromatic is $\frac{k(k-1)(k-2)}{3} / \frac{2k(2k-1)(2k-2)}{6} = \frac{k(k-1)(k-2)}{3} \cdot \frac{6}{2k(2k-1)(2k-2)} = \frac{k-2}{4k-2}$. This goes to 0.25 as $k \to \infty$.

18. $f(x) = mx - x^2$. Then $f'(x) = m - 2x$. So $f'(x) = 0$ when $x = m/2$. Since $f'' = -2$, this yields a maximum.

19. We find the partial derivatives $f_x = my - 2xy - y^2$ and $f_y = mx - x^2 - 2xy$ and then set them each equal to zero and solve simultaneously. By doing so, we obtain $x = y$ or $x + y = m$. Together with the original equation $x + y + z = m$, these give, respectively, $x = y = z = m/3$, and $x = m/2$, $y = m/2$, $z = 0$. Using the second partials test, we get $f_{xx} = -2y$, $f_{yy} = -2x$, and $f_{xy} = m - 2x - 2y$. So $D = f_{xx} \cdot f_{yy} - [f_{xy}]^2 =$

$4xy - (m^2 + 4x^2 + 4y^2 - 4mx - 4my + 8xy) = -m^2 - 4x^2 - 4y^2 + 4mx + 4my - 4xy$ evaluated at $x = y = z = m/3$ gives $m^2/3 > 0$, yielding either a relative max or relative min. $f_{xx} = -2y = -2m/3 < 0$ confirms it is a relative maximum. On the other hand, at $x = m/2$, $y = m/2$, $z = 0$, we find $D = 0$, so the test fails. However, $x = y = z = m/3$, the product $xyz = m^3/27$; whereas at $x = m/2$, $y = m/2$, $z = 0$, the product $xyz = 0$, so a maximum occurs only when $x = y = z = m/3$.

Exercises 6.3 (page 188)

1. The wombats would be safest in the third enclosure with the goats, rabbits, and shrews.

3. Take advantage of the fact that a unique 3-coloring of H is displayed in Figure 6.21. We can therefore determine that adding in edge fk would increase the chromatic number to 4 since f and k have the same color. On the other hand, adding in edge ah, each of whose vertices have a distinct color, would not alter the chromatic number.

5. **a.**

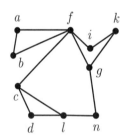

   b. No

   c. a, e, g, i in cabinet 1; d, f, k in cabinet 2; and b, c, n in cabinet 3 is one possible safe arrangement.

7. **a.**

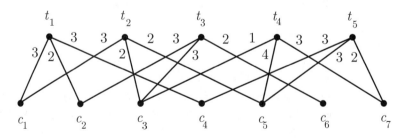

   b. Nine periods are needed since $\Delta(G) = 9$

Answers/Solutions to Selected Exercises   335

c.

Period \ Teacher	t_1	t_2	t_3	t_4	t_5
1	c_1	c_3	c_2	c_5	c_4
2	c_2	c_5	c_3	c_7	–
3	c_4	c_1	c_6	c_5	c_7
4	c_1	c_3	c_2	c_7	c_5
5	c_2	c_1	c_3	c_5	c_4
6	–	–	c_2	c_3	c_5
7	c_4	c_1	–	c_7	c_5
8	c_1	c_5	–	–	c_4
9	c_4	–	c_3	c_5	c_7

Chapter 7

Exercises 7.1 (page 198)

1. $AB = \begin{bmatrix} 13 & 15 & 13 \\ 2 & 19 & 19 \\ 28 & 17 & 11 \end{bmatrix}$, $AC = \begin{bmatrix} 5 & 26 & 18 \\ 5 & 15 & 4 \\ 13 & 21 & 32 \end{bmatrix}$, $BC = \begin{bmatrix} 9 & 1 & 16 \\ 3 & 18 & 6 \\ 4 & 12 & 8 \end{bmatrix}$

2. $BA = \begin{bmatrix} 8 & 9 & 22 \\ 18 & 18 & 18 \\ 14 & 13 & 17 \end{bmatrix} \neq AB$

3. $(AB)^t = \begin{bmatrix} 13 & 2 & 28 \\ 15 & 19 & 17 \\ 13 & 19 & 11 \end{bmatrix}$

4. $(AB)C = \begin{bmatrix} 41 & 80 & 78 \\ 23 & 114 & 46 \\ 73 & 72 & 134 \end{bmatrix}$

5. $Av = \begin{bmatrix} 33 \\ 20 \\ 40 \end{bmatrix}$. D is a 3×2 matrix and v is 3×1, so we can't find Dv.

6. $(A+B)v = \begin{bmatrix} 6 & 2 & 5 \\ 0 & 8 & 5 \\ 7 & 3 & 6 \end{bmatrix} \begin{bmatrix} 3 \\ 2 \\ 5 \end{bmatrix} = \begin{bmatrix} 47 \\ 41 \\ 57 \end{bmatrix}$. On the other hand, $Av + Bv = \begin{bmatrix} 33 \\ 20 \\ 40 \end{bmatrix} + \begin{bmatrix} 14 \\ 21 \\ 17 \end{bmatrix} = \begin{bmatrix} 47 \\ 41 \\ 57 \end{bmatrix}$.

7. $v^t = \begin{bmatrix} 3 & 2 & 5 \end{bmatrix}$

8. $I + A + 10B = \begin{bmatrix} 43 & 11 & 5 \\ 0 & 36 & 32 \\ 16 & 21 & 25 \end{bmatrix}$

9. DA, $C+D$, $A+v$, D^2

10. $(A+B)^2 = (A+B)(A+B) = A(A+B)+B(A+B) = A^2+AB+BA+B^2 \neq A^2 + AB + AB + C^2$. This is because $AB \neq BA$.
11. Show $[(MN)^t]_{ij} = (N^t M^t)_{ij}$ for all i,j. (a) $[(MN)^t]_{ij} = (MN)_{ji}$ This is the dot product of row j of M and column i of N. (b) $(N^t M^t)_{ij}$ is the dot product of column j of M^t and row i of N^t. This is the dot product of row j of M and column i of N.
12. The i,j entry of D^2 is the dot product of row i and column j of D. This is equivalent to the dot product of rows i and j of D, which is 1 if $i = j$, and is 0 otherwise. This yields a diagonal matrix. Now compute $D^2 D$ and then generalize.
13. $(A^k)^t = (A^t)^k = A^k$
14. $[(A+B)^t]_{ij} = (A+B)_{ji} = A_{ji} + B_{ji} = (A^t)_{ij} + (B^t)_{ij} = [A^t + B^t]_{ij}$

Exercises 7.2 (page 206)
1. There are two cities that require at least three flights.
2. Dot the ith row of A^2 and the jth column of A (same as jth row of A) and apply the multiplication principle whenever two nonzero entries are multiplied.
3. There are 7! permutations of the numbers $1, 2, \ldots, 7$.
5. The adjacency matrix is $A(K_4 - e) = \begin{bmatrix} 0 & 1 & 1 & 1 \\ 1 & 0 & 1 & 1 \\ 1 & 1 & 0 & 0 \\ 1 & 1 & 0 & 0 \end{bmatrix}$.
6. They have the same neighborhood.

Exercises 7.3 (page 211)
1. The first row is $0\ 1\ 2\ 3 \cdots n\ (n-1)\ (n-2) \cdots 1$. Each successive row is obtained by cyclically shifting the previous row. So row 2 is $1\ 0\ 1\ 2\ 3 \cdots n\ (n-1)\ (n-2) \cdots 2$, row 3 is $2\ 1\ 0\ 1\ 2\ 3 \cdots n\ (n-1)\ (n-2) \cdots 3$, and so on.
3. The distance matrix has 0's on the diagonal, all 1's in row 1 and column 1 except for $d_{1,1}$, and for all other entries, $d_{i,j} = 1$ if $|i-j| = 1$ or $|i-j| = n-1$, and $d_{i,j} = 2$ otherwise.
5. The order is 6, $e(v_2) = 3$, $e(v_5) = 2$, $r(G) = 2$, $d(G) = 3$, and $C(G) = \{v_1, v_3, v_5\}$.
7. The order is 7, $e(v_3) = 2$, $e(v_6) = 4$, $r(G) = 2$, $d(G) = 4$. $C(G) = \{v_3\}$.
9.

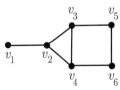

11. K_n has the desired property.

13. The matrix fails the triangle inequality at v_2 because $d(v_2, v_4) = 4$ but $d(v_2, v_1) + d(v_1, v_4) = 1 + 2 = 3$.

Chapter 8

Exercises 8.1 (page 221)

1. Here is one of them. The root is labeled.

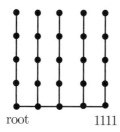

2. A BFS spanning tree whose root is the center vertex of the odd 2-mesh will generate the desired spanning tree. The BFS preserves the radius, and, since the diameter of the mesh is twice the radius, must also preserve the diameter.

3.

4. $\mathrm{diam}(Q_n) = \mathrm{rad}(Q_n) = n$. If T is a spanning tree of Q_n, $\mathrm{diam}(T) \geq 2\,\mathrm{rad}(T) - 1 = 2n - 1 > n$.

5. If the root is the center, we obtain the star $K_{1,7}$. Otherwise, we obtain either of the two graphs depicted below.

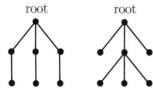

6. Include all the vertical edges of the first column. (This is a copy of P_6.) Now delete all other vertical edges.

7. The geodesics from v and w to the vertices (i, k) and (h, j) forms a cycle.

8. By the previous exercise, all reach-preserving vertices must lie in a single column or row. Choose the larger of these two options.

9. The edges of the tree are generated in the order: $ac, ah, cb, cg, be, ed, ef$. Then draw the tree.

11. The algorithm proceeds as follows: ac, cb, be, ed, backtrack to e, ef, backtrack to e, backtrack to b, bg, backtrack to b, backtrack to c, backtrack to a, ah, backtrack to a. The resulting tree therefore contains the edges: $ac, cb, be, ed, ef, bg, ah$. Draw the tree.

13. The algorithm proceeds as follows: cb, bh, ha, backtrack to h, backtrack to b, be, ef, backtrack to e, ed, backtrack to e, backtrack to b, bg, backtrack to b, backtrack to c. The resulting tree therefore contains the edges: $cb, bh, ha, be, ef, ed, bg$. Draw the tree.

15. First, include an additional labeling in BFS to find the eccentricity of the root. The root r gets label $d(r) = 0$. Each subsequent vertex v receives a label $d(v)$ that is one more that the label $d(u)$ of the vertex u that caused v to be labeled. The label $d(v)$ gives the distance from vertex v to root r. The maximum of the $d(v)$'s is the eccentricity of r. We can successively repeat the process, with each vertex of G serving as the root and thereby find each eccentricity. We can then easily determine the center and the periphery.

Exercises 8.2 (page 227)

1. Assign the following colors: a, 1; b, 2; c, 1; d, 2; e, 3; f, 2; and g, 3.
2. Assign the following colors: a, 1; e, 2; f, 3; d, 1; c, 2; g, 3; and b, 2.
3. First we color a by 1. Then b, e, f, and g all have color degree 1, but g has maximum degree among those, so vertex g gets colored next; it gets color 2. Next to be colored is vertex f, which gets color 3, then e gets color 2, d gets color 1, g gets color 3, and finally b gets color 2.
4. Vertices a and g were the only vertices for which color degree was maximum and maximum degree was equal at the point they were candidates for coloring, so those were the only vertices where vertex ordering was used as the final tie breaker. Vertex a was still earlier in the ordering, so the same coloring resulted.
5. Assign the following colors: a, 1; b, 1; c, 2; d, 1; e, 2; f, 1; g, 3; and h, 2.
6. First we color b, which has degree 4, by 1. Then c, e, g, and h all have color degree 1, but c and e have maximum degree among those, so vertex c gets colored next since it is earlier in the (alphabetical) vertex ordering among tied vertices in the maximum degree listing; it gets color 2. Next to be colored is vertex g, which gets color 3. Then a, e, and h are tied with color degree 1, but e has maximum degree among those, so e gets colored next by color 2. Now all remaining uncolored vertices have color degree 1 and a and h have maximum degree among those, so a gets colored next by 1, then h gets color 2, and finally d and f each get color 1.
7. Consider P_4 with consecutive vertices labeled a, c, d, b and suppose that vertices are ordered alphabetically. Then vertex a gets color 1, b gets

color 1, c gets color 2, and d gets color 3. Thus a 3-coloring is produced rather than a 2-coloring.

8. At each step in the maximum color-degree algorithm, the next vertex to be colored is adjacent to a vertex that has already been colored. Since neighbors in a bipartite graph are in opposite parts, the newly colored vertex will always receive the opposite color of its neighbor that is causing it to be colored. Thus only two colors will be used for a bipartite graph having at least one edge.

9. Vertices a, b, e, h, i, j, and k will have color 1, c and d get color 2, f gets color 3, and g gets color 4.

10. Suppose that consecutive vertices along the outer cycle are labeled a, e, i, b, d, and on the inner cycle suppose that the neighbors of a, e, i, b, d are, respectively, g, h, j, f, c. Then a, b, and c get color 1; then d, e, and f get color 2; g, h, and i get color 3, and j gets color 4. Thus a 4-coloring is produced. We can see that $\chi(G)$ is actually 3, by coloring the consecutive vertices a, e, i, b, d by $1, 2, 1, 2, 3$ and their respective neighbors g, h, j, f, c by $2, 1, 3, 3, 2$. The fact that G has an odd cycle implies that $\chi(G) \geq 3$ and our achieving a 3-coloring then shows that $\chi(G) = 3$.

11. There are two basic cases to consider. Since there are many ties along the way, the first four vertices that get colored will either induce a $K_{1,3}$ with the center vertex colored, say, 1 and its three neighbors colored 2, or those vertices will induce a P_4 with consecutive vertices colored $1, 2, 1, 2$. The next vertex colored will force the issue. There are still a couple of cases to consider, but it is not difficult from here to determine that a 3-coloring will necessarily be produced.

Exercises 8.3 (page 233)

1. $(2, 3, 4, 5)$ and $(2, 3, 4, 5, 6)$
2. $(2, 3, \ldots, n-1)$
3. $(1, 1, \ldots, 1)$
4.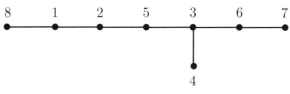
5.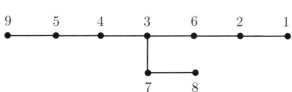

6. 125
7. Three trees: $K_{1,4}$, P_5, and a 3-branched starlike tree
8. We delete $n-2$ end vertices
9. The vertex n can never be deleted. At each stage, we delete the end vertex with lowest number.
10. Without loss of generality, the center is $(0,0,\ldots,0)$. Now each end vertex must have exactly one 1, requiring $n-1$ places in the addressing.
11. Label the vertices consecutively, $(0,0,0,\ldots)$, $(1,0,0,\ldots)$, $(1,1,0,\ldots)$, $(1,1,1,0,\ldots)$, and so on.
12. Here's one.

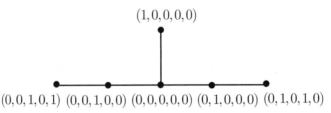

13. The Prüfer code is d,b,e,d,e.
14. We begin by labeling vertex 1 by 0 and its neighbor 5 by 1. Then consider edge $(2,5)$. Alter 1's label to $(0,0)$, 5's label to $(0,1)$, and then vertex 2 gets label $(1,1)$. Next consider edge $(5,4)$. Alter 1's label to $(0,0,0)$, 5's label to $(0,0,1)$, 2's label to $(0,1,1)$, and then since vertex 4 is adjacent to 5, it gets label $(1,0,1)$. Continuing in this way, if we consider next edges $(4,3)$, $(4,6)$, $(4,7)$, and then $(6,8)$, the binary addressing will make vertices $1,5,2,4,3,6,7$, and 8 receive labels $(0,0,0,0,0,0,0)$, $(0,0,0,0,0,0,1)$, $(0,0,0,0,0,1,1)$, $(0,0,0,0,1,0,1)$, $(0,0,0,1,1,0,1)$, $(0,0,1,0,1,0,1)$, $(0,1,0,0,1,0,1)$, and $(1,0,1,0,1,0,1)$, respectively.

Chapter 9

Exercises 9.1 (page 243)

1. $K_{3,4}$ with two edge crossings:

2. See the chart.

Answers/Solutions to Selected Exercises

polyhedron	n	q	f	$n - q + f$
tetrahedron	4	6	4	2
cube	8	12	6	2
pyramid	5	8	5	2
octahedron	6	12	8	2
dodecahedron	20	30	12	2
icosahedron	12	30	20	2

3. See the chart.

4. See the chart.

5. Since each edge is subdivided exactly once, each odd cycle will become an even cycle, and each even cycle will remain an even cycle. Thus, for all graphs G, $S(G)$ contains no odd cycles, so by Theorem 3.4, $S(G)$ is bipartite.

6. The cycle in G becomes an even cycle C in $S(G)$. As we travel around C the vertices will be alternately colored red and blue, so there is the same number of red vertices as blue vertices on C. Attached to C in $S(G)$ are various trees with the property that as we travel away from the vertex on C to which a given tree T is attached toward an endpoint of T, the path we traverse will necessarily have even length because of the way $S(G)$ is formed. Thus, on each such path we will encounter the same number of red vertices as blue ones. Thus $S(G)$ is equitable for any connected unicyclic graph G.

7. $\bar{C}_n$ is regular of degree $n - 3$. So when $n > 8$, $\delta(\bar{C}_n) \geq 6$, which means that G is nonplanar according to Corollary 9.3a. Thus there is one exceptional case we must handle separately, namely, when $n = 8$. In this case, the number of edges in $\bar{C}_8$ is $\binom{8}{2} - 8 = 28 - 8 = 20$. Using Theorem 9.3, we find that $q = 20 > 3(8) - 6 = 18$, so $\bar{C}_8$ is nonplanar.

8. The number of edges in $\bar{C}_7$ is $\binom{7}{2} - 7 = 21 - 7 = 14$, and $q = 14 \leq 3(7) - 6 = 15$. Thus $\bar{C}_7$ does satisfy the inequality in Theorem 9.3. We can show that $\bar{C}_7$ is nonplanar by using Theorem 9.4. $\bar{C}_7$ contains a subgraph that is a subdivision of $K_{3,3}$. To see this, let the vertices of $\bar{C}_7$ be labeled consecutively $1, 2, 3, 4, 5, 6, 7$ so that $1, 2, 3, 4, 5, 6, 7, 1$ is the cycle in the complementary graph. Then in $\bar{C}_7$, vertices 1 and 2 are adjacent to 4, 5, and 6, and vertex 3 is directly adjacent to 5 and 6 and adjacent by way of 7 to vertex 4. This describes a subgraph that is a subdivision of $K_{3,3}$, so by Theorem 9.4, $\bar{C}_7$ is nonplanar. This does not contradict Theorem 9.3 because that theorem says that a planar graph must satisfy the inequality; it does not say that satisfying the inequality is sufficient to guarantee that the graph is planar.

9. The redrawing is shown below.

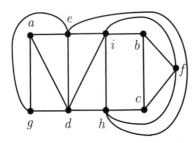

10. a. For the graph in Figure 9.12, $n = 8$, $q = 14$, and $f = 8$. Thus $n - q + f = 8 - 14 + 8 = 2$.

    b.

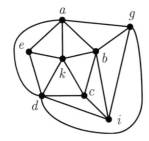

11. The subgraph is shown here with it redrawn so that the subdivision of $K_{3,3}$ is apparent.

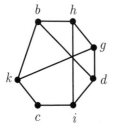 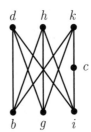

12. The standard drawing of $W_{1,n}$ has one vertex in the interior region adjacent to each vertex of the cycle C_n. This shows that $W_{1,n}$ is planar. It is also easy to draw $W_{2,n}$ as a plane graph by inserting one new vertex v in the exterior region of $W_{1,n}$ and joining v to each vertex of the cycle C_n. Finally, to show that $W_{m,n}$ is nonplanar when $m \geq 3$, just verify that $W_{m,n}$ has a subgraph that is a subdivision of $K_{3,3}$ (this can be done by designating three vertices of $\bar{K}_m$ to be in one part of the $K_{3,3}$ and selecting the other three vertices from the cycle).

13.

14. If G has seven faces and n vertices, all of degree 4, then $q = (\sum \deg)/2 = 4n/2 = 2n$. So by Euler's formula, $n - q + f = n - 2n + 7 = 2$. So $n = 5$ and $q = 2n = 10$, but this implies that G is K_5, which is nonplanar. Thus, there is no such plane graph.

15. By Theorem 9.3, we know that $q \leq 3n - 6$. Our graph has $q = (\sum \deg)/2 = (10(4) + 8(5) + (n-18)(7))/2 = (7n - 46)/2$. So $(7n - 46)/2 \leq 3n - 6$ implies that $n \leq 28$. Since there are $n - 18$ vertices of degree 7, there are at most 10 such vertices.

Exercises 9.2 (page 250)

1. It is clear that a graph G that can be embedded in the plane can be embedded on a sphere. Simply use a large enough sphere so that the plane drawing of G can be "pasted onto" the sphere. On the other hand, suppose that a graph H is embedded on a sphere. Let F be one of the faces in the embedding. Make a hole in F (and thus in the sphere) and then stretch the resulting punctured sphere out to a portion of the plane. This shows that H can be embedded in the plane. Note that F will correspond to the outer face in the resulting planar embedding.

2.

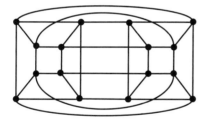

3. Assuming a standard binary labeling of Q_4 (see Section 2.4), let the vertices of the first part of the subdivision graph of $K_{3,3}$ be 0000, 0101, and 1010. Let those in the second part be 0001, 0010, and 0100. Then 0000 is directly adjacent to each vertex in the second part. Vertex 0101 is directly adjacent to 0100 and 0001 and is adjacent to 0010 by way of a path through 0111 and 0110. Finally, vertex 1010 is directly adjacent to 0010, is adjacent to 0001 by a path through 1011 and 0011, and is adjacent to 0100 by a path through 1110 and 1100. This shows that Q_4 contains a subdivision of $K_{3,3}$, so Q_4 is nonplanar by Kuratowski's theorem.

4. View Q_4 as the cartesian product $C_4 \times C_4$. Draw four C_4's around four "parallel" small circular cross sections of the torus. Then add in the edges of the C_4's that go in the perpendicular direction around the torus.

5. Q_4 has no odd cycles, so Q_4 is bipartite. Thus $\chi(Q_4) = 2$.

8. We set $\lceil (3-2)(n-2)/4 \rceil \geq 2$ and solve to find $n \geq 7$.

9. We set $\lceil (4-2)(n-2)/4 \rceil \geq 2$ and solve to find $n \geq 5$.

10. The genus of $K_{6,n}$ is $n-2$ for $n \geq 2$, and $\lim_{n \to \infty} (n-2) = \infty$, so there are bipartite graphs of arbitrarily large genus.

11. ($\Rightarrow$) If every planar graph is 4-colorable, then since a maximal plane graph is also planar, it too is 4-colorable. ($\Leftarrow$) Suppose that every maximal plane graph is 4-colorable. Let G be a planar graph and let G' be a plane drawing of G. We can obtain a maximal plane graph H' containing G' by adding additional edges to G' if G' is not already a maximal plane graph. Then H' is 4-colorable. Using the same coloring on the vertices in G' shows that G', and therefore G is also 4-colorable.

12. Consider, for example, the states California, Oregon, Idaho, Nevada, Utah, and Arizona. The graph modeling their adjacencies is $W_{1,5}$ with Nevada as the center vertex. We know that the odd cycle C_5 requires three colors. The center vertex requires a fourth color since it is adjacent to each vertex of the cycle. Thus, $\chi(W_{1,5}) = 4$ and a map of the United States is not 3-colorable.

14. We need to find all n for which $\lceil (n-3)(n-4)/12 \rceil = n$. Then we must have $(n-3)(n-4) = 12n - k$, where $0 < k < n$. Multiplying out and combining, we obtain the quadratic equation $n^2 - 19n + (12+k) = 0$, which has solution $n = \dfrac{19 \pm \sqrt{313-4k}}{2}$. Since n must be an integer, $313 - 4k$ must be an odd perfect square where $0 < k < n$. This occurs precisely when $k = 6$. Thus the resulting values of n are 18 and 1.

Exercises 9.3 (page 256)

1. Draw K_4 like this:

Answers/Solutions to Selected Exercises

2. The dual is shown using 12 dashed lines and 6 diamond-shaped vertices.

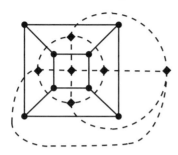

3. Draw w below, and z above, the plane of the remaining 4-cycle. This is an octahedron.

4. Place a vertex of the dual in each of the $(a-1)(b-1)$ 4-cycles of $M(a,b)$.

5. It is self-dual.

6. See the figure and observe that $(G^*)^* \cong G$.

8. $(G^*)^* \cong G$

9. a. It is easy to verify that the mapping $f(a) = u$, $f(b) = y$, $f(c) = z$, $f(d) = x$, $f(e) = w$, and $f(f) = v$ preserves adjacencies. Thus $G \cong H$.

b.

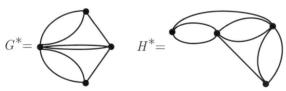

c. G^* has a vertex of degree six whereas H^* does not. Thus $G^* \not\cong H^*$.

10. The vertices of G correspond to the faces of G^*, and $\deg(v)$ equals the number of edges in the corresponding face of G^*. Thus, if G is eulerian, then by Theorem 5.1, every vertex of G has even degree. But this means that every face of G^* contains an even number of edges; that is, every cycle of G^* is even. Since G^* contains no odd cycles, by Theorem 3.4, G^* is bipartite.

11.

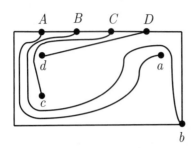

12. The number of vertices of G equals the number of faces of G^*. But then G being self-dual implies that $n = f^* = n^* = f$. So in G, $n - q + f = 2$ becomes $n - q + n = 2$. Thus $2n = q + 2$.

13. a.

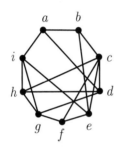 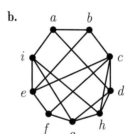

b. $C = a, b, c, d, h, g, f, e, i, a$ is a hamiltonian cycle for G. There are other possible hamiltonian cycles. Graph G is redrawn with C on the outside in the figure above.

c.

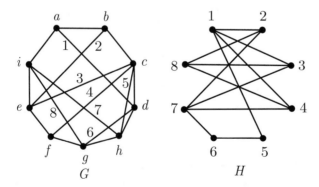

d. Color vertices 1, 7, and 8 red; color vertices 2, 3, 4, and 6 blue; and color vertex 5 green. Put the red edges inside and the blue edges outside of the cycle C. Edge 5 must cross at least one edge, so put it inside, crossing edge 1.

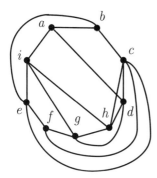

e. We can satisfy all but one of the desired adjacencies as indicated by the one pair of crossing edges. A possible optimal arrangements of the stores is shown in the figure.

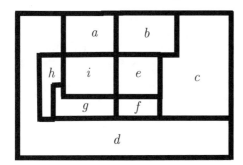

Chapter 10

Exercises 10.1 (page 269)

1. Here is a strongly connected digraph with underlying graph Q_3.

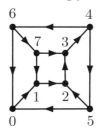

3. Label the vertices a, b, and c. Without loss of generality, let ab be a directed edge. Now if bc is a directed edge, then a, b, c is a directed spanning path. On the other hand, if cb is a directed edge, then consider undirected edge ac. If ac is a directed edge, then a, c, b is a directed spanning path. If ca is a directed edge, then c, a, b is a directed spanning path.

4. See the figure. The indegrees and outdegrees are as follows: $id(a) = 0$, $id(b) = 2$, $id(c) = 2$, $id(d) = 1$, $id(e) = 0$, $id(f) = 4$, $od(a) = 1$, $od(b) = 3$, $od(c) = 2$, $od(d) = 3$, and $od(e) = od(f) = 0$. $\sum id(v_i) = \sum od(v_i) = 9 = q$.

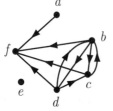

5. It is none of those. The digraph is disconnected since vertex e is isolated.
6. In the first digraph, vertex a is a transmitter and vertex c is a receiver. In the second digraph, vertex f is a transmitter and both vertices d and g are receivers.
7. **a.** $\subseteq$ is reflexive since each set within A is equal to itself. $\subseteq$ is antisymmetric because if $X \subseteq Y$ and $Y \subseteq X$, then $X = Y$. Finally, $\subseteq$ is transitive because if $X \subseteq Y$ and $Y \subseteq Z$, then $X \subseteq Z$. Note that $\subseteq$ acts as a partial order relation on any set, not just on set A.

b.

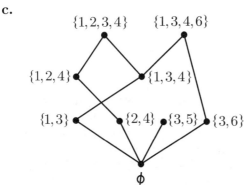

c.

Answers/Solutions to Selected Exercises 349

8.

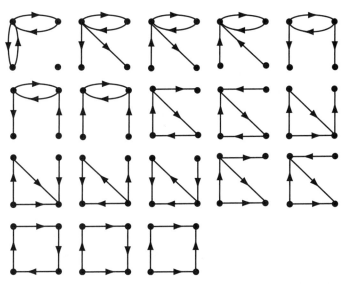

9. If we find the complement of a digraph with four vertices and four arcs, we obtain a digraph with four vertices and eight arcs. Since a digraph has precisely one complement, there is a one-to-one correspondence between the number of digraphs having four vertices and four arcs with those having four vertices and eight arcs. Thus there are the same number of each.

10. All we need is $0 \le r < n$. This is related to the fact that the complete graph K_n can be decomposed into $(n-1)/2$ edge-disjoint cycles when n is odd and into $(n-2)/2$ cycles and a perfect matching when n is even.

11. Begin at any vertex v_0. Since $od(v_0) > 0$, we may leave by some arc, say $v_0 v_1$. Each time we arrive at a new vertex v_i, we may leave by a new arc, say $v_i v_{i+1}$ since $od(v_i) > 0$. Since $|V(D)|$ is finite, we will eventually reach a vertex v_k already visited and will travel there from a vertex, say v_j. Then $v_k, v_{k+1}, v_{k+1}, \ldots, v_j, v_k$ is a directed cycle in D.

14. ($\Rightarrow$) Let T be a transitive tournament. Suppose $C = v_0, v_1, v_2, \ldots, v_k, v_0$ is a cycle in T. Then since T is transitive, the presence of arcs $v_i v_j$ and $v_j v_{j+1}$ implies the existence of $v_i v_{j+1}$. So $v_0 v_2$ is present. Then $v_0 v_2$ and $v_2 v_3$ imply that $v_0 v_3$ is present. Similarly, $v_0 v_3$ and $v_3 v_4$ imply that $v_0 v_4$ is present. Continuing in this way, we eventually find that $v_0 v_k$ is present. But this then prevents the existence of the arc $v_k v_0$. Hence the cycle C cannot exist in T. Therefore, T is acyclic. ($\Leftarrow$). If T is acyclic, then if $uv \in T$ and $vw \in T$, then $wu \notin T$. But then uw must be in T. Thus $uv \in T$ and $vw \in T$ implies that $uw \in T$. Thus, T is a transitive tournament.

15. Let T be a transitive tournament of order n, and let $x, y \in V(T)$. Assume without loss of generality that xy is an arc of T. Let $N(y)$ be the set of out-neighbors of y. So $N(y) = \{z : yz \in T\}$. Then since T is transitive, $xy \in T$ and $yz \in T$ implies that $xz \in T$. Thus $N(y) \subset N(x)$, the out-neighborhood of x, so $|N(x)| > |N(y)|$ and no two vertices can have the same outdegree. Hence there is a unique vertex of maximum outdegree. That vertex should be declared the winner in the round-robin tournament.

16. By Exercise 15, we know that in a transitive tournament, every vertex has distinct outdegree. If there are n vertices, then the outdegree sequence must be $n-1, n-2, n-3, \ldots, 1, 0$. On the other hand, if T is a tournament with outdegree sequence $n-1, n-2, n-3, \ldots, 1, 0$, then we may label the vertices v_i so that $\text{od}(v_i) = i$. Then $v_i v_j \in T$ for $i > j$. But this implies that T is transitive. So T is transitive if and only if the outdegree sequence is $n-1, n-2, n-3, \ldots, 1, 0$. Since there is only one unlabeled transitive tournament on n vertices, only one unlabeled tournament has outdegree sequence $n-1, n-2, n-3, \ldots, 1, 0$.

17. $\sim$ is not a partial order on Z because $a \not\sim a$ when $a \neq 0$. That is, $a \neq 3a$ unless $a = 0$. Thus $\sim$ is not reflexive.

18. $\sim$ will be a partial order on Z only if $k = 1$. See Exercise 17.

19. First we obtain the DFS tree T and direct the edges of T toward the higher labeled vertices. Edges not in T are then directed toward vertices with lower labels. Here is the tree T.

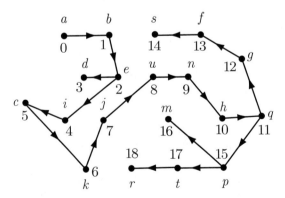

Here is the resulting strong digraph that can provide good traffic flow for Graphland Valley.

Answers/Solutions to Selected Exercises   351

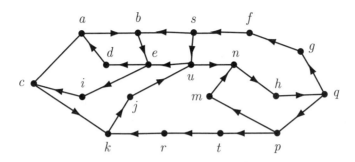

20. No. $1 \not\sim 1$ since $1 + 2(1) \neq 0$.

22. There are only $n - 1$ arcs incident with each vertex. If $od(u) = n - 1$, every arc incident with u points away from u. Thus there is no vertex v with an arc pointing toward u. Thus only u can have outdegree $n - 1$.

Exercises 10.2 (page 284)

1. After the first labeling pass, our network looks as in (a). We increment by 5 along the path s, a, b, t to obtain a network with positive flow, which we then label. We obtain the network in (b).

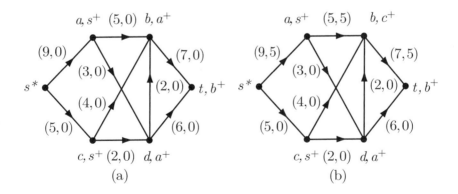

We increment by 2 along the path s, c, b, t and relabel. We obtain the network in (c). We then increment by 3 along the path s, a, d, t and relabel and obtain the network in (d).

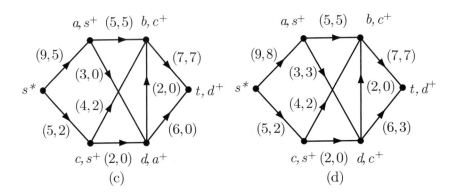

We increment by 2 along the path s, c, d, t and relabel to obtain the network (e). Since t is unlabeled, we stop. We have achieved maximum flow.

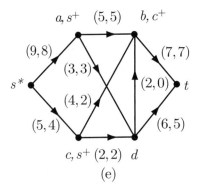

2. The cut value of $\{dt, et\}$ is $15 + 6 = 21$; however, $\{ce, be, bd, ad\}$ has a cut value of $4 + 4 + 5 + 4 = 17$. Thus $\{dt, et\}$ is not a minimum cut. It is, however, a minimal cut.

3. We set up the network that puts an arc from each applicant to each job for which he or she is qualified. Then we put an arc from the source s to each applicant, and an arc from each job to the sink t. Each arc is assigned a capacity of 1. Then we maximize flow. Here is the initial network (without capacities, which are all 1, or flows, which are all zero, listed).

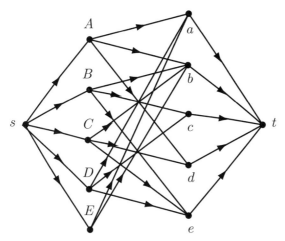

4. The first increment is 1 along s, A, a, t, then 1 along s, B, b, t. Then we increment by 1 along s, C, d, t, followed by 1 along s, D, c, t. Finally, we increment by 1 along $s, E, a, A, b, B, c, D, e, t$. Note that flow is diverted along the reverse arcs a, A, b, B, and c, D in the final flow-augmenting path.

6. First, note that vertex a would have to be the source because it is the only vertex with indegree zero. Similarly, vertex d would have to be the sink, because it is the only vertex with outdegree zero. But the flow is not a legal flow because it is not conserved at vertex e. For conservation of flow, we must have the inflow minus the outflow equal to zero at all vertices other than the source and the sink. But at vertex e, we have inflow − outflow $= 2 - 1 = 1$.

7. Assign all arcs flow value of 1 except arc db, which gets a flow value of 2. The resulting flow is legal.

Index

A
activity digraph, 286–293
acyclic digraph, 264–269, 292
acyclic graph, 85
address, 191
adjacent to, 262–263, 272
adjacent from, 263, 272
adjacency list, 215–217
adjacency matrix, 199–207, 216
adjacent vertices, 48, 55, 77, 79, 140, 203, 205
algorithm, 68, 97–101, 199, 207, 215–234
    breadth-first search, 99, 216–218, 221, 272, 279–280
    binary addressing, 230–234
    bipartite matching, 107–111
    depth-first search, 219–222, 264
    determining if G is bipartite, 104, 106
    Dijkstra's, 272–273, 285
    eulerian trail, 145–148
    graphical degree sequence, 68, 87
    greedy, 96
    Kruskal's, 97–100
    maximum-color–degree coloring, 224–227
    maximum flow, 280–283
    orienting a bridgeless graph, 264
    Prim's, 98–101, 216, 272
    Prüfer code, 228–230, 233–234
    sequential coloring, 223–224, 227
    smallest last, 226
alternating path, 109
antipode, 121–124
antisymmetric, 265
arc, 261, 269, 273
augmenting flow, 279
augmenting path, 109–110, 114, 279–281

B
backtrack, 219–221, 281–283
Berge's matching theorem, 109
binary, 79
    addressing, 231–235
    Hamming graph, 232
    matrix, 200–202, 207
    tree, 92
    vector, 33
binomial coefficient, 37, 40–45
binomial theorem, 40
bipartite graph, 52, 102–114, 116, 159, 174–176, 204–206, 239–240, 243, 250–251, 257
bipartition set, 102
block, 132–133, 135, 138, 204–205, 210, 221
branch, 123, 125–126, 150
breadth-first search, 99, 216–218, 272, 279–280

bridge, 85, 93, 130, 135, 144, 252–253, 264
Brooks' theorem, 221

C

capacity, 274–276, 277, 286
cardinality, 7, 21, 109
cartesian product
    of graphs, 77–78, 82, 106–107, 135, 141, 149, 157
    of sets, 10, 14, 34
Cayley's Theorem, 231
ceiling function, 2, 14, 179, 249–250, 304
center, 121–124, 128, 132, 136, 211–212, 221–222
    central vertex, 123, 136–139, 140–141
    self-centered graph, 127, 138
    central block, 132–133, 138
    central distance set, 137, 141
centroid, 125-127, 129
chemistry, 50, 90-93, 186, 188
Chinese postman problem, 148, 159-161, 163
chromatic number, 165–172, 184–185, 223–224, 227, 249–250
circuit, 57
clique, 169
closed, 12, 149
closed trail, 57
codes
    binary addressing, 230–234
    Prüfer, 228–230, 233–234
color degree, 223–227
coloring, 246–251, 255, 258
    algorithms, 223–227
    chromatic number, 165–172, 184–185, 223–224, 227, 249–250
    edge, 173–181, 298–304
    maximum color-degree, 224–227

k–colorable, 165, 167, 169–172, 248–251
    proper, 165, 171, 180
    sequential, 223–224, 227
    uniquely k–colorable, 170–172, 185–186, 188
    uniquely k–edge colorable, 181
    vertex-, 165–172
column, 191–197
combination, 28–45
combinatorial argument, 38–39, 44–45
committee, 30–31, 34, 38–39, 43–44
complement of a graph, 75, 82, 106, 243–244
complement of a set, 8–10, 13, 21–22, 75
complete
    bipartite graph, 61, 174, 205, 239–240, 242, 245, 250–251, 304, 309
    graph, 37, 61, 175–181, 204, 239, 242, 296–304
    matching, 107–108, 112
component, 60, 71, 85, 93, 132, 140, 220, 241, 245, 248
connected, 58, 62, 91, 129, 131–132, 208–209, 241, 245
    digraph, 263
    hamiltonian, 155
    homogeneous, 154
    k–connected, 130
    minimally, 63
connectivity, 129–136, 283–284
    vertex, 130, 135–136
    edge, 130, 135–136
conservation of flow, 274
constructive proof, 138, 147
contraction of an edge, 81–82
contrapositive, 113
converse, 113
converse digraph, 270

cover, 309
covering number, 309–310
critical,
    activity, 288, 290–293
    path, 288, 290–293
    path method, 286–293
crossing graph, 255–259
crossing number, 238
cube, 78–79, 116, 243–244
cubic graph, 135
cut, 277-278
    minimal, 277, 284
    minimum, 135, 277–279, 283–286
    value, 278
cutset, 130–131, 134–135
cut edge (see bridge)
cut vertex, 129–132, 135, 138
cycle, 57–58, 61, 105, 116, 203, 211, 241
    girth, 169, 242
    hamiltonian, 151–159, 161–163, 257–258
    unicyclic graph, 237, 243

D

deficiency, 94, 100
degree, 48, 55, 66, 145, 202, 245
    color, 224, 226–227
    minimum, 48–49, 130–131, 135, 241, 247
    maximum, 49, 173–176, 187, 223–225, 307
    maximum color, 224–225
    of a face, 251
    -preserving vertex, 94
    sequence, 67–72, 87, 90, 159
deletion, 75, 93–94, 228–229
DeMorgan's laws, 9, 14, 22, 24, 247
depth-first search, 219–222, 264
diagonal, 194–195, 199–202
diagonal matrix, 199

diameter, 121, 123–125, 128, 135, 172, 208–209, 211–212, 221
    -preserving spanning tree, 219–221
diametral pair of vertices,
diametral path, 121, 124, 127–128
digraph, 56, 261–273
    activity, 286–293
    acyclic, 264–269, 292
    connected, 263–265
    converse, 270
    r–regular, 270
    strong, 263–265, 269
    unilateral, 263, 269
    weakly connected, 263, 269
Dijkstra's algorithm, 160, 272–273, 285
dimension, 191, 196–197
Dirac's theorem, 153–154
directed graph, 261–273
directed path, 262–263, 268
disconnected graph, 58, 62, 73, 85–86, 209
distance, 52, 119–125, 132, 136, 256, 269, 271–273, 285
    Hamming, 228–230
    matrix, 207–212
    matrix realizability, 210–212
distinct representatives,, 113–114, 117
dodecahedron, 244, 257
dominating set, 305–310
domination number, 307–310
dual, 251–253, 256–257
Dudeney's puzzle, 257

E

eccentric
    mutually, 120
    vertex, 120–123, 207
eccentricity, 120–122, 127, 132, 211–212, 221–222
edge, 37, 44, 55, 244

edge (continued)
- -chromatic number, 173–176, 180–182, 187
- coloring, 173–181, 183–187, 298–304
- connectivity, 130, 135–136
- contraction, 81–82
- cutset, 130–131, 134
- deletion, 93–94
- induced subgraph, 97
- multiple, 143

efficiency, 280
element, 4
elimination tournament, 267–268
embedding
- planar, 249
- toroidal, 249–251

empty set, 6
end vertex, 48, 62, 82, 93, 100, 123, 129, 228–229, 232
entry, 191–196, 201–203, 206, 208
equitable, 103, 111, 244
Erdős, 19
error-correcting codes, 231
Euler, 48, 144
eulerian
- circuit, 144–150
- graph, 53, 143–150, 153, 157, 163, 253
- semi-, 147, 153, 157
- trail, 147

walk, 150
Euler's formula, 241, 243–246
event, 140

F

F-graph, 136–138, 141
face, 240–241, 251–253
facility layout, 254–258
facility location, 121-122, 137
factorial, 26, 33, 44
finite, 6, 13
five-color theorem, 247–249

floor plan, 255–256
flow, 273–286
Ford-Fulkerson, 277–278
forest, 91
four-color problem, 237, 247
four-color theorem, 247
function, 11–12, 14–15, 35
- ceiling, 2, 12, 14, 179, 249–250, 310
- floor, 2, 179
- one-to-one, 11, 14–15, 35, 64–65, 252
- onto, 11, 14–15, 64–65

G

Gauss, 16
generalized ramsey number, 300–304
generalized wheel, 135, 245
genus, 249–251
geodesic, 58, 67, 71, 86, 119, 128, 135
girth, 169, 242
graph, 37, 48, 55
- acyclic, 85
- algorithms, 97–101, 104, 106, 215–236
- binary, Hamming, 232
- bipartite, 52, 102–114, 116, 159, 174–176, 204–206, 243, 250–251, 257
- coloring algorithms, 222–227
- complement, 75, 82–83, 243–244
- complete, 37, 61, 73, 177–181, 204, 239, 242, 299–304
- complete bipartite, 61–62, 174, 205, 239–240, 242, 245, 250–251, 304, 309
- connected, 58, 62, 105, 208–209, 241, 245
- crossing, 255, 259
- cubic, 135

Index   359

graph (continued)
    cycle, 57–58, 61, 203, 211, 245, 304
    directed, 56, 257–269
    disconnected, 58, 62, 73, 85–86, 92
    dodecahedron, 244, 257
    eulerian, 53, 143–150, 153, 157, 163, 257
    F–, 136–138
    generalized wheel, 135, 245
    hamilton laceable, 156–158
    hamiltonian, 53, 151–159, 161, 254
    hamiltonian connected, 155
    homogeneous connected, 154
    hypercube, 78–79, 82–83, 149, 206, 219, 250, 256
    icosahedron, 244, 257
    invariant, 71
    isomorphic, 64–73, 82, 87, 23, 257
    k–colorable, 165–166, 169–172, 247–251
    k–connected, 130, 135
    k–critical, 167
    , labeled, 60, 93, 117, 228, 231
    line, 80–81, 106, 135, 149
    maximal planar, 245
    mesh, 79–80, 82, 100, 116, 141, 155, 172, 221, 256, 269
    minimally connected, 64
    nonisomorphic, 65–72, 91–92
    nonplanar, 238–242, 246
    octahedron, 243–244, 256
    operations, 74–83
    order, 48
    oriented, 263–265
    path, 61, 203, 300–304
    permutation, 159
    Petersen, 163, 227, 242–243, 246
    planar, 237–259
    plane, 238–239, 241
    pseudograph, 253
    regular, 49, 56, 62–63, 82, 91
    searching, 215–222
    self-centered, 127, 138
    self-complementary, 76, 124–125
    self-dual, 258
    semi-eulerian, 147, 153
    semi-hamiltonian, 151, 153
    semiregular bipartite, 106–107
    sequential join, 75, 128–129, 135, 301, 304
    simple, 143
    size, 48
    star, 62, 90-91, 231
    subdivision, 244
    subgraph, 58–60, 62–63, 82–83, 127
    supergraph, 58
    tetrahedron, 243
    traceable, 151, 153, 156–157
    traversable, 147, 152
    tripartite, 170
    trivial, 61
    underlying, 262–263, 269
    unicyclic, 237, 243
    unlabeled 60, 76
    union, 74, 82
    uniquely k–colorable, 171–173, 185–186, 188
    uniquely k–edge colorable, 181
    weighted, 52, 54, 100, 114, 216
    wheel, 74, 82, 166, 211, 219
graphical sequence, 67–72, 87
greedy, 96
grid, 79

H
Hall's marriage theorem, 112
Hall's matching theorem, 112
hamilton laceable, 156, 158
Hamilton, 152

hamiltonian
  connected, 155, 157
  cycle, 151–159, 161–163, 257–258
  graph, 53, 151–159, 161, 254
  path, 151, 155–157, 268
  semi-, 151, 153, 157
Hamming distance, 231–233
Hasse diagram, 267, 269
height of a tree, 89
hockey-stick theorem, 42
homogeneous connected, 154
hydrocarbon, 50, 90, 92–93
hypercube, 78–79, 82, 149, 156, 206, 221, 250, 256

I
icosahedron, 244, 257
Icosian game, 152
identity matrix, 194, 198
incidence matrix, 202–203, 206
incident, 48, 102
indegree, 262, 266, 269, 292
independent set, 168, 172, 305–306, 308–310
independent dominating set, 306
independence number, 168–169, 172, 308–310
induced subgraph, 60, 141
induction, 15–27, 44–45, 86, 180
  inductive hypothesis, 17, 21, 40
  strong, 23
infinite set, 6, 13
integer, 6, 44, 72
internally disjoint paths, 133
intersection, 8–10, 13, 21–22
invariant, 71
isolated vertex, 48
isomer, 90–92
isomorphic graphs, 64–73, 82-83, 87, 253, 257
isomorphism, 65–67, 73

J
job assignment, 107–108, 110–111, 116, 284
join, 74, 82-83, 105
  sequential, 75, 128–129, 135

K
k–colorable, 165-166, 169–172, 243–247
k–connected, 130, 135
k–critical, 167, 174
k–deficient, 94
K–terminal reliability, 139
king-chicken theorem, 269
knight's tour, 150
König, 176
Königsberg bridge problem, 144
Kruskal's algorithm, 97–99
Kuratowski's theorem, 242

L
lattice, 79
leaf, 48, 88–90
length, 57–58, 188, 201, 206, 208
line graph, 80–81, 106, 135, 149
loop, 265, 267, 269

M
map, 246–247, 250
matching, 174–176
  Berge's matching theorem, 109
  bipartite, 107–111, 283
  complete, 107–108, 111
  maximal, 108
  maximum, 107, 115, 283
  perfect, 107, 111, 113–116, 175–176
mathematical induction, 15–27, 44–45, 86, 180
matrix, 191–212
  addition, 192–193, 198–199
  adjacency, 199–207

Index   361

matrix (continued)
    binary, 200–202, 207
    block of, 204–205, 210
    column of, 191–197
    diagonal, 199
    diagonal of, 193–195, 199–202, 207
    distance, 207–212
    identity, 193
    incidence, 202–203, 206
    multiplication, 195–199
    operations, 192–199
    power, 198, 201, 206, 208
    row of, 191–197
    scalar multiplication, 193, 198
    square, 191, 198
    submatrix, 204–205
    subtraction, 193, 198
    symmetric, 194–195, 199, 205, 207, 209–210
    trace, 202
    transpose, 193, 198–199, 207
    upper-triangular, 270
max-flow min-cut theorem, 278–280
maximal, 60, 131
    matching, 55, 108
    planar, 245
maximum, 60
    color degree, 224–225
    degree, 49, 173–176, 187, 223–225, 308
    flow algorithm, 280–283
    matching, 108–110, 115, 282–283
Menger's theorem, 133, 136, 283
mesh, 79–80, 82, 100, 116, 141, 155, 172, 221, 252, 269, 307–308, 310
metric, 119, 207, 209
minimal
    connector problem, 95
    cut, 277, 284
    dominating set, 305–306
minimally connected, 63
minimum
    cut, 135, 277–279, 283–286
    degree, 48–49, 131, 135, 237
    dominating set, 247, 306–310
    spanning tree, 93–101
monochromatic triangle, 177–181, 298
multigraph, 50, 143–145, 148, 160
multiple edge, 143
multiplication principle, 25, 27, 34, 226
mutually eccentric vertices, 120, 127

N

n–cube, 206, 250, 256
n factorial, 25
n-mesh, 80
n queens problem, 305
natural number, 6
nearly equitable, 103
neighbor, 111–112, 139, 218, 220–221, 223, 228–229, 233, 247
network, 130, 139–141, 144, 271–286
network flow, 273–286
nonadjacent, 48
nonisomorphic, 64–72, 91–92 134, 234, 252, 269–270, 308
nonplanar, 238–244, 246
NP-complete, 220
number
    chromatic, 165–172, 184–185, 223–224, 226, 249–250
    covering, 308–310
    crossing, 238
    domination, 307–310
    edge-chromatic, 173–176, 180–182, 187
    generalized ramsey, 294–298

number (continued)
    independence, 168–169, 172, 308–310
    integer, 6, 44
    natural, 6
    prime, 5–6, 35
    ramsey, 298–300, 303–304
    rational, 6, 154, 279
    real, 6
    tetrahedral, 43
    triangular, 15, 23

O

octohedron, 135, 149, 243–244, 256
one-to-one function, 11, 14-15, 34, 64–65, 252
onto function, 11, 14–15, 64–65
operation
    cartesian product, 10, 14, 34, 106–107, 135, 141, 149, 151
    complement of a graph, 75
    complement of a set, 8–10, 13, 21–22
    intersection of sets, 8–10, 13, 21–22
    join, 74, 81
    line graph, 80–81
    mesh, 79–80, 82
    sequential join, 75
    union of graphs, 74, 81
    union of sets, 8–10, 21–22
order
    of a graph, 48, 77, 211–212
    partial, 264
orientation, 263–265
outdegree, 262, 266, 269–270, 292
outer face, 240

P

parity, 3–4, 127
part, 102
partial order, 264
partially ordered set, 264–267, 269–270
partition, 102, 169–170, 173, 278
Pascal, 36
Pascal's triangle, 36–39, 41, 44–45
path, 57–58, 61, 89–90, 203, 300–304
    alternating, 109
    augmenting, 275–277
    critical, 284, 286–289
    critical path method, 282–289
    diametral, 121, 124, 138
    directed, 262–263
    geodesic, 58, 67, 71, 86, 161, 208
    hamiltonian, 151, 155–157, 268
    internally disjoint, 133–134
    longest, 88, 219
    radial, 121
    shortest (see geodesic)
    spanning, 63, 151, 155
perfect matching, 107, 111, 113–116
periphery, 121, 221–222
permutation, 25–35, 44, 159
PERT, 286
Petersen graph, 163, 225, 242–243, 246
pigeonhole principle, 297
planar
    embedding, 248–249
    graph, 236–259
    planarity, 236–248
plane graph, 238–239, 241
Platonic solid, 243
poset, 264–267, 269–270
power of a matrix, 198, 201, 206, 208
power set, 7, 14, 21–22
precedence relation, 286
predecessor, 287–290, 292–293
predecessor label, 286

prime number, 5–6, 35
Prim's algorithm, 216, 272
principle
    multiplication, 25, 27, 34
    mathematical induction, 14–15
    pigeonhole, 297
probability, 139–141, 290
product
    cartesian, 10, 14, 34, 77–78, 106–107, 135, 141, 149, 157
    matrix, 195–199
proper coloring, 165, 171, 180
proper subset, 7
property format, 5–7, 13
Prüfer code, 228–230, 233–234
pruning, 123
pseudograph, 253

R

radial path, 121
radius, 121, 136–138, 141, 209, 211–212, 221
    -preserving spanning tree, 219, 221
Ramsey, 299
    generalized ramsey number, 300–304
    ramsey number, 298–300, 303–304
Ramsey's theorem, 291–292
rational number, 6, 154, 279
reach-preserving, 216, 222
receiver, 262, 269, 273
recursion, 38–39
redundant arc, 277
reflexive, 265
regular graph, 49, 56, 73, 82, 91, 112–113
reliability, 139–142
relation
    on a set, 11, 34, 265, 270

precedence, 291
reverse arc, 276, 282, 282
root, 216, 219–220, 272–273
rooted tree, 89
rounding, 1–2, 12, 179
round-robin tournament, 267–268
row, 191–197

S

saturated arc, 275–279, 281–282
saturated hydrocarbon, 90, 92–93
scan, 279
scheduling, 54, 184–189, 286–293
searching, 215–222
self-centered graph, 127, 138
self-complementary graph, 76, 124–125, 128
self-dual graph, 258
semi-eulerian graph, 147, 153, 157
semi-hamiltonian graph, 151, 153, 157
semipath, 275–276, 279
semiregular bipartite graph, 106–107
separating set, 133
sequence, 229
    degree, 67–72, 87, 90
    graphical, 67–72, 87
sequential coloring, 223–224, 227
sequential join, 75, 128–129, 135, 301, 304
set, 4–14, 261
    central distance, 137
    complement, 8–10, 13, 21–22, 75
    cutset, 130–131, 134–135
    dominating, 305–310
    edge, 44
    empty, 6
    finite, 6, 13
    independent, 168, 172, 305–306, 308–310
    infinite, 6, 13

set (continued)
  intersection, 8–10, 13, 21–22
  partially ordered, 264–267, 269–270
  poset, 265
  power, 7, 14, 21–22
  relations, 11, 34, 265, 270
  separating, 133
  subset, 7, 13, 112, 169
  union, 8–10, 13, 22, 24
  universal, 8, 76
  vertex, 44
skewed, 103
simple graph, 143
sink, 274
size, 48
slack, 275–276
slack time, 288-290, 292
source, 273–274
spanning, 60–61, 64
  cycle (see hamiltonian cycle)
  path, 63, 151, 155
  subgraph, 60, 82, 127, 239–240
  tree, 93–101, 128–129, 216–222, 232
  tree algorithm, 96–100
square matrix, 191
star, 62, 90–91, 234
strong digraph, 263–265, 269
strong induction, 23
strongly connected, 263–265
subdivision, 243–245
subgraph, 58–60, 62–63, 82–83, 106, 127, 242
  clique, 169
  edge-induced, 97
  induced, 60, 63, 141
  spanning, 60–61, 93–101, 127, 239–240
subset, 7, 13, 44, 112, 169
submatrix, 204–205
successor, 287

supergraph, 58
symmetric
  arcs, 262
  matrix, 194–195, 199, 205, 207, 209–210
system of distinct representatives, 113–114, 117

T
target vertex, 272
terminal, 273
tetrahedral number, 43
tetrahedron, 243–244
timetabling problem, 186–189
toroidal, 249–251
torus, 249–250
tournament, 267–271
trace, 202
traceable graph, 151, 153, 156–157
traversable graph, 147, 153
traffic flow, 264, 270–271
trail, 57–58, 63
transitive tournament, 270
transitive, 265
transmitter, 262, 269, 273
transpose, 193, 198–199, 207
traveling salesman problem, 161–163
tree, 50–51, 85–103, 105, 115–116, 123–127, 150, 216–222, 232
  binary, 92
  codes, 226–232
  height, 89
  minimum spanning, 93–101
  rooted, 89
  spanning, 93–101, 128–129, 216–222, 230
triangle, 177–181, 206
  inequality, 119, 122, 177, 209
  Pascal's, 36–39, 44–45
triangular number, 15, 23, 42
tripartite, 170
trivial graph, 61

U

underlying graph, 262–263, 269
unicyclic graph, 237, 245
unilateral digraph, 263, 269
union
    of sets, 8–10, 13, 22, 24
    of graphs, 74, 82
uniquely k–colorable, 170–171, 173, 186–187, 188
unique k–edge colorable, 181
unmatched edge, 109
upper-triangular matrix, 270

V

vector, 194
vertex, 37, 48, 55
    cover, 309–310
    cutset, 130–131
Vizing's theorem, 176
vulnerability, 130

W

walk, 57–58, 63, 201, 206, 208
weakly connected digraph, 263–269
weight, 93, 126, 128–129
weighted digraph, 272
weighted graph, 52, 54, 100, 114, 144, 187, 214, 284
wheel, 74, 82, 166, 211, 256
while loop, 219–220
Whitney's theorem, 130